Quantitative Analysis

Problem Solutions

Selected Chapters from:

Applied Management Science
A Computer-Integrated Approach for Decision-Making
John A. Lawrence, Jr.
Barry A. Pasternak

Business Statistics
Decision Making With Data
Richard A. Johnson
Dean W. Wichern

Printed in the United States of America.

ISBN 0-471-32645-3

Contents

Materials Selected from Business Statistics

Richard A. Johnson
Dean W. Wichern

1.1. Proportion = Number of companies / Total

Country	Number of companies	Proportion
United States	135	0.270
Japan	128	0.256
Germany	45	0.090
Britain	42	0.084
France	33	0.066
Canada	18	0.036
Spain	16	0.032
Italy	15	0.030
Switzerland	15	0.030
Other	53	0.106
Total	500	1.000

1.2.

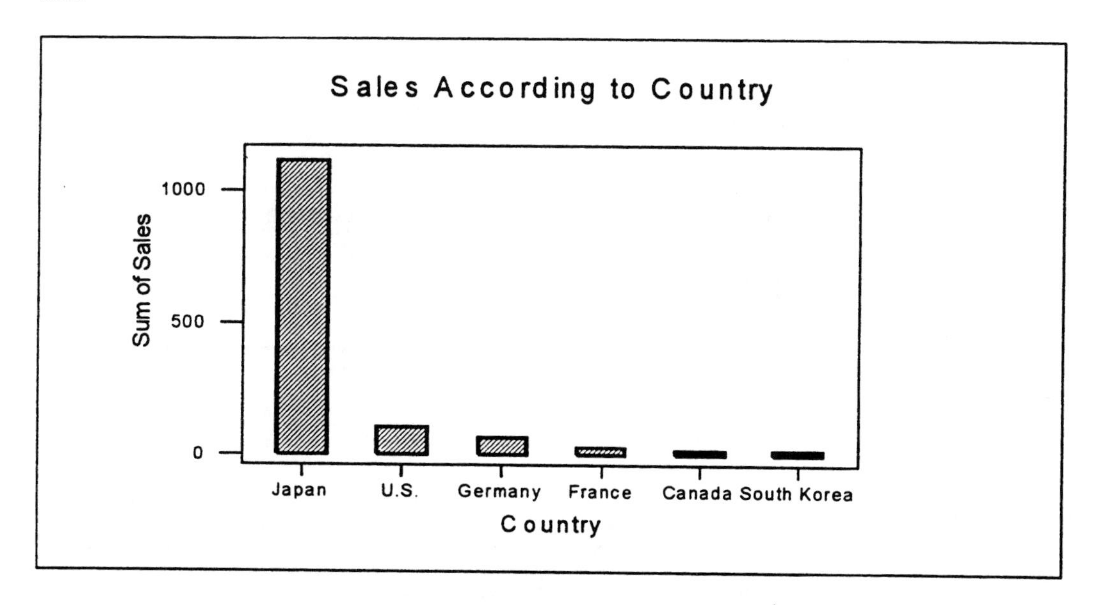

As compared to Figure 1.1, the United States now has many fewer companies (4 as compared to Japan's 16) instead of having slightly more companies (135 to Japan's 128). Figure 1.1 shows numbers of companies whereas, this bar chart shows sales according to country. Several countries are represented on the Figure 1.1 bar chart that are not on this bar chart.

1.3. Bar chart showing the number of banks from each country:

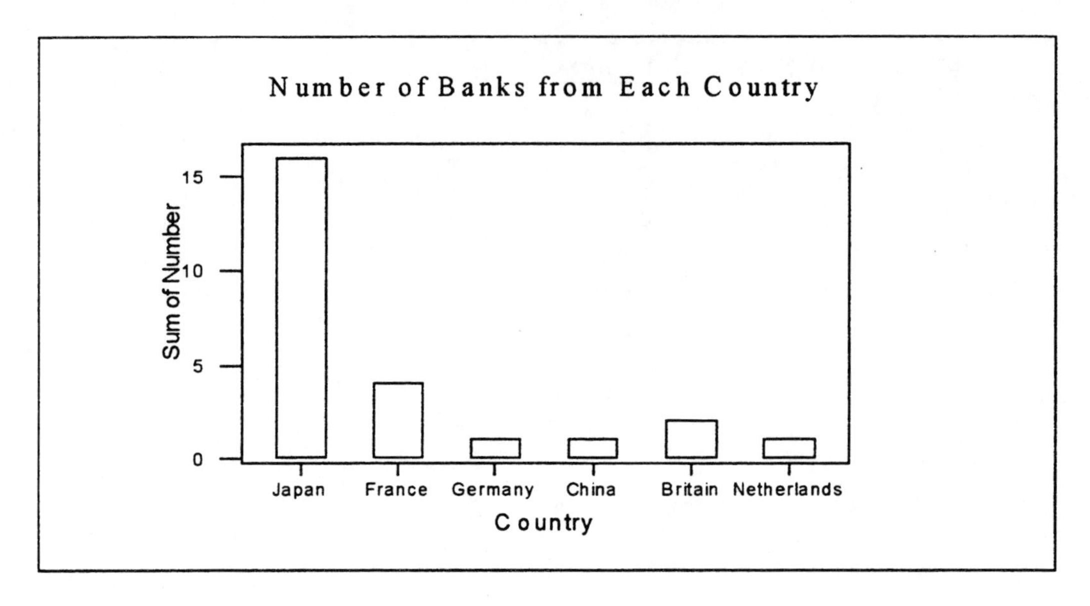

The proportion of banks from Japan is calculated as 16/25 = 0.64.

1.4. Total assets = $8305.10 (in billions)

Country	Assets per Country	Proportion of total assets
Japan	5796.00	0.698
France	1189.40	0.143
United States	0	0

1.5. Assets per bank dot plot:

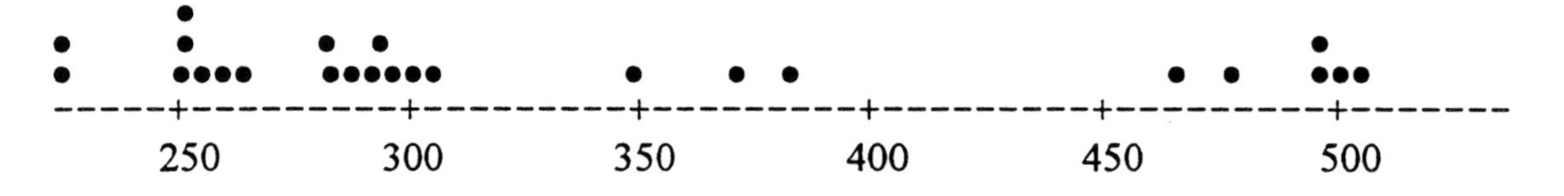

There are gaps in the asset pattern. The group of largest banks, greater than $450 billion, includes 6 banks. Japan dominates this group.

1.6. Population per city (millions):

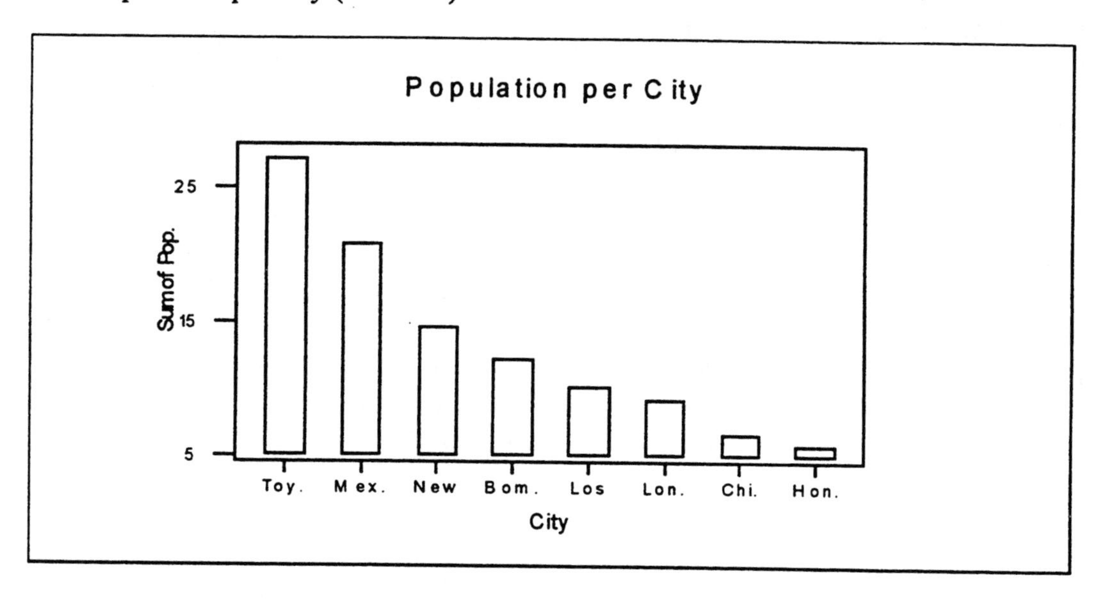

1.7. The proportion of persons in these cities who live in the United States = (6.529 + 10.130 + 14.625) / 106.337 = 0.294

1.8. Population Density by City:

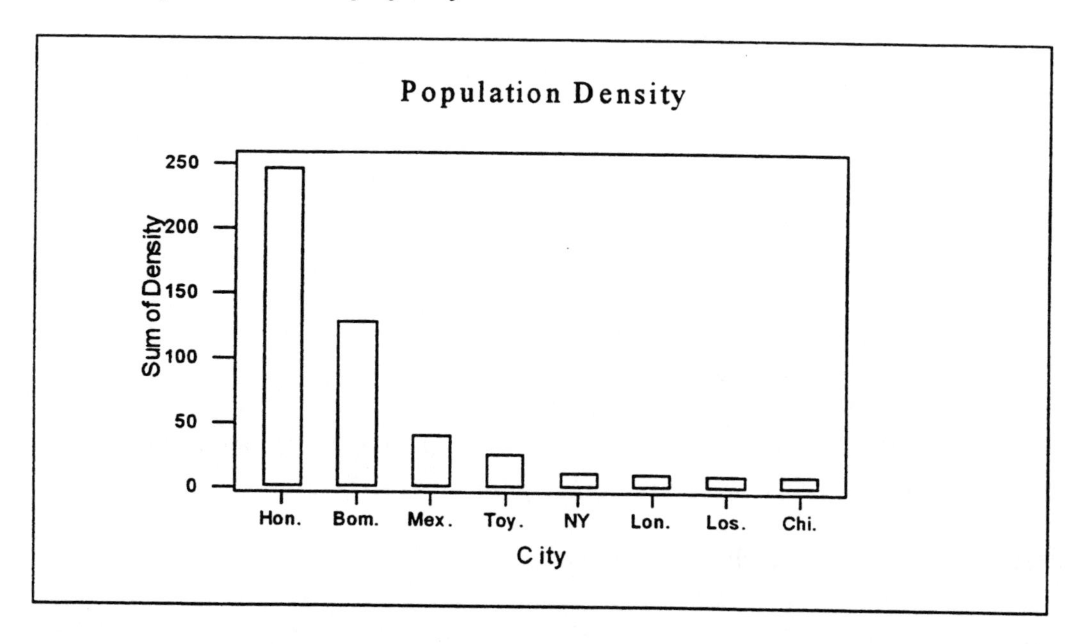

The data in this bar chart shows little relation to that in 1.6. However, note that Hong Kong, with the smallest population, has the largest density since its area is so small.

1.9. Academic processes which need improvement and steps that might be taken to improve them using the scientific method.

Registration process:

Step 1: Observe the current methods being used, identifying existing problems. This may be a computerized system that allows students to register by telephone. Problems that exist are related to the time that it takes to get into the system since the lines are generally busy. Gather data and analyze the data using methods introduced in this chapter.
Step 2: Hypothesize that there are not enough lines into the computerized registration system.
Step 3: Deduce that more telephone lines are needed, or perhaps students could use the computers in the computer labs as another registration method .
Step 4: Verify that more telephone lines solves the problem.

Method of instruction:

Step 1: Observe that some students sleep during class. Gather and analyze data based on the number of students who appear inattentive or are sleeping during class lectures and the teaching method used at the time.
Step 2: Hypothesize that the lecture method of presenting classes is a major portion of the problem.
Step 3: Deduce that classes should consist of a variety of presentation techniques: lecture, discussion, short answer, films, outside speakers.
Step 4: Verify that variety in presentation will better keep the attention of the students.

Grading process:

Step 1: Observe that there is too long a wait between testing and posting of grades. Gather and analyze data related to the time frame between testing, grading, and posting of grades as related to the type of exam (multiple choice, essay, fill in the blank).
Step 2: Hypothesize that multiple choice exams with answers recorded on machine-graded forms can be graded more rapidly than other methods and that a true picture of the students' knowledge can be gained using this testing method.
Step 3: Deduce that machine-graded tests are better.
Step 4: Verify this by using two groups of students, and two testing methods, comparing the knowledge of the two groups.

1.10. Financial transactions which need improvement:

Student loans:

Step 1: Observe that student loans occasionally arrive after the beginning of the semester, leaving those students without money for tuition, books, or living expenses. Gather data from students about the number of loans that arrive late and the reason in each case.
Step 2: Hypothesize that the main problem is that students apply for the loans too late, not leaving enough time for processing.
Step 3: Deduce that the application deadline needs to be moved back.
Step 4: Verify that fewer late loans occur for the next semester.

Check cashing:
Step 1: Observe that there is occasionally a line at the check cashing counter. Collect and analyze data related to the line size and time of day.
Step 2: Hypothesize that the line is longer at certain times of the day such as the lunch hour.
Step 3: Deduce that an extra employee needs to work at the check cashing counter during some parts of the day.
Step 4: Verify that the extra employee solves the problem.

Parking ticket payments:
Step 1: Observe that there is a large number of unpaid parking tickets. Gather data related to unpaid tickets, the classification of students, and the penalty for non-payment.
Step 2: Hypothesize that students would pay for tickets if the payment was a pre-requisite for graduation.
Step 3: Deduce that students should be required to pay for their parking tickets prior to being allowed to graduate.
Step 4: Verify that grading seniors' tickets are now paid.

1.11. Definitions of quality for these items:
A winter coat is a quality coat if it is warm, fits well, is becoming to the wearer, is made of good quality fabric, and is liked by the wearer.

A swimsuit is a quality swimsuit if it is made of good quality fabric, fits the wearer well, is becoming to the wearer, does not fade in the sun or chlorinated water, and is liked by the wearer.

Sunglasses are a quality product if they provide UV protection, are becoming to the wearer, are scratch resistance, light weight, have the correct amount of light that shows through the lens, and not breakable.

Tennis shoes are a quality product if they are comfortable, look nice, are made of quality materials and workmanship, and have good support for the wearer's arch and foot in general.

1.12. Definitions of quality for these services:
Airline service is a quality product if the airplanes are dependable and ride well, the flights run on schedule, the stewardesses are polite and helpful during the trip, the luggage arrives with the customer, the check in process is short and streamlined, and meals or snacks are served .

Public transportation is a quality service if it is dependable and comfortable, the schedules are convenient and run on time, and the driver is polite and helpful.

Drive through banking services are quality if they have either no lines or short lines, they handle your banking needs, and they are courteous and friendly.

Hotel service is quality service if it is done properly and on time, and the personnel are courteous and helpful.

Maintenance service is a quality service if it makes repairs quickly and dependably, and the personnel are courteous and helpful.

The common factors of general service quality include: keeping up with a schedule, quick and dependable service, and courteous and helpful personnel.

1.13.

Factors determining the quality of sunglasses:	Type of Data to collect
becoming to the wearer	numeric scale
scratch resistance	yes/no
light weight	numeric
correct amount of light that shows through the lens	numeric
not breakable	yes/no

1.14. In order to determine the quality of airline service, data should be collected for a period of time on each of these factors: airplanes are dependable and ride well, the flights run on schedule, the stewardesses are polite and helpful during the trip, the luggage arrives with the customer, the checkin process is short and streamlined, and meals or snacks are served .

1.15. Sample mean = $(-3 + 1 + 0 + 3 + 4)/5 = 1$
Sample variance = $[(-3 - 1)^2 + (1 - 1)^2 + (0 - 1)^2 + (3 - 1)^2 + (4 - 1)^2] / (5 - 1) = 7.5$
Sample standard deviation = $(7.5)^{1/2} = 2.74$

1.16. Sample mean =$(7 + 3 + 4 + 10)/4=6$
Sample variance = $[(7 - 6)^2 + (3 - 6)^2 + (4 - 6)^2 + (10 - 6)^2] / (4 - 1) = 10$
Sample standard deviation = $(10)^{1/2} \cong 3.1622$

1.17. Sample mean =$(3.5 + 2.9 + 3.4 + 4.3)/4 = 3.525$
Sample variance =$[(3.5 - 3.525)^2 + (2.9 - 3.525)^2 + (3.4 - 3.525)^2 + (4.3 - 3.525)^2] \cong$ 0.3358
Sample standard deviation = $(0.3358)^{1/2} \cong 0.580$

1.18. Since the mean is smaller on Saturday as compared to Wednesday, the plant closure on Saturday coincides with the direction of the difference.

1.19. Mean for five plants = $(530 + 1135 + 1125 + 778 + 476) / 5 = 808.8$

1.20. Mean for the U.S. = 99,624 / 111 ≅ 897.51
Two of the five plants were above the U.S. mean.

a.

$$\sum_{i=1}^{n}(x_i - \bar{x}) = \sum_{i=1}^{n} x_i - (n * \bar{x})$$
$$= \sum_{i=1}^{n}(x_i) - n*\left(\frac{1}{n}\right)\sum_{i=1}^{n}(x_i)$$
$$= \sum_{i=1}^{n}(x_i) - \sum_{i=1}^{n}(x_i)$$
$$= 0$$

b. If all of the numbers are the same, for any i:

$$\bar{x} = \left(\frac{1}{n}\right)\sum_{i=1}^{n} x_i = \frac{1}{n}(n)x_i = x_i$$

1.22.a. Sample mean = (34.1 + 34.0 + 33.8 + 33.8 + 33.9)/5 = 33.92

b. Sum of the differences of the observations from the sample mean:
(34.1 - 33.92) + (34.0 - 33.92) + (33.8 - 33.92) + (33.8 - 33.92) + (33.9 - 33.92)= 0

c. Sample variance = [(34.1 - 33.92) + (34.0 - 33.92) + (33.8 - 33.92) + (33.8 - 33.92) + (33.9 - 33.92)] / 4 = 0.17
Sample Standard deviation = $(0.17)^{1/2} \cong 0.13038$

d. The mean in Example 1.7 was 34 and the standard deviation was 0.16. The two plants are getting equally reliable determinations of moisture content since the means and standard deviations are nearly equal.

1.23.a. Mean = 19.115
Standard deviation = 7.014

b.

Range	Actual Data
Within one standard deviation of the mean	20 data items = 20/26 = 0.77
Within two standard deviations of the mean	25 data items = 25/26 = 0.96

1.24. A unit is a single flight.

The population of units is the complete collection of all flights of U.S. based airlines.

The statistical population is the collection of measurements of all the flights.

The sample is the set of 158 measurements taken.

1.25. The population is the collection of responses from all readers.

The sample is the collection of responses received from the readers.

Since readers who have not purchased any new products are not as likely to respond, the sample may not be representative of the population.

1.26. The population is the collection of all readers.

The sample is the collection of readers who respond.

The sample may not be representative of the population since readers are not randomly selected to respond.

1.27. Anecdotal data are given in *a* and *c* while *b* is based on a sample.

1.28. Anecdotal data are given in *a* and sample data are given in *b* and *c*.

1.29. Bar chart for total exports arranged from largest to smallest:

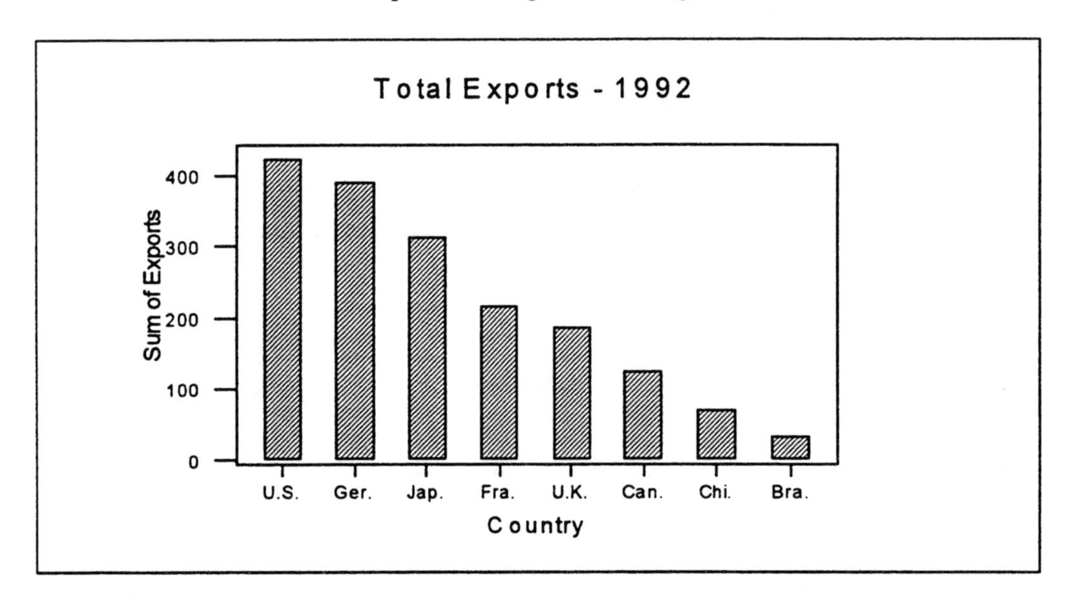

1.30.

Sample mean = 219.13

Sample variance = 21078.98

1.31.a. Bar graph of exports per capita:

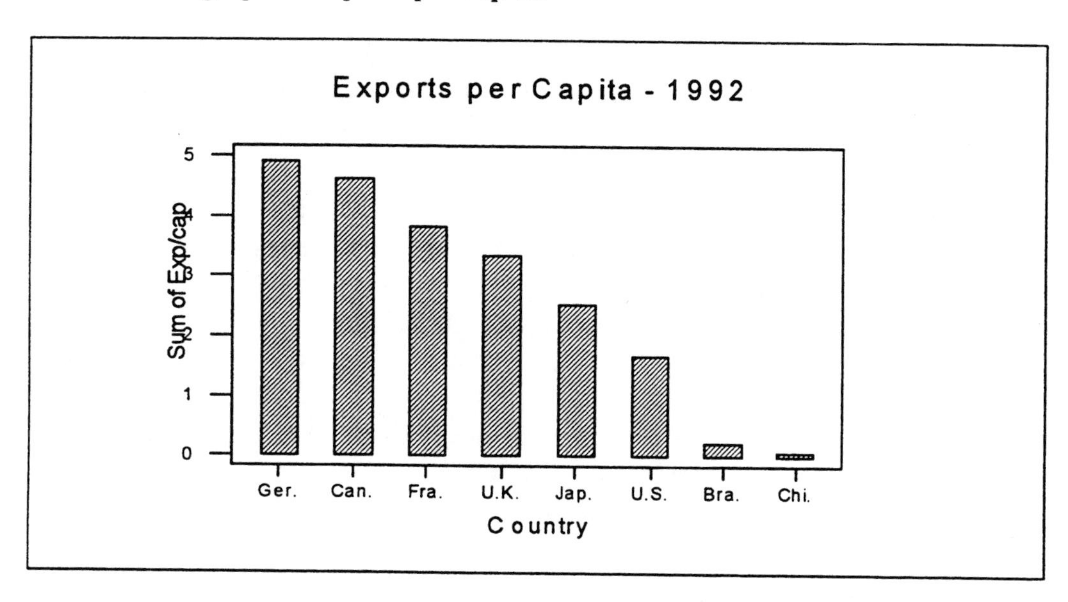

b. The U.S., Canada, and Japan changed the most as far as order of nations on the two graphs.

1.32.a. Bar graph showing the percent of total energy consumption by fuel type:

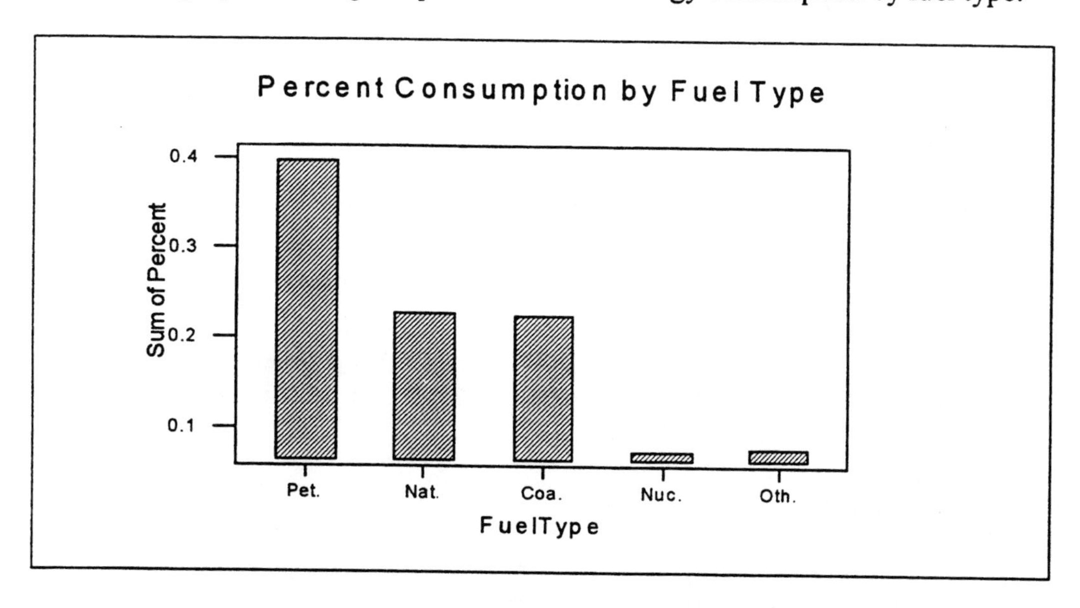

b. In order to track the changes in the dependence on natural gas relative to other fuels as an energy source, the consumption percents should be used since the overall fuel usage may change.

1.33. Processes associated with daily living which need improvement:

Room cleanup:
Step 1: Observe the level of clutter in the room on each of various days, rated on a numerical scale from 1 to 10. Analyze the data to determine the frequency and quality of the cleaning process.
Step 2: Hypothesize that room cleanup is done on a very irregular basis.
Step 3: Deduct that a regular cleanup schedule, say weekly for some tasks and daily for other tasks, might improve the room.
Step 4: Verify that the new schedule decreases the room's clutter.

Grocery shopping:
Step 1: Observe that grocery items are not obtained even though there has been a recent trip to the grocery store. Also observe that multiple trips are made to the grocery store with just a few items purchased each time. Gather and analyze data to determine the extent of the problem.
Step 2: Hypothesize that the use of a grocery list and a weekly pre-planned menu should cut down on items forgotten at the grocery store and cut down on the number of trips.
Step 3: Deduct that a weekly menu and shopping list should remedy the situation.
Step 4: Verify that the new system solves the problem.

1.34. Quality for each of these foods:

Pizza: good crust with proper thickness, the proper amount of tomato sauce, cheese, spices, and other toppings, temperature
Frozen yogurt: proper temperature, good ingredients, flavor
Orange juice: good flavor, correct amount of pulp, added calcium
Potato chips: low fat but does not taste like low fat, good flavor, crinkles
Chocolate chip cookies: lots of chocolate chips, good flavor and consistency

1.35. Collect data on pizzas by taking the temperature, measuring the amount of tomato sauce, cheese, and other toppings, and measure the thickness of the crust.

1.36.a. Mean = (12 + 6 + 8 + 0 + 14)/ 5 = 8
b. Variance = 29.9997
c. Standard deviation = 5.4772

1.37.a.

```
               • •  •
       •   • ••••   ••                                  •
 +---------+---------+---------+---------+---------+-------
1.5       3.0       4.5       6.0       7.5       9.0
```

b. Sample mean = 4.192
Sample standard deviation = 1.819

c. With the outlier removed, sample mean = 3.7 and standard deviation = 0.671.

d. Removing the outlier decreased both the mean and standard deviation. In general, the mean will be decreased when large outliers are removed and increased when small outliers are removed. The standard deviation will decrease in either case.

1.38. The population is the collection of all people. The sample is the collection of 100 people who responded to the question. This is not a random sample, but rather a response from those who are interested in the topics covered by the consumer magazine and possibly shopping in particular.

1.39.a.

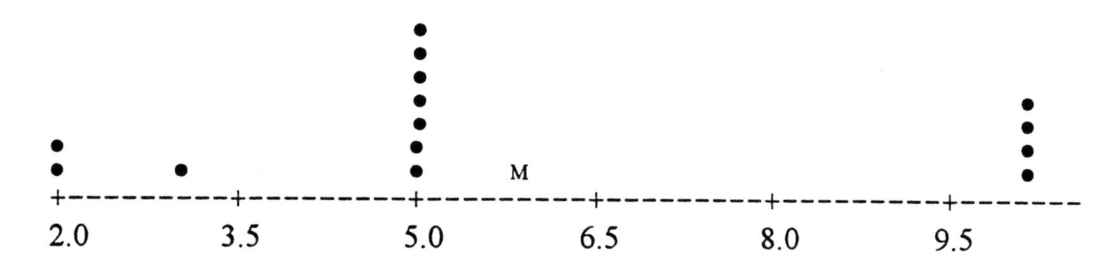

b. Mean = 5.857 ; Standard deviation = 2.931 The mean is marked with an "M" on the above dotplot. Yes, it appears that the dot diagram "balances" at the mean, which it should.

1.40.a. New values after 3 added to each: 13 8 8 13 8 8 13 8 6 8 8 13 5 5
New mean = 8.857 is exactly 3 units larger than the old mean.

b. In general, adding *c* to each number in a data set will always result in the new mean being equal to the old mean plus *c*.

1.41.a. New values: 30 15 15 30 15 15 30 15 9 15 15 30 6 6
Both the standard deviation and the variance would be expected to increase. The new standard deviation is 8.794 which is 3 times the old standard deviation. The new variance is 9 times the old variance.

b. In general, multiplying each number in a data set by a positive constant *c* will also increase the standard deviation by a multiple of 3. No, the constant need not be positive. The standard deviation is multiplied by the absolute value of the constant.

c. Adding a constant, *c*, to each number in a data set does not change the standard deviation.

1.42.a. Mean = 5

New values: 14 16 22 28 20

New mean = 20 is found by taking the old mean and multiplying it by 2 then adding 10. Both the additive and multiplicative constants affect the mean. The same would be true for different values of the constants.

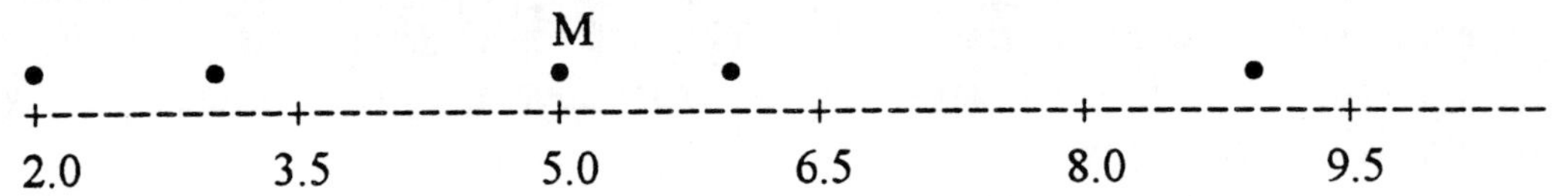

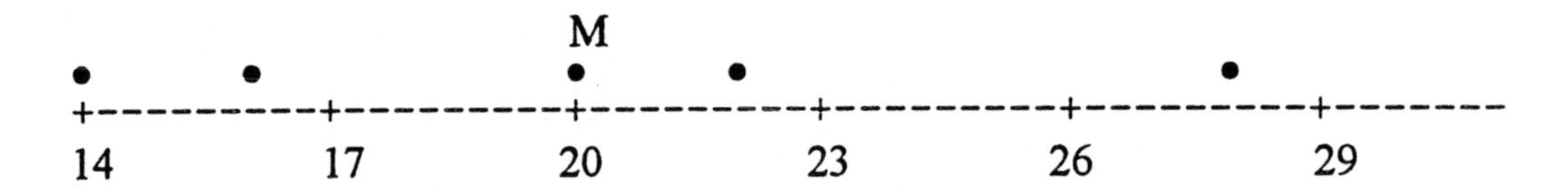

b. Standard deviation = 2.7386

New standard deviation = 5.477 is found by multiplying the old standard deviation by 2. The additive constant does not affect the standard deviation. Different values would not change this conclusion.

2.1.

a. The number of working telephones is enumerative since it is a finite collection of units.

b. The number of on-time arrivals yesterday is enumerative since it is an unchanging constant.

c. The amount of supervisory time to be allocated is analytical since it is a prediction.

d. The prediction of number of freshman is analytical since it is a prediction.

2.2.

a. The number watching a television show is enumerative since it is a constant.

b. The number of defective golf balls is enumerative since it is a constant.

c. The effect of advertising on sales is analytical since it is a prediction.

d. The shipping time is analytical since it is a prediction.

2.3.

a. The number of working telephones is quantitative and discrete.

b. The number of on-time arrivals yesterday is quantitative and discrete.

c. The amount of supervisory time to be allocated is quantitative and continuous.

d. The prediction of number of freshman is quantitative and discrete.

2.4.

a. The number watching a television show is quantitative and discrete.

b. The number of defective golf balls is quantitative and discrete.

c. The effect of advertising on sales is quantitative and discrete.

d. The shipping time is quantitative and continuous.

2.5.

a. Enumerative studies for which the students in the class may be regarded as a sample from a larger population include:

Determine the grade point average of all business students.

Determine the average age of all students who take this statistics class.

b. Enumerative studies for which the students in the class may be regarded as the population:

Determine the average grade in the class.

Determine the average age of students in this class.

2.6.

a. Six out of fifteen people in the sample favored high speed rail transportation. This may be interpreted as forty percent of the population favors high speed rail transportation.

b. The sample mean is 0.40. For coded data, this means that 40 percent of the data is coded as 1 as opposed to 0.

2.7.

a. Seven out of ten recent accounting graduates were satisfied with their job. This can be interpreted as seventy percent of new accounting graduates are satisfied with their job.

b. The sample mean is 0.70, meaning that seventy percent of binary data is recorded as 1 instead of 0.

2.8.

a.

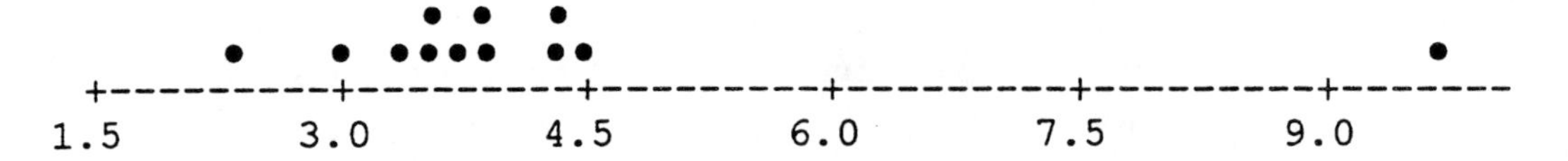

b. The dot diagram is consistent with the stem-and-leaf diagram since they both represent the same data. Most of the data is between 3.0 and 4.5.

2.9.

a. The choice of intervals is poor since the intervals are not of equal width and the range from 50 to 55 is not included in any interval.

b. The choice of intervals is poor since the intervals are not of equal width and the range from 75 to 80 is included in two intervals.

2.10.

a.

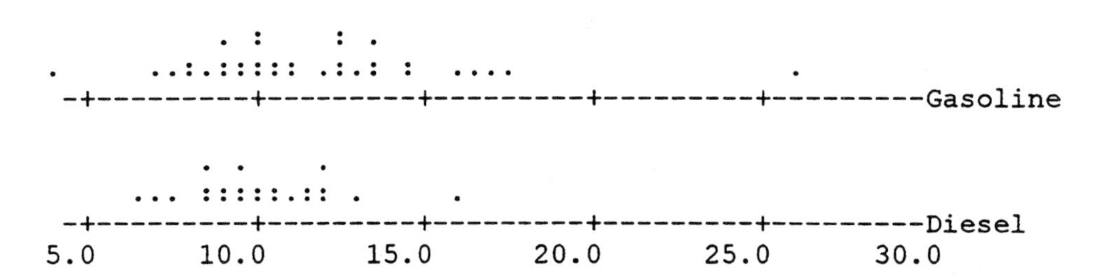

There are not any outliers with the diesel fuel cost as compared to the gasoline which has two outliers. Also, the diesel figures are more concentrated in the 6.00 to 12.00 range than are the gasoline figures.

b. Character Stem-and-Leaf Display

```
Stem-and-leaf of Gasoline  N  = 36
Leaf Unit = 0.10


    1     4 2
    1     5
    1     6
    3     7 15
    8     8 22589
   13     9 14799
   17    10 1223
   (2)   11 12
   17    12 134679
   11    13 357
    8    14 27
    6    15 8
    5    16 49
    3    17 3
    0    18
    0    19
    0    20
    0    21
    0    22
    0    23
    0    24
    0    25
    1    26 1
    0    27
    0    28
    1    29 1
```

```
Stem-and-leaf of Diesel     N  = 23
Leaf Unit = 0.10
    1      6 4
    1      6
    3      7 14
    3      7
    4      8 2
    6      8 55
    8      9 01
   (4)     9 5677
   11     10 124
    8     10 8
    7     11 3
    6     11 689
    3     12
    2     12 7
    0     13
    0     14
    1     15 9
```

c. Stem and leaf for the gasoline fuel cost (leaf unit = hundredths):

```
42  4
71  9
72
73
74
75  1
76
77
78
79
80
81
82  1
82  2
83
84
85  1
86
87
88  8
89  8
90
91  8
92
93
94  9
95
96
97  0
```

```
 98
 99  0
 99  2
100
101  8
102  4
102  5
103  2
104
105
106
107
108
109
110
111  1
112  0
113
114
115
116
117
118
119
120
121  7
123  4
124  9
125
126  8
127  2
128
129  5
130
131
132
133  2
134
135  0
136
137  0
138
139
140
141
142  5
143
144
145
146
```

```
147  0
148
149
150
151
152
153
154
155
156
157
158  6
159
160
161
162
163
164  4
165
166
167
168
169  3
170
171
172
173  2
261  6 (Outlier)
291  1 (Outlier)
```

d. Back-to-back stem-and-leaf showing fuel costs for diesel (left) and gasoline (right):

```
        4 2
        5
4       6
14      7 15
255     8 22589
01567   9 14799
1248   10 1223
3689   11 12
07     12 134679
       13 357
       14 27
9      15 8
       16 49
       17 3
       18
       19
       20
       21
       22
       23
       24
       25
       26 1
       27
       28
       29 1
```

The differences are consistent with those of the dot diagrams in part a.

2.11.

a. Frequency Distribution for the Fuel Costs of Gasoline Trucks:

Class Interval	Frequency	Relative Frequency
[1.5,4.5)	1	0.02778
[4.5,7.5)	1	0.02778
[7.5,10.5)	15	0.41667
[10.5,13.5)	9	0.25000
[13.5,16.5)	6	0.16667
[16.5,19.5)	2	0.05556
[19.5,22.5)	0	0.00000
[22.5,25.5)	0	0.00000
[25.5,28.5)	1	0.02778
[28.5,31.5)	1	0.02778
Totals:	36	1.00000

b. Frequency Distribution for the Fuel Costs of Diesel Trucks:

Class Interval	**Frequency**	**Relative Frequency**
[1.5,4.5)	0	0.0000
[4.5,7.5)	3	0.1304
[7.5,10.5)	12	0.5217
[10.5,13.5)	7	0.3043
[13.5,16.5)	1	0.0435
[16.5,19.5)	0	0.0000
[19.5,22.5)	0	0.0000
[22.5,25.5)	0	0.0000
[25.5,28.5)	0	0.0000
[28.5,31.5)	0	0.0000
Totals:	23	1.0000

c. Relative frequencies for Gasoline and Diesel costs shown back-to-back .

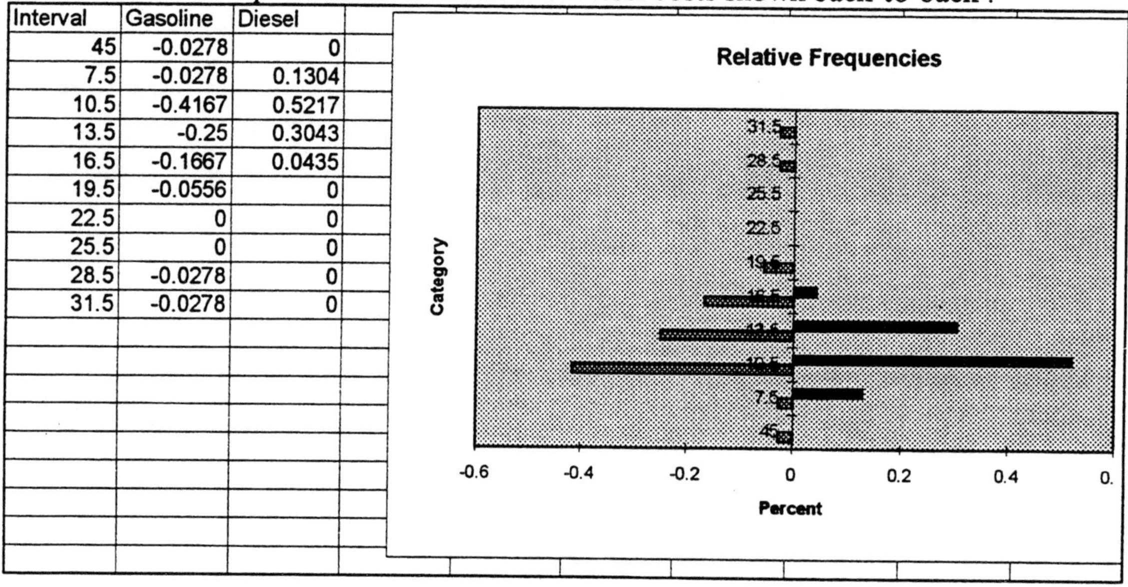

Interval	Gasoline	Diesel
45	-0.0278	0
7.5	-0.0278	0.1304
10.5	-0.4167	0.5217
13.5	-0.25	0.3043
16.5	-0.1667	0.0435
19.5	-0.0556	0
22.5	0	0
25.5	0	0
28.5	-0.0278	0
31.5	-0.0278	0

Diesel costs vary less than gasoline costs.

d. Since there are more gasoline trucks than diesel trucks, the histogram would change if densities rather than relative frequencies were used.

2.12.

a.

Character Dotplot of bankrupt and non-bankrupt companies:

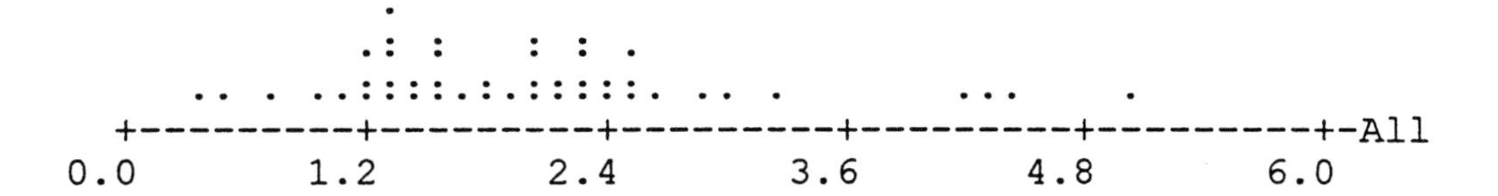

This is a multi-modal set of data.

b. Character Dotplot of non-bankrupt vs. bankrupt companies:

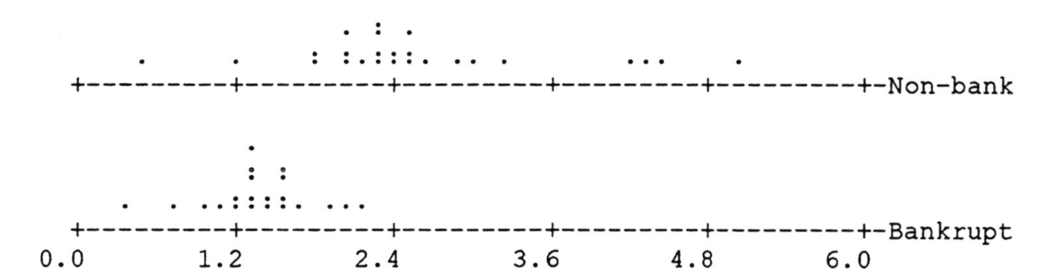

Non-bankrupt firms have a larger CA/CL ratio than do bankrupt firms. This variable may be useful in distinguishing bankrupt from non-bankrupt firms.

c. Back-to-back Stem-and-Leaf Display of bankrupt firms and non-bankrupt firms:

Leaf Unit = 0.10

```
                 37  0 4
0011223334455556 89  1 2889
                  1  2 00122333445569
                     3 02
                     4 224
                     5 0
```

This result is consistent with the dot diagrams in part b.

2.13.

a. Frequency distribution for bankrupt firms:

Class Intervals	Frequencies	Relative Frequencies
[0.25 - 0.75)	2	2/21 = 0.095
[0.75 - 1.25)	4	4/21 = 0.190
[1.25 - 1.75)	12	12/21= 0.571
[1.75 - 2.25)	3	3/21 = 0.143
Total:	21	0.999*

* This entry is 1.000 within rounding error.

b. Frequency distribution for non-bankrupt firms:

Class Intervals	Frequencies	Relative Frequencies
[0.25 - 0.75)	1	1/25 = 0.04
[0.75 - 1.25)	1	1/25 = 0.04
[1.25 - 1.75)	0	0/25 = 0
[1.75 - 2.25)	7	7/25 = 0.28
[2.25 - 2.75)	9	9/25 = 0.36
[2.75 - 3.25)	2	2/25 = 0.08
[3.25 - 3.75)	1	1/25 = 0.04
[3.75 - 4.25)	1	1/25 = 0.04
[4.25 - 4.75)	2	2/25 = 0.08
[4.75 - 5.25)	1	1/25 = 0.04
Total:	25	1.00

c.

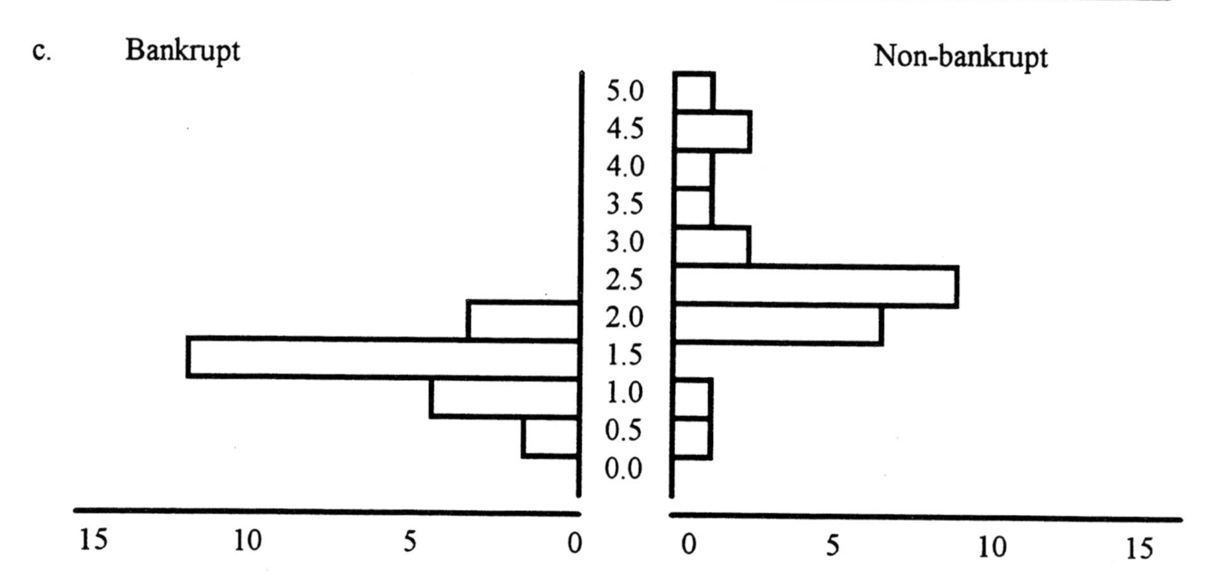

The average CA/CL for bankrupt firms is lower than the average for non-bankrupt firms. Non-bankrupt firms have a larger range of values.

d. Densities would look similar to these frequencies since the interval widths are equal.

2.14.

a.

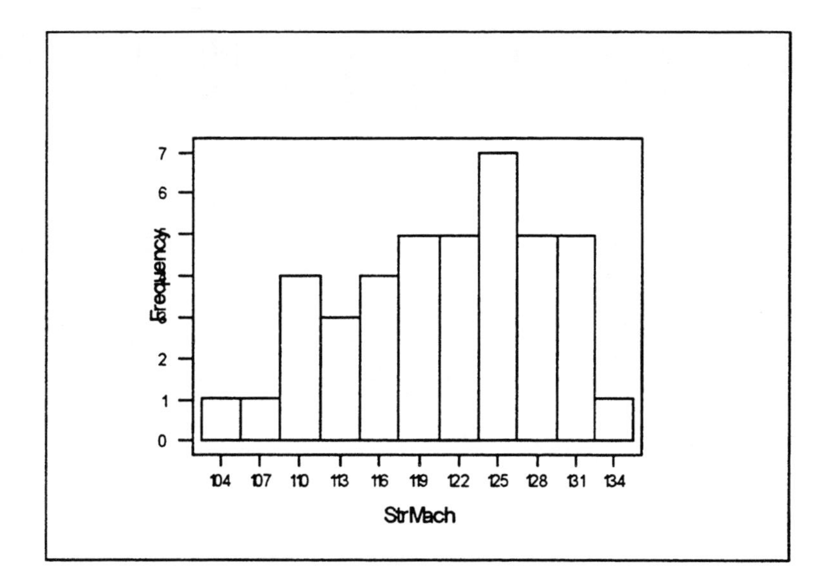

As compared to the histogram of strengths in the machine direction in Example 2.6, the peaks are not as clearly defined. The changes appear to be more irregular. The highest point is in the same range.

b.

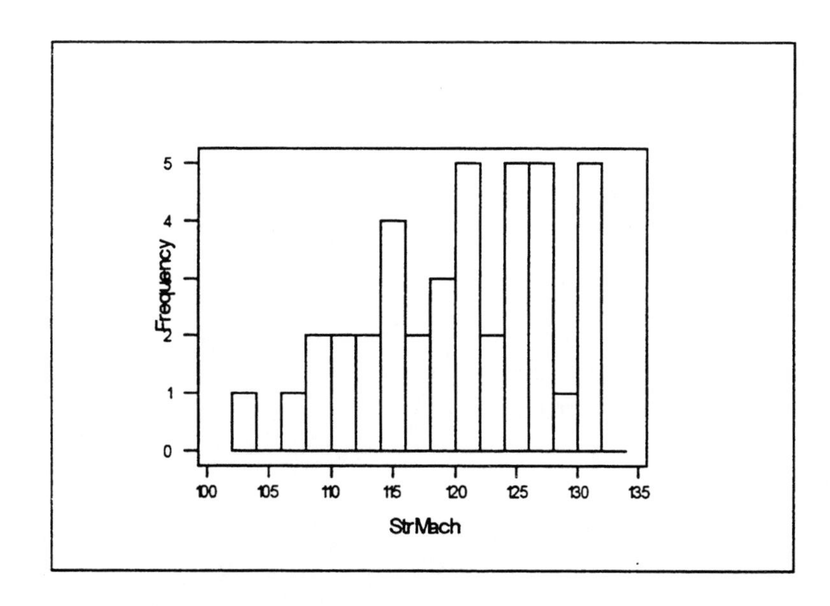

In changing the interval to a smaller unit, there are even more irregularities in the histogram. The old paper tends to have a smaller strength in the machine direction, but the division is not clear cut.

2.15.

a.

Size of Raise	Frequency	Relative Frequency	Density
[-.1, .1)	602	602/4145 = 0.145	0.145/0.2 = 0.725
[.1, 3.1)	715	715/4145 = 0.172	0.172/ 3 = 0.057
[3.1, 5.1)	1405	1405/4145 = 0.339	0.339/ 2 = 0.170
[5.1, 7.1)	805	805/4145 = 0.194	0.194/ 2 = 0.097
[7.1, 10.1)	386	386/4145 = 0.093	0.093/ 3 = 0.031
[10.1, 15.1)	178	178/4145 = 0.043	0.043/ 5 = 0.009
[15.1, 20.1)	54	54/4145 = 0.013	0.013/ 5 = 0.003
Total:	4145	0.999*	

* This entry is 1.000 within rounding error.

b. Density histogram for sizes of raise:

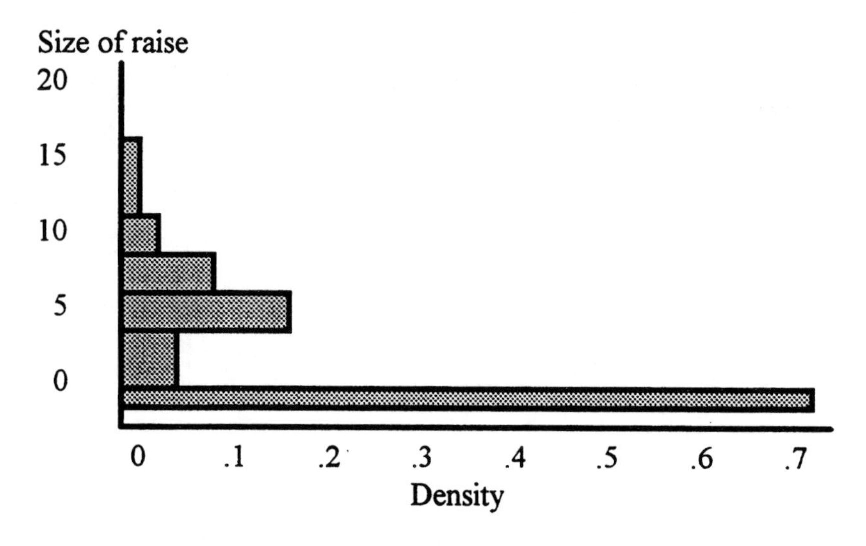

This density histogram shows a high density of raises that are in the interval between -.1 and .1. The second highest interval is from 3.1 to 5.1.

2.16.

a. Frequency Distribution:

Midpoint	Class limits	Frequency	Relative Frequency	Density
2.99	[2.98,3.00)	1	0.01	0.01/.02=.5
3.01	[3.00,3.02)	4	0.04	0.04/.02 = 2
3.03	[3.02,3.04)	4	0.04	0.04/.02 = 2
3.05	[3.04,3.06)	4	0.04	0.04/.02 = 2
3.07	[3.06,3.08)	7	0.07	0.07/.02 = 3.5
3.09	[3.08,3.10)	17	0.17	0.17/.02 = 8.5
3.11	[3.10,3.12)	24	0.24	0.24/.02 = 12
3.13	[3.12,3.14)	17	0.17	0.17/.02 = 8.5
3.15	[3.14,3.16)	13	0.13	0.13/.02 = 6.5
3.17	[3.16,3.18)	6	0.06	0.06/.02 = 3
3.19	[3.18,3.20)	2	0.02	0.02 /.02 = 1
3.21	[3.20,3.22)	1	0.01	0.02/.02 = 1

b.
Relative frequency histogram for weight of pennies:

Relative Frequencies (in hundredths)

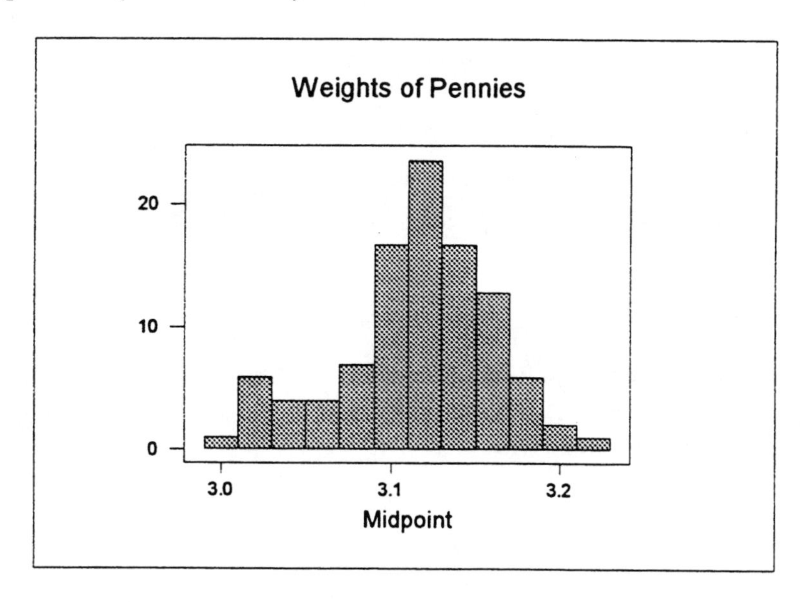

The configuration of the density histogram would look like this relative frequency histogram since the intervals are of equal widths.

2.17.

a.

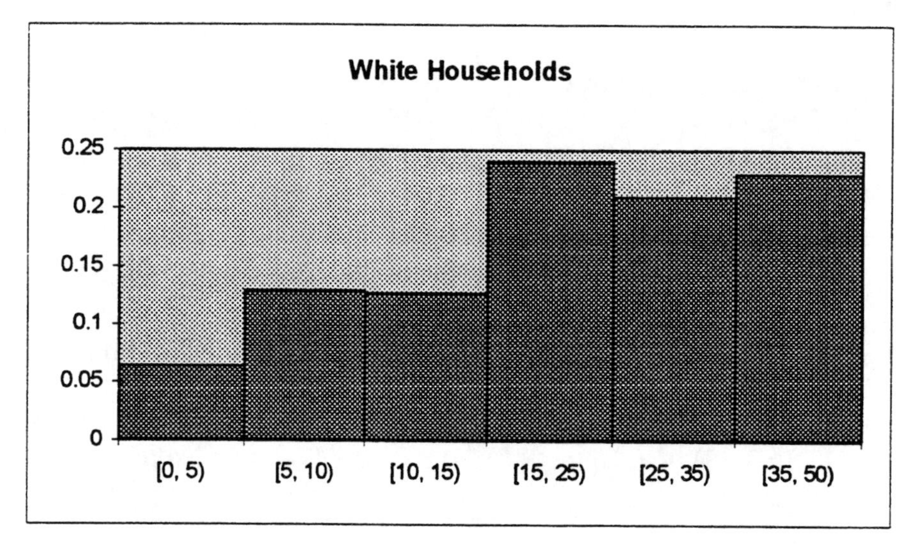

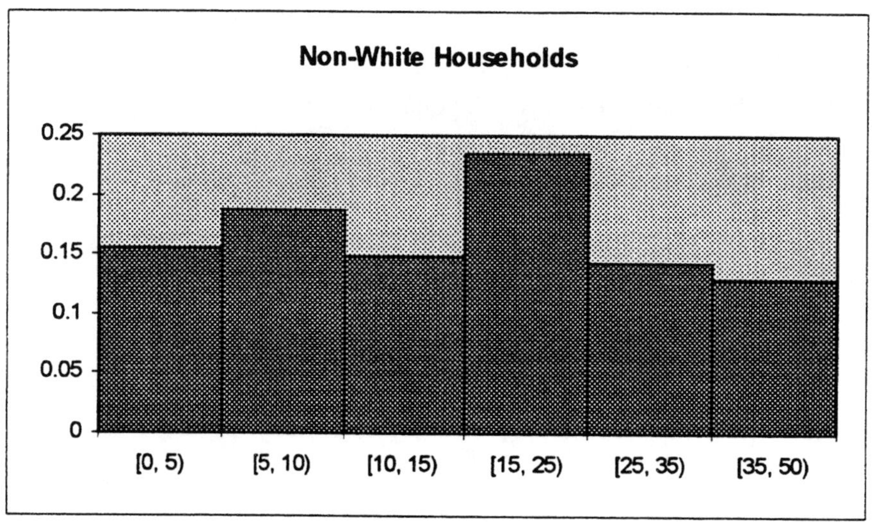

White household incomes are larger than non-white household incomes. Non-white household incomes have larger relative frequencies in the smaller income ranges as compared to white household incomes.

b.

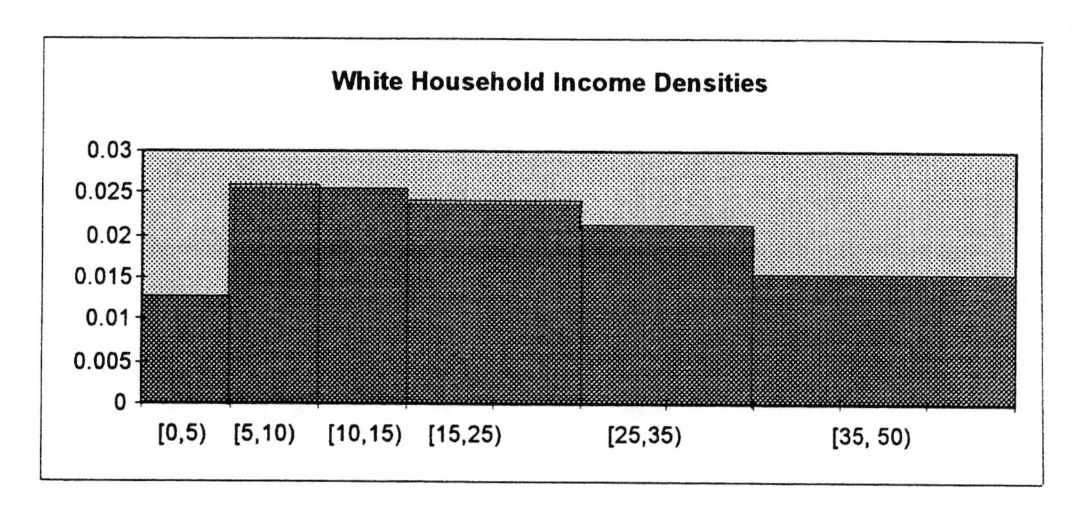

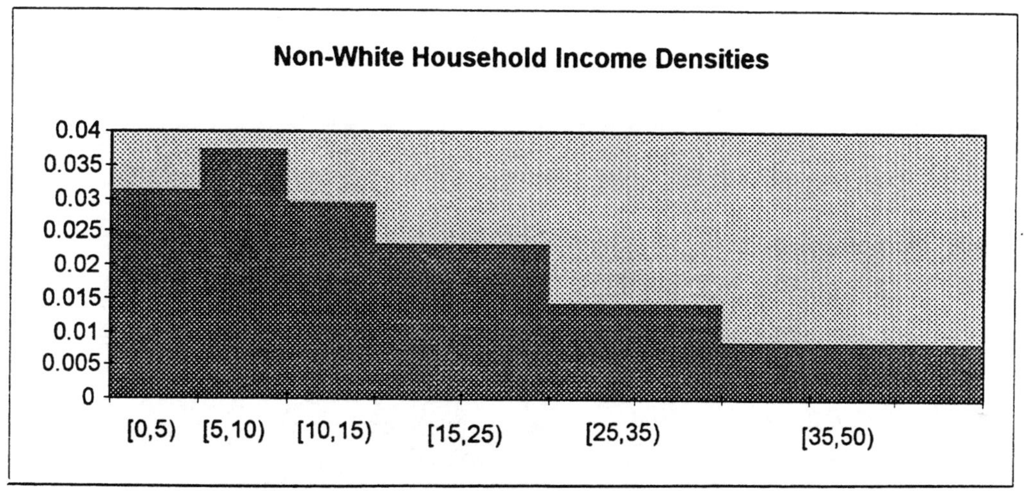

White household incomes are higher than non-white household incomes based on these density histograms. It is much easier to interpret the distribution of incomes using the density histograms when the intervals are of uneven widths.

c. It better to use density histograms instead of relative frequency histograms when the intervals are of unequal width.

2.18.

Sample variation of {3,1,0,4} = 3.33318
Sample standard deviation = 1.8257
Range = 4.0000
Interquartile range: 3 - 1 = 2

2.19.

	Sample Mean	Variance	Standard deviation
a.	9	9.5	3.082
b.	25	10	3.162
c.	0.8	2.415	1.554

2.20.

Mean = 284.00
Median = 240.00

The median is a more appropriate measure of central tendency since there are a few very high salaries.

2.21.

Sorted data: 150 175 190 210 230 250 260 300 425 650
First quartile: np = 10 * .25 = 2.5; round up to 3rd number: 190
Median: Average 2 middle numbers: (230 + 250) / 2 = 240
Third quartile: np = 10 * .75 = 7.5; round up to 8th number: 300

2.22.

$\bar{x}$ = Mean; M = median

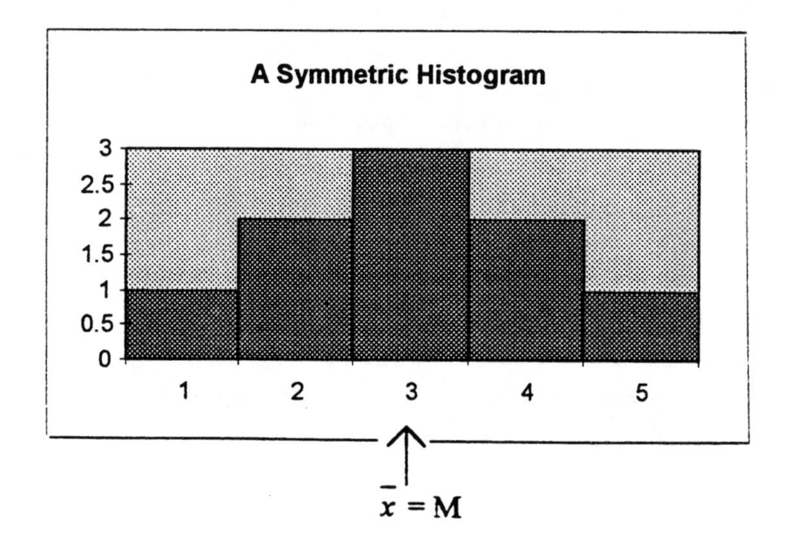

$\bar{x} = M$

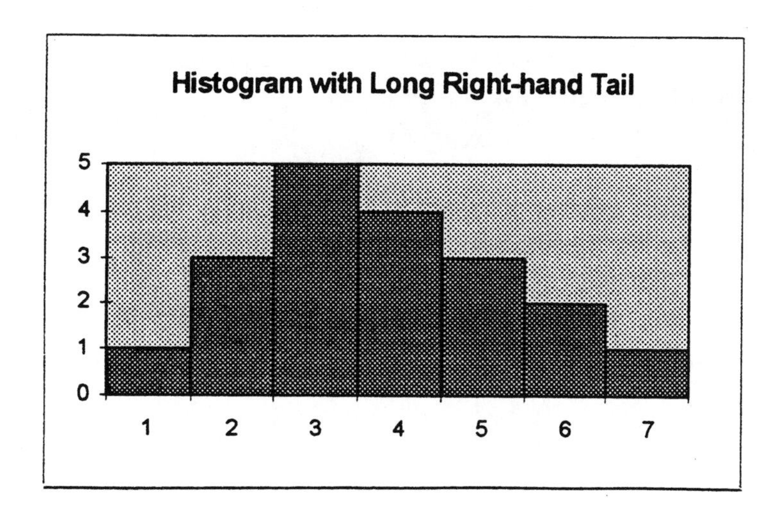

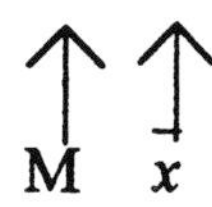

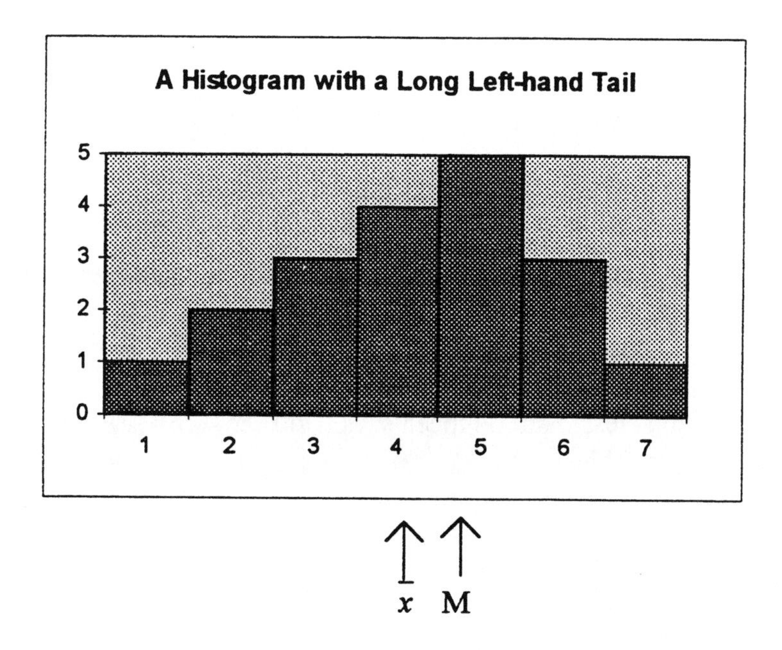

↑ ↑

$\bar{x}$ M

2.23.

a. Sample median = 3.2
First quartile = 0.25 (15 numbers) = 3.75; Round 3.75 up to 4.
Fourth element in the sorted list = 1.0
Third quartile = 0.75 (15 numbers) = 11.25; Round 11.25 up to 12.
The twelfth element in the sorted list = 7.6

b. Range = 27 - 0.1 = 26.9
Interquartile range = 7.6 - 1.0 = 6.6

c. 90th percentile = .90 (15 numbers) = 13.5; Round 13.5 up to 14.
14th element of the sorted list = 14.9

2.24.

a. Company A offers a better prospect to a machinist having superior ability since the mean being greater than the median indicates that there are a few high salaries for the best employees.

b. A medium quality machinist can expect to earn more with company B since the median salary is higher.

2.25.

a.

Dot diagram of workers per vehicle:

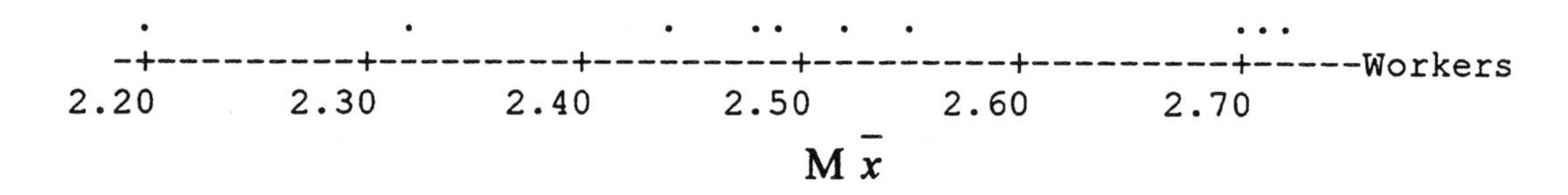

Mean of Workers = $\bar{x}$ = 2.513

b. The 5% trimmed mean equals the mean without the first and last numbers
= 2.5262

c. Median of Workers = M = 2.505
The fact that the median is less than the mean says that there are several large numbers that make the mean larger. The data is skewed to the right; there is a long tail to the right.

2.26.

a. 5% trimmed mean = 3.55

b. Mean of R&D = 4.1917 is much larger than the trimmed mean due to the one comparatively large number.

c. Median of R&D = 3.8000

d. The outlier has an effect on the sample mean (making it larger), but not on the trimmed mean or the median.

2.27.

a. Mean of Claims = 16,653.23
5% trimmed mean = 12,094.44

b. First quartile = 31 (.25) = 7.75; Round 7.75 up to 8. The eighth element is 4,375.
Third quartile = 31 (.75) = 23.25; Round 23.25 up to 24. The 24th element is 13,125.
Interquartile range = 13,125 - 4,375 = 8,750.

c. Boxplot:

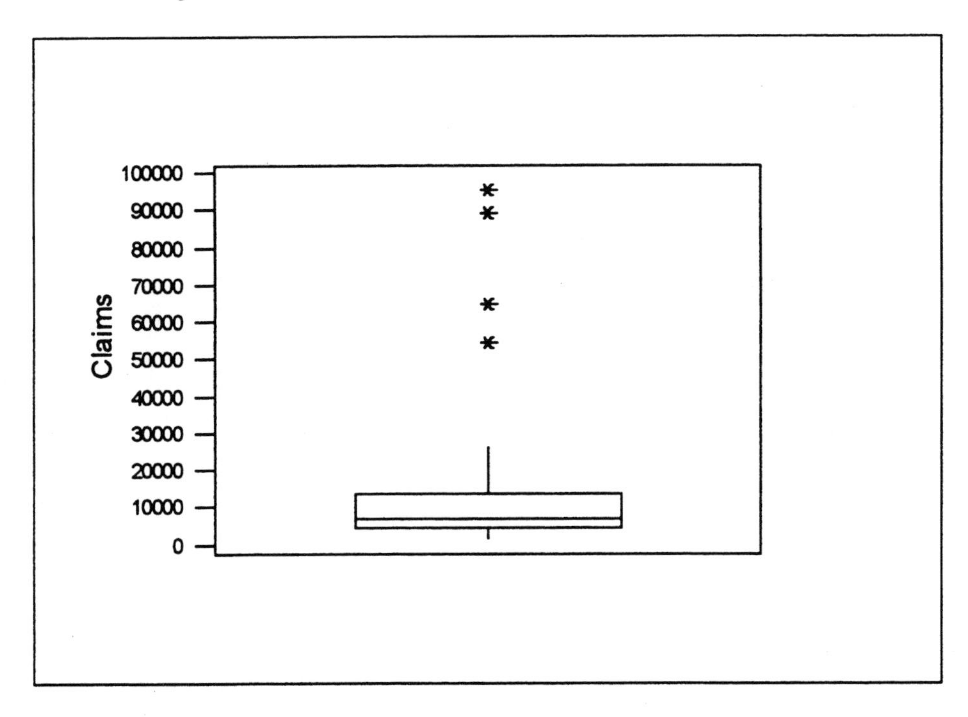

2.28. Sample variance using the computing formula = (.3333) [3* 1,000,000,000,000 + 1,000,002,000,000 - (.25)*16,000,008,000,000] =0

Sample variance using the definitional formula = 0.25

Yes, inaccurate results were obtained using the computing formula due to round off error on the calculator.

2.29.

a. The boxplot for education shows these values for GRE quantitative score:

Summary Number	Value
Minimum	375
First Quartile	450
Median	495
Third Quartile	525
Maximum	600

b. The highest quantitative scores are for the natural science department followed by the engineering department. The departments with the most highly concentrated quantitative scores about the median are the engineering department followed by the humanities and arts department. The departments with the largest range of scores are the natural sciences department followed by the social sciences department.

c. The distribution for all departments is symmetric since the box is centered over the entire range and the median is in the center of the box. The engineering department is the most heavily skewed; it is skewed to the left. The humanities and arts department is skewed to the right.

2.30.

a. The mean of a data set with the outliers eliminated will be **smaller or larger than** the average of the data set with the outliers included.

b. The standard deviation of a data set with the outliers eliminated will be **smaller than** the standard deviation of the data set with the outliers included.

c. The median of a data set with the outliers eliminated will be **smaller or larger than** the median of the data set with the outliers included.

2.31.

a. Stem-and-Leaf Display

Leaf Unit = 1.0

```
  2  4 23
  2  4
  5  4 667
  8  4 899
 14  5 001111
 16  5 22
(8)  5 44445555
 18  5 6667777
 11  5 8
 10  6 0111
  6  6 2
  5  6 445
  2  6
  2  6 89
```

b. Median of Age = 55.

First quartile: .25(42) = 10.5; Round up to 11. Eleventh element is 51.

Third quartile: .75(42) = 31.5; Round up to 32. 32nd element is 58.

2.32.

a. Stem-and-Leaf Display

Leaf Unit = 0.10

```
   1    3 8
   2    4 7
   8    5 112666
  25    6 01244455567889999
(15)    7 011345566677899
  20    8 01237788
  12    9 24
  10   10 23358
   5   11 26
   3   12 08
   1   13
   1   14
   1   15 8
```

b. The cities with an unusually large percentage of workers with graduate degrees are: Washington, D.C., San Francisco, and San Jose.

2.33.

Back-to-back stem-and-leaf diagram of the first 10 cities as compared to the last 50 cities:

```
First 10    Leaf   Last 50
               3 8
               4 7
               5 112666
               6 01244455567889999
97             7 0113455666789
873            8 01278
               9 24
53            10 238
62            11
0             12 8
              13
              14
              15 8
```

There is a difference between the two groups of cities with respect to the percent of workers with graduate degrees, the average being higher in the first 10 cities.

2.34.

	z value	Area under the standard normal curve to the left of z
a.	1.16	0.8770
b.	.24	0.5948
c.	-.57	0.2843
d.	-2.1	0.0179

2.35.

	z value	Area under the standard normal curve to the left of z
a.	.77	0.7794
b.	1.68	0.9535
c.	-.21	0.4168
d.	-1.39	0.0823

2.36.

	z value	Area under the standard normal curve to the right of z
a.	.84	0.2005
b.	2.25	0.0122
c.	-.10	0.5398
d.	-1.60	0.9446

2.37.

	z value	Area under the standard normal curve to the right of z
a.	.21	0.4168
b.	2.03	0.0212
c.	-.67	0.7486
d.	-1.115	0.8676

2.38.

	interval	Area under the standard normal curve over the given interval
a.	0 to .37	0.1443
b.	-.42 to 1.06	0.5182
c.	-1.62 to .09	0.4832
d.	.25 to 1.97	0.3766

2.39.

	interval	Area under the standard normal curve over the given interval
a.	-2.07 to .04	0.4968
b.	-1.12 to -.35	0.2318
c.	-.77 to 0	0.2794
d.	.69 to 1.893	0.2159

2.40.

a. The area to the left of z is 0.2643, so z = -0.63.
b. The area to the right of z is 0.20, so z =0.84.
c. The area between 0 and z is 0.35. The area to the left of 0 is 0.5, so the area to the left of z is 0.5 - 0.35 = 0.15, giving a z value of 1.04.
d. The area between -1 and z is 0.756. From the table, the area to the left of -1 is 0.1587, so the area to the left of z is 0.756 + 0.1587 = 0.9147, giving a z value from the table of 1.37.
e. The area between - z and z is 0.6528, so the area between 0 and z is 0.3264. Therefore the area to the left of z is 0.3264 + 0.5 = 0.8264, giving a z value of 0.94.
f. The area between z and 1.82 is 0.59. The area to the right of 1.82 is 0.0344. The area to the right of z is 0.59 + 0.0344 = 0.6244, so z = -0.317.

2.41.

a. The probability of being to the left of z is 0.20, so z = -0.8418.
b. The probability of being to the right of z is 0.125, so z = 1.1503.
c. The probability of being between z and - z is 0.668, so the probability of being between 0 and z is 0.668/2 =0 .334. The probability of being to the left of z is 0.334 + 0.5 = 0.834, so z = 0.97.
d. The probability of being between z and 2.0 is 0.888. The probability of being to the right of 2.0 is 0.0228, so the probability of being to the right of z is 0.888 + 0.0228 = 0.9108. z = -1.3457.

2.42.

Quartile	Probability of being to the left of z	z
First Quartile	0.25	-0.67449
Median = Second Quartile	0	0
Third Quartile	0.75	0.67449

2.43.

a. The area under the standard normal to the left of z = 0.35 is 0.6368.
b. The 35th percentile of the standard normal distribution is -0.3853.
c. The area under the standard normal to the left of z = 0.6 is 0.7257.
d. The 60th percentile of the standard normal distribution is 0.2533.

2.44.

a. The area under the standard normal to the left of $z = 0.15$ is 0.5596.
b. The 15th percentile of the standard normal distribution is -1.0364.
c. The area under the standard normal to the left of $z = 0.99$ is 0.8389.
d. The 99th percentile of the standard normal distribution is 2.3263.

2.45. For x distributed as N(50, 20), in general, $y = a + bx$ is distributed with mean = $a + b\mu$ and standard deviation = $|b|\sigma$.
a. If $y = 10 + 3x$, the mean = 10 + 3(50) = 160 and the standard deviation = 3(20) = 60.
b. If $y = -50 + 0.2x$, the mean = -50 + 0.2(50) = -40 and the standard deviation = 0.2(20) = 4.
c. If $y = (x-25)/2 = -12.5 + 0.5x$, the mean = -12.5 + 0.5(50) = 12.5 and the standard deviation = 0.5(20) = 10.
d. If $y = -4x$, the mean = 0 + -4(50) = -200 and the standard deviation = 4(20) = 80.

2.46. Using the results of Exercise 1.21 of chapter 1,

$$\text{Sample Mean} = \bar{z} = \frac{\sum z_i}{n} = \frac{\sum \frac{(x_i - \bar{x})}{s}}{n} = \frac{0}{n} = 0$$

Sample standard deviation =

$$\sqrt{\frac{\sum z_i^2 - \frac{\left(\sum z_i\right)^2}{n}}{n-1}} = \sqrt{\frac{\sum\left(\frac{x_i - \bar{x}}{s}\right)^2 - \frac{0^2}{n}}{n-1}} = \sqrt{\frac{\sum \frac{\left(x_i - \bar{x}\right)^2}{s^2}}{n-1}} = \frac{1}{s}\sqrt{\frac{\sum\left(x_i - \bar{x}\right)^2}{n-1}} = \frac{1}{s}\sqrt{s^2} = 1$$

2.47.

	x	z	Area
a.	43	(43-40)/3=1	Area to the left of x = 0.8413
b.	45	(45-40)/3=1.667	Area to the left of x = 0.9522
c.	38	(38-40)/3=-0.6667	Area to the left of x = 0.7475
d.	40	(40-40)/3=0	Area to the right of x = 0.5
e.	36	(36-40)/3=-1.333	Area to the right of x =0.9087
f.	37 to 41	(37-40)/3=-1 (41-40)/3=.3333	Area over the interval = 0.4719

2.48.

	x	z	Area
a.	16.5	(16.5-15)/6= 0.25	Area to the left of x = 0.5987
b.	7	(7-15)/6= -1.3333	Area to the left of x = 0.0912
c.	22	(22-15)/6= 1.1667	Area to the right of x = 0.1217
d.	11	(11-15)/6= -0.6667	Area to the right of x = 0.7475
e.	17 to 27	(17-15)/6= 0.3333 (27-15)/6= 2	Area over the interval = 0.3467
f.	-1 to 19	(-1-15)/6= -2.6667 (19-15)/6= 0.6667	Area over the interval = 0.7437

2.49.

	Area		c
a.	0.8461	to left of c	204.0794
b.	0.5897	to right of c	199.0929
c.	0.0116	to left of c	190.9195
d.	0.2297	to right of c	202.9593

2.50.

	Area		c
a.	0.7995	to left of c	-8.3203
b.	0.9429	to right of c	-13.1592
c.	0.6826	Over the interval from c to -c	-9.30165 to -10.6983
d.	0.9544	Over the interval from c to -c	-9.00331 to -10.9967

2.51.

a. Probability(> 25) = 0.09
b. Probability(< 20) = 0.37
c. Probability of a score being between 19 and 27 = 0.9772 - 0.2514 = 0.7258

2.52.

a. Probability(score over 24) = 0.84
b. Probability is 50% at the mean = 21.
c. Probability of being greater than x is 20%; x = 24.

2.53.

a. Estimating the number of two day letters that arrive within 48 hours is analytical.
b. Knowing the number of employees who were absent last month is enumerative.

2.54.

a. The variable of interest is the estimated number of two day letters, which is quantitative and discrete.

b. The variable of interest is the number of employees who were absent, which is quantitative and discrete.

2.55.

a.

Frequency	Relative Frequency	Density
8	0.133	0.013
17	0.283	0.028
14	0.233	0.023
40	0.667	0.067
11	0.183	0.009

b.

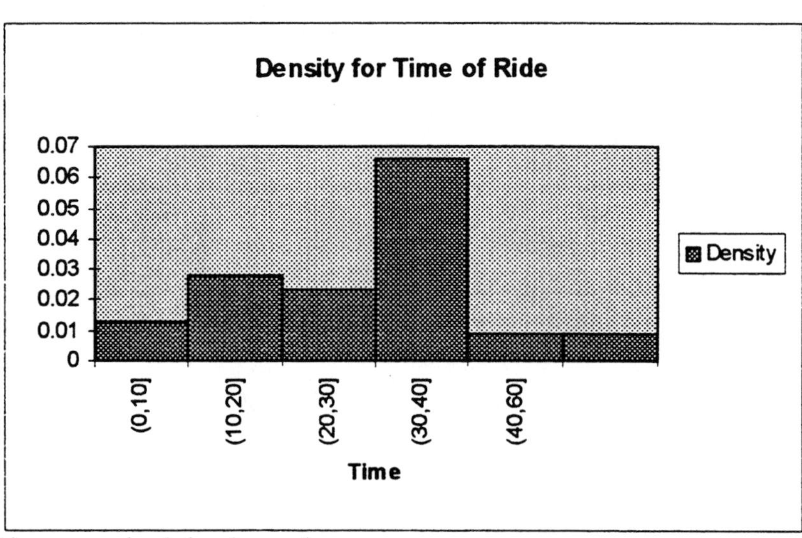

The density histogram is right skewed.

2.56.

The mean is the average of all numbers, not just the highest and lowest numbers. Only if the data is symmetrical will the mean be halfway between the highest and lowest amounts.

2.57.

a. Sample mean = 2; median = 3.

b. s^2 using the definitional formula = $(2.828)^2 = 8$

c. s^2 using the computing formula = 1/(4-1)(40-.25*64) = 8

2.58.

a. Sample mean = 13.4

b. Sample first quartile = .25(10) = 2.5; round up to 3; result is 8.
Sample third quartile = .75(10) = 7.5; round up to 8; result is 18.

c. Range = 25 - 3 = 22.

d. Interquartile range = 18-8 = 10.

2.59.

Answers vary depending on the data collected.

2.60.

a. Mean = 0.56426
Standard deviation = 0.083920

b.

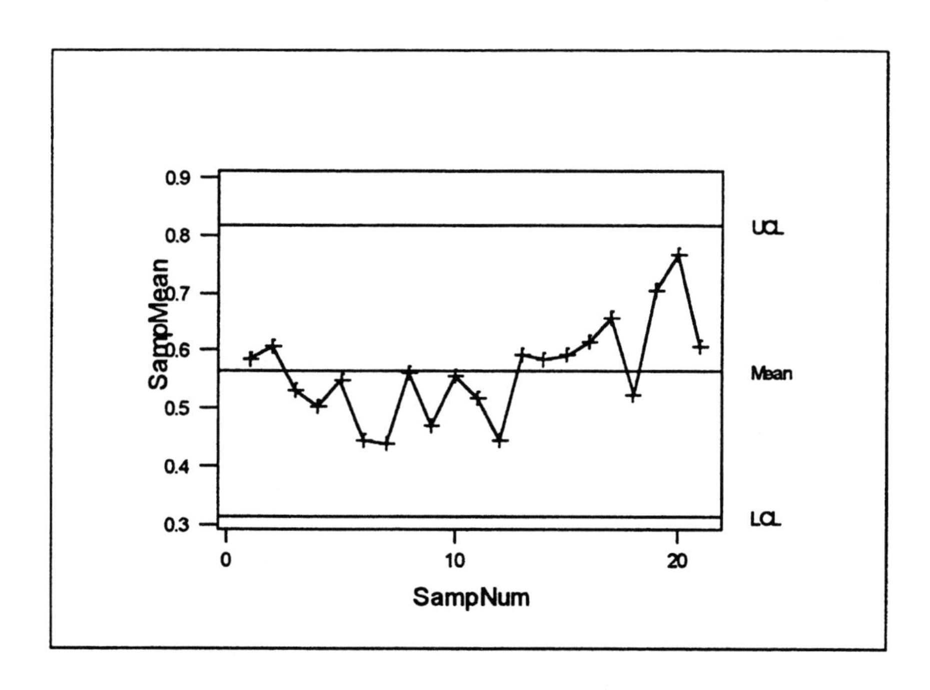

c. The radiation measurements are not stable over time; they seem to be increasing. All transformed measurements are within three standard deviations of the mean. Higher recent measurements indicate that there may be a problem in recently manufactured ovens. Additional monitoring may be required.

2.61.

a.

N = 42

Mean	=	0.16381
Median	=	0.11000
Standard deviation	=	0.12724
Minimum	=	0.010000
Maximum	=	0.60000
First quartile	=	0.09

Third quartile = 0.25

Stem-and-leaf N = 42
Leaf Unit = 0.010

```
   3     0 114
  12     0 555799999
 (15)    1 000000000222222
  15     1 55
  13     2 00
  11     2 58
   9     3 000023
   3     3
   3     4 0
   2     4 5
   1     5
   1     5
   1     6 0
```

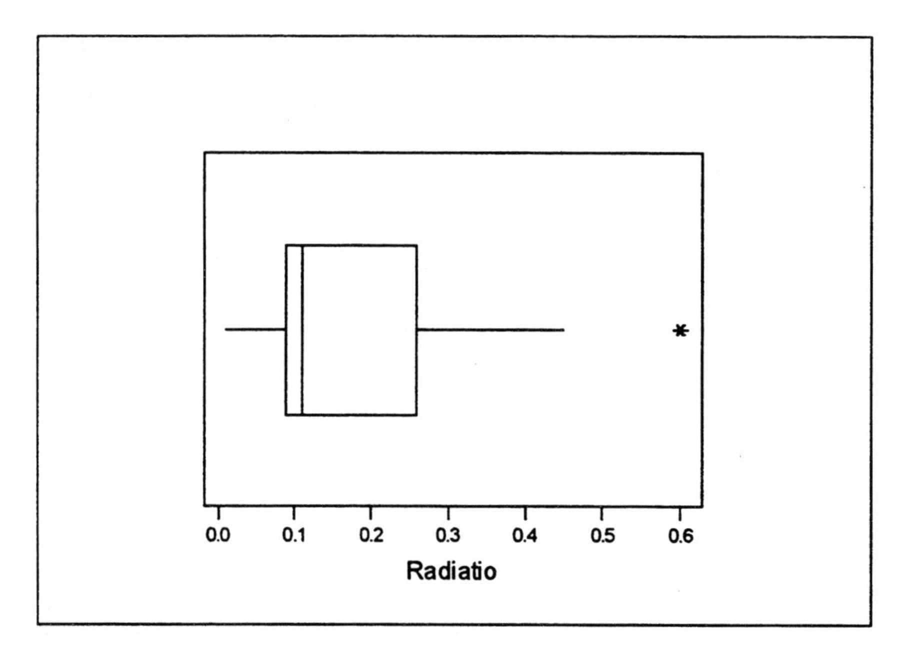

These measurements are not normal.

b.

Mean	=	0.60298
Median	=	0.57545
Standard deviation	=	0.12060
Minimum	=	0.31623
Maximum	=	0.88011
First quartile	=	0.547723
Third quartile	=	0.707107

Stem-and-leaf N = 42
Leaf Unit = 0.010

```
   2   3 11
   2   3
   3   4 4
   6   4 777
  12   5 144444
 (15)  5 666666666888888
  15   6 22
  13   6 66
  11   7 024444
   5   7 559
   2   8 1
   1   8 8
```

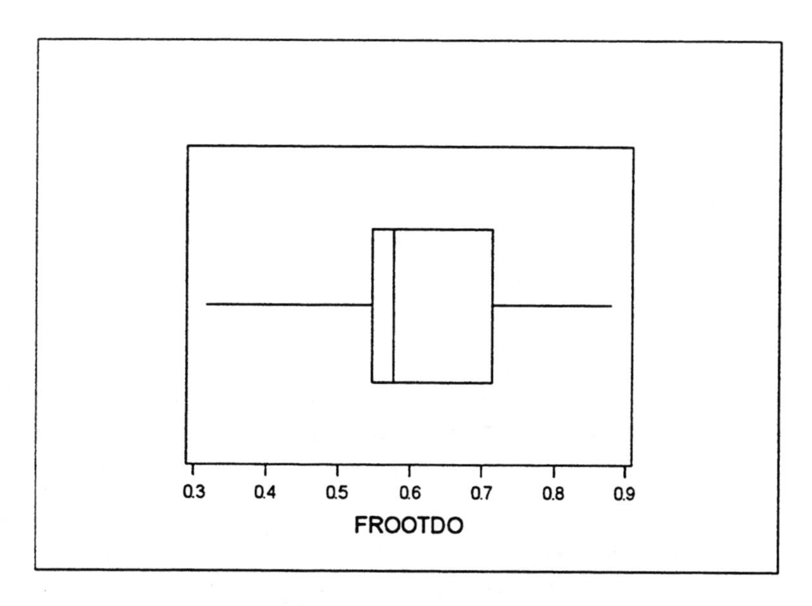

These measurements are not strictly speaking normal since the mean is not equal to the median. The measurements are not symmetric. However, the transformed measurements are more nearly normal than the original data.

2.62.

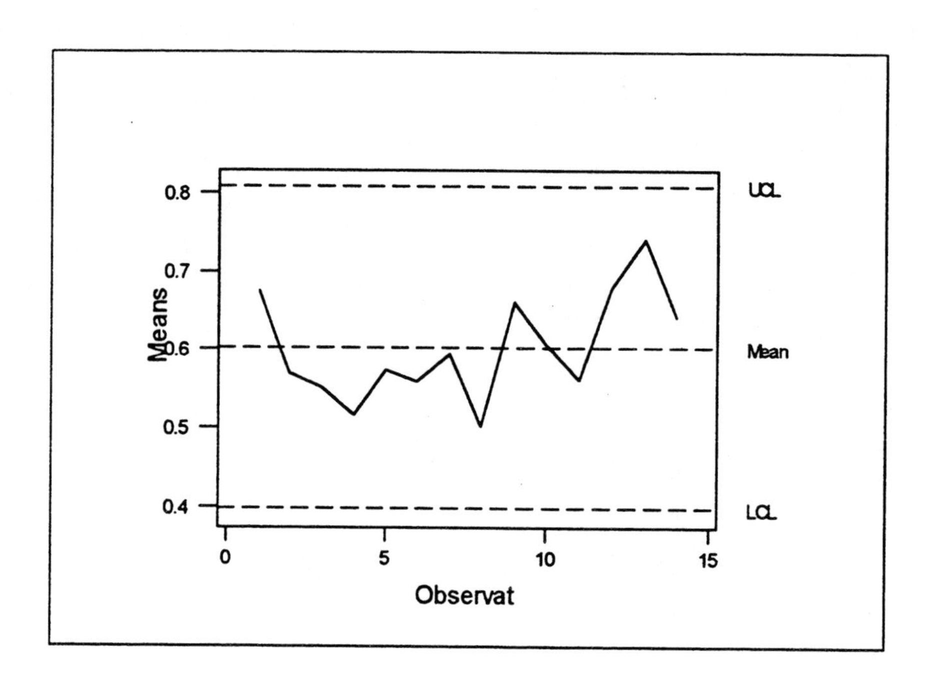

2.63.

a. Yes, the X-bar chart in Exercise 2.62 is similar in appearance to the one in Figure 2.21 in that there is a drift in the open door radiation measurements over time with the older measurements below the center line and the more recent measurements above. Yes, the two charts would be expected to be similar. If the observation pairs are not from the same ovens, the charts are not as likely to be similar.

b. In general, the radiation measurements through the open doors of the ovens should be larger than the corresponding measurements through closed doors. The two X-bar charts are consistent with this answer. The variation is slightly larger for the open door measurements.

2.64.

a.

Stem-and-leaf N = 60
Leaf Unit = 1.0

```
1 33
1 777777888888899999999
2 0001111222222233333344
2 55566668999
3 0124
3 7
```

b. The cities with an unusually high percentage of knowledge workers are: Washington, D.C., San Francisco, San Jose and Raleigh/Durham.
The cities with an unusually low percentage of knowledge workers are: Las Vegas and Scranton.

2.65.

Back-to-Back Stem-and-leaf
Leaf Unit = 1.0

```
         1 33
         1
         1 777777
         1 8888888999999
         2 0001111
     2   2 222222333333
   554   2 45
     6   2 666
    98   2 99
    10   3
     2   3
         3 4
```

The first ten cities appear to have a larger percentage of workers with college degrees.

2.66.

a. Mean of %BA = 22.657
Standard deviation of %BA = 4.8690

b. XBAR - $2s$ = 22.657 - 2*4.8690 = 12.9190
XBAR + $2s$ = 22.657 + 2*4.8690 = 32.3950
57/60 = 95 % of data is within 2 standard deviations of the mean.
The empirical rule states that 95% of the data is within 2 standard deviations of the mean.

2.67.

a. The minimum percentage is 13.3.
The maximum percentage is 37.0.
The first quartile is 0.25 * 60 = 15. The fifteenth element is 18.8.
The third quartile is 0.75 * 60 = 45. The forty-fifth element is 25.0.

b. The interquartile range is 25.0 - 18.8 = 6.2.

c. Character Boxplot

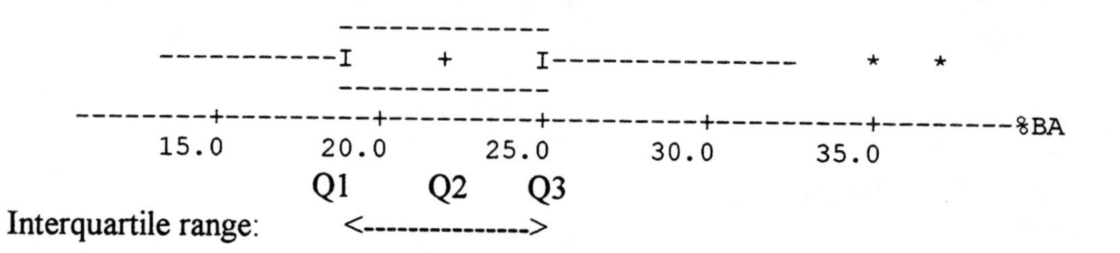

2.68. Answers vary depending on data collected.

2.69.

a.

Descriptive Statistics

```
Variable        N     Mean    Median    TrMean    StDev    SEMean
OpIncome      102    5.480     5.100     5.097    7.250     0.718

Variable      Min      Max       Q1        Q3
OpIncome  -11.400   30.400    0.275     9.000
```

Character Stem-and-Leaf Display

```
Stem-and-leaf of OpIncome   N  = 102
Leaf Unit = 1.0


     1    -1 1
     3    -0 77
    23    -0 44444332211111111000
    49     0 00000111222233333334444444
  (29)     0 55555555566666666667777778889
    24     1 0000001112223444
     8     1 5568
     4     2 0
     3     2 6
     2     3 00
```

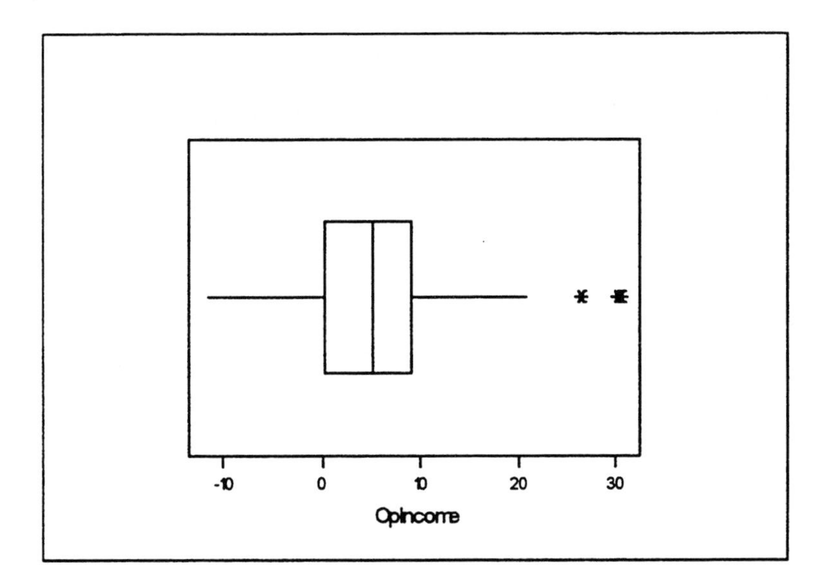

These data are not normal and are skewed to the right. Since the mean is greater than the median and the data has a long right tail, it is skewed to the right.

b.

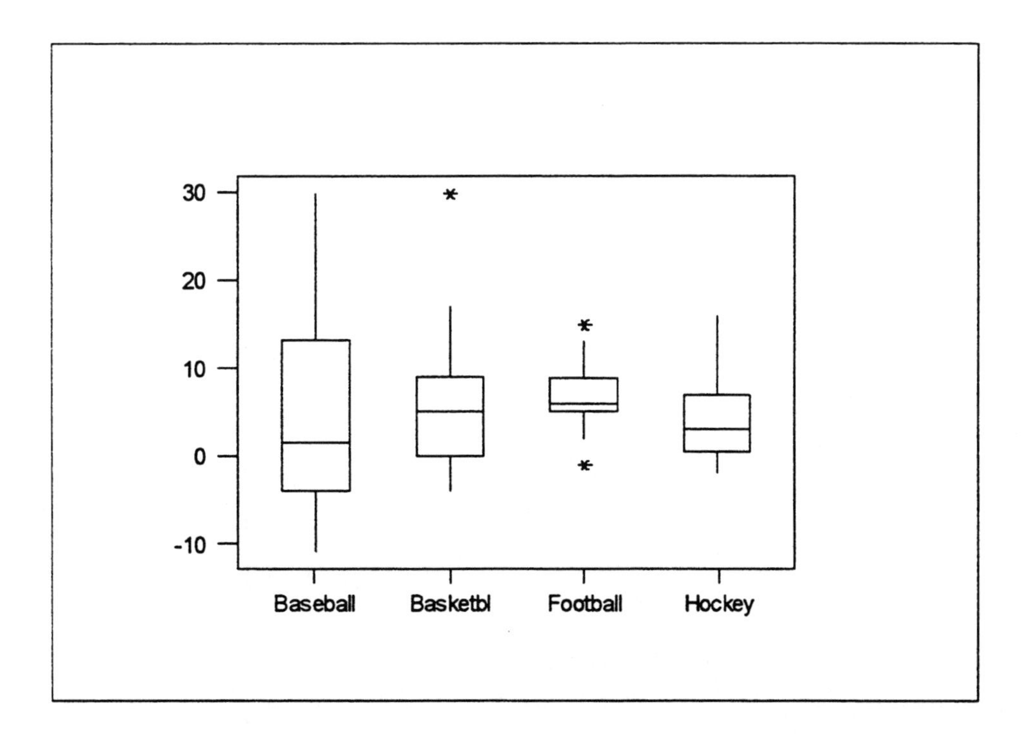

There are outliers for basketball and football only, as identified by the asterisks. Baseball has the largest range as compared to the other sports; football has the smallest range. Also, baseball has the smallest mean; football has the largest mean.

c.
As shown with the following information, the mean is larger than the median for all sports. Therefore, they all have right tails; they are skewed to the right.

Baseball:
N = 26
Mean of Baseball = 4.9231
Median of Baseball = 1.5000
Standard deviation of Baseball = 10.976
Minimum of Baseball = -11.000
Maximum of Baseball = 30.000
First quartile: 26 * 0.25 = 6.5; round up to 7th element: -4
Third quartile: 26 * 0.75 = 19.5; round up to 20th element: 13

Stem-and-leaf of Baseball N = 26
Leaf Unit = 1.0

```
  1    -1 1
  4    -0 875
 12    -0 44443210
(2)     0 03
 12     0 577
  9     1 1134
  5     1 58
  3     2 1
  2     2 6
  1     3 0
```

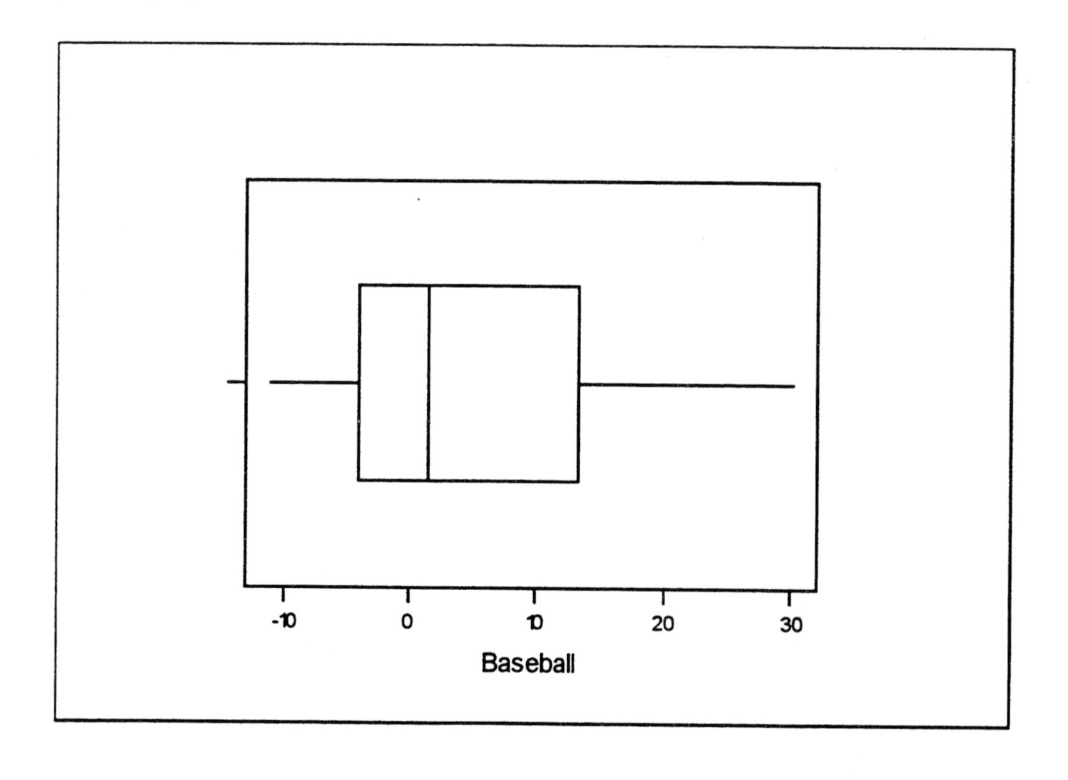

Basketball:

Total number of observations in Basketbl= 27
Mean of Basketbl= 5.7778
Median of Basketbl= 5.0000
Standard deviation of Basketbl= 7.4438
Minimum of Basketbl= -4.0000
Maximum of Basketbl= 30.000
First quartile: 27 * 0.25 = 6.75000; round up to 7th element: 0
Third quartile: 27 * 0.75 = 20.2500; round up to 21st element: 9

Stem-and-leaf of Basketbl N = 27
Leaf Unit = 1.0

```
  6  -0 433111
 13   0 0133334
(8)   0 55566899
  6   1 034
  3   1 57
  1   2
  1   2
  1   3 0
```

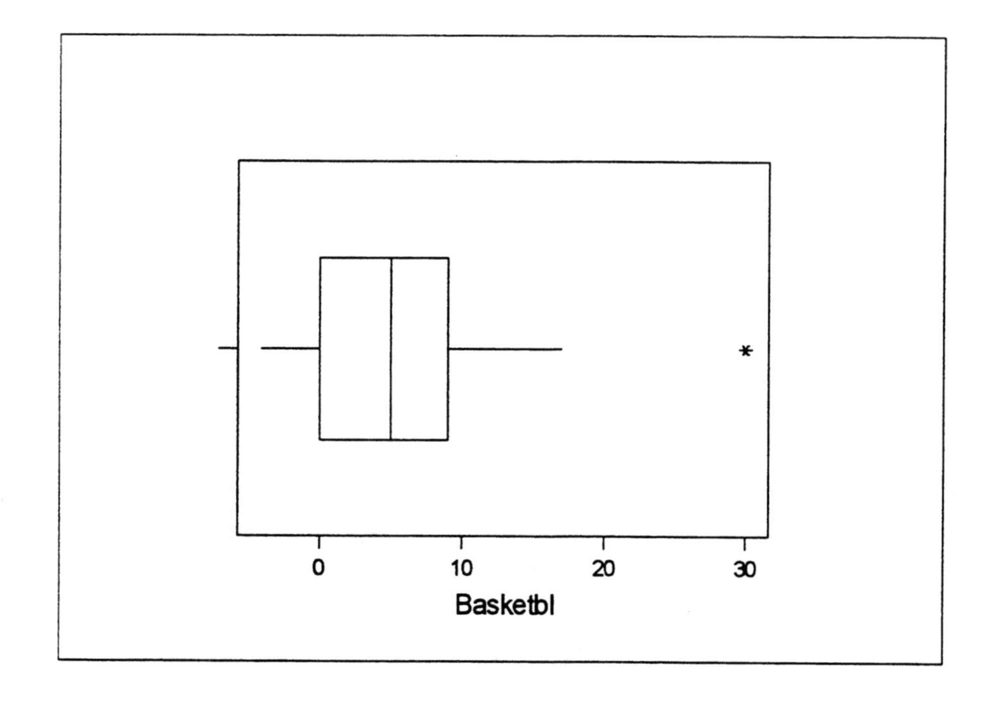

Football:

Total number of observations in Football = 28
Mean of Football = 6.7857
Median of Football = 6.0000
Standard deviation of Football = 3.5209
Minimum of Football = -1.0000
Maximum of Football = 15.000
First quartile: 28 * 0.25 = 7th element: 5
Third quartile: 28 * 0.75 = 821st element: 8

Character Stem-and-Leaf Display

Stem-and-leaf of Football N = 28
Leaf Unit = 1.0

```
  1    -0 1
  1     0
  4     0 223
 10     0 445555
 (7)    0 6666677
 11     0 88889
  6     1 001
  3     1 23
  1     1 5
```

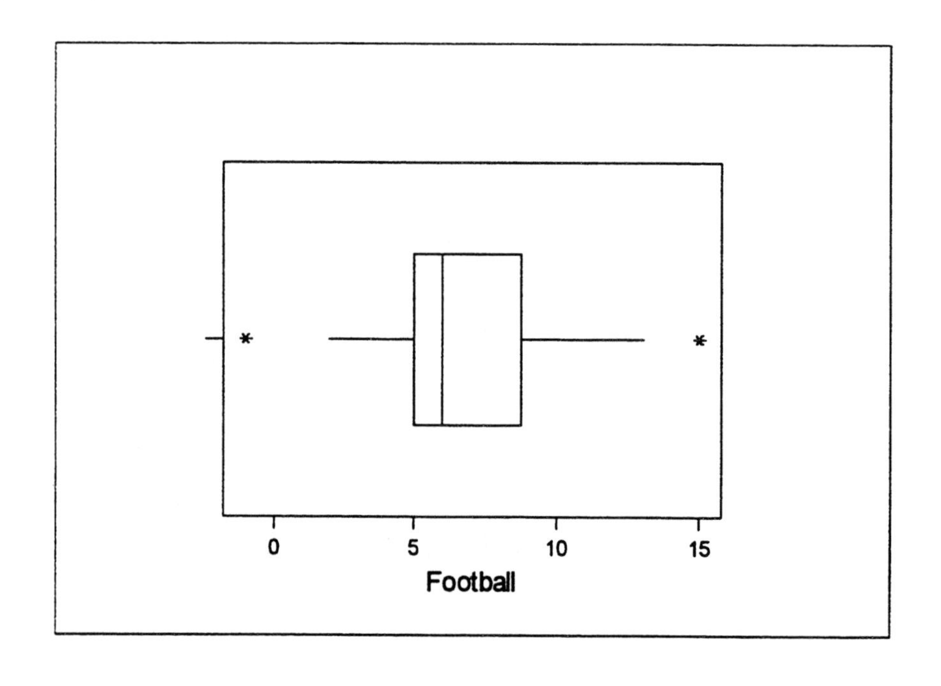

Hockey:

Total number of observations in Hockey = 21
Mean of Hockey = 4.1905
Median of Hockey = 3.0000
Standard deviation of Hockey = 4.7394
Minimum of Hockey = -2.0000
Maximum of Hockey = 16.000
First quartile: 21 * 0.25 = 5.25000; round up to 6th element: 1
Third quartile: 21 * 0.75 = 15.7500; round up to 16th element: 7

Stem-and-leaf of Hockey N = 21
Leaf Unit = 1.0

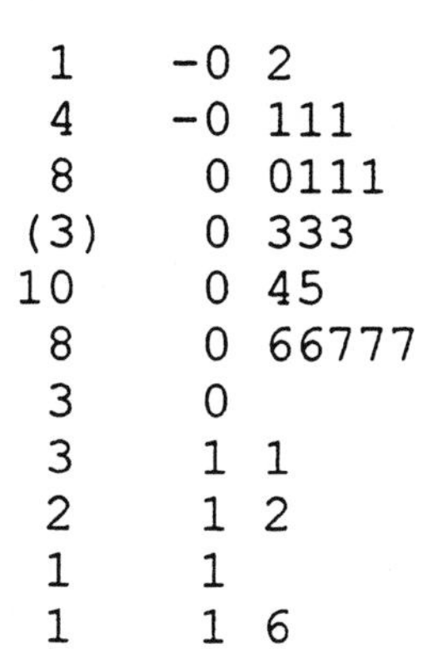

```
  1    -0 2
  4    -0 111
  8     0 0111
 (3)    0 333
 10     0 45
  8     0 66777
  3     0
  3     1 1
  2     1 2
  1     1
  1     1 6
```

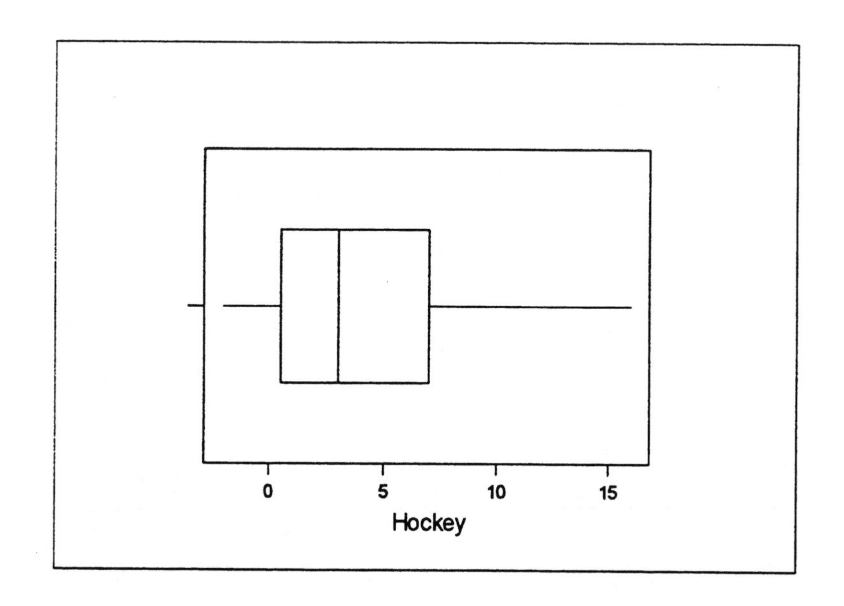

2.70.
Descriptive Statistics

```
Variable          N      Mean    Median    TrMean     StDev    SEMean
GurAcc10        182     11293     11329     11312      1457       108

Variable        Min       Max        Q1        Q3
GurAcc10       7172     17059     10543     12199
```

Character Stem-and-Leaf Display

```
Stem-and-leaf of GurAcc10  N  = 182
Leaf Unit = 100

    5    7 13347
   13    8 55578899
   31    9 012233556667888999
   70   10 011122333334455555566666677788889999999
  (53)  11 00000000000122222222333334444445666677777788888888999
   59   12 000000000111112222234444556666677777888889999
   14   13 0002335666
    4   14 258
    1   15
    1   16
    1   17 0
```

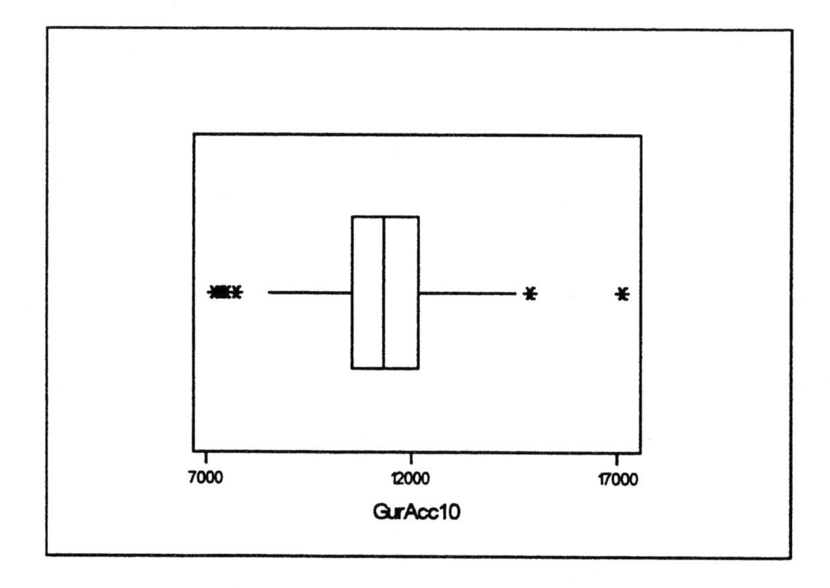

Since the mean is slightly less than the median, the data has a left-tail; it is skewed slightly to the left.

3.1.

a. Since the numbers in both import cars and business school graduates are both increasing, there would be a positive correlation between the numbers, even though the categories may not otherwise be related.

b. The number of luxury car sales and the number of certified public accountants for cities of differing sizes in a given year would have a positive correlation since more CPAs would be needed for a city with higher individual incomes and higher income individuals, including CPAs, would be more likely to buy luxury cars.

c. The number of sales persons and the dollar amount of real estate sold in a year would give a positive correlation since the larger sales force should result in higher sales.

d. The number of felonies and the number of persons in upper management positions may give a positive or negative correlation. If more individuals in upper management positions also means that the population of the city is higher, then more felonies might also be involved.

3.2.

a. As the population increases, so also would import car sales (and domestic car sales). Business graduates increase in numbers over the years since the population increases and also since more people are seeking college degrees in this decade as compared to previous decades. The lurking variable is the population growth.

b. The lurking variable is the income level in the city since individuals with higher incomes hire more CPAs and are more likely to buy luxury cars.

c. The number of sales people causes a higher sale of real estate. This is a causal effect.

d. A lurking variable of population size of the city can effect the number of felonies and the number of upper management positions. The upper management positions can also vary from city to city. Upper management positions could be decreasing as a result of downsizing but, in general, the larger the city, the larger the upper management positions.

3.3.

a.

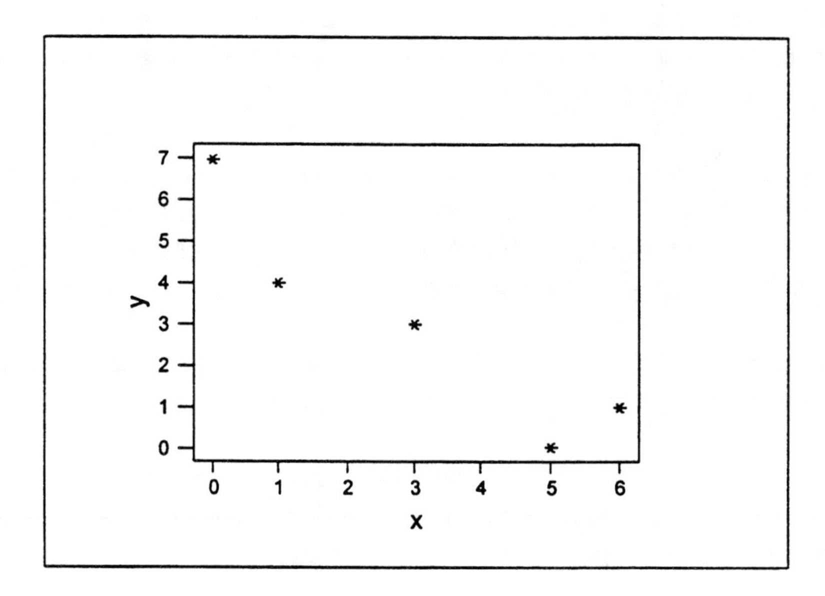

b. The correlation coefficient is negative. My guess: -0.9. (Answers will vary.)

c.

x	y	x-xbar	y-ybar	$(x\text{-}xbar)^2$	$(y\text{-}ybar)^2$	c3*c4
0	7	-3	4	9	16	-12
1	4	-2	1	4	1	-2
6	1	3	-2	9	4	-6
3	3	0	0	0	0	0
5	0	2	-3	4	9	-6
Totals:						
15	15	0	0	26	30	-26
Means:						
3	3			S_{xx}	S_{yy}	S_{xy}
Square roots:				5.099	5.477	

r = -26/(5.099)(5.477) = -0.931

3.4.

a.

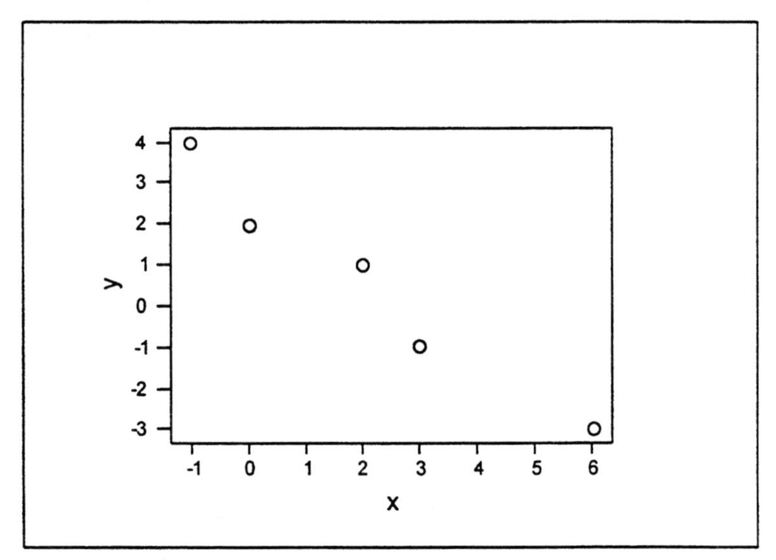

b.

My guess at the correlation coefficient: -0.9. (Answers may vary.)

c.

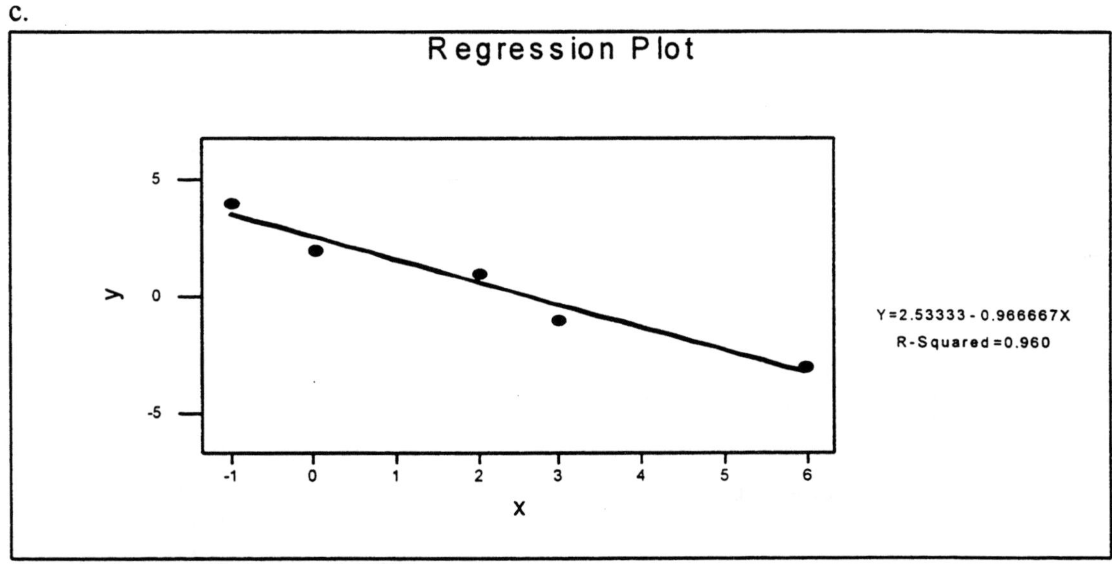

Correlation coefficient = -0.9798.

3.5.

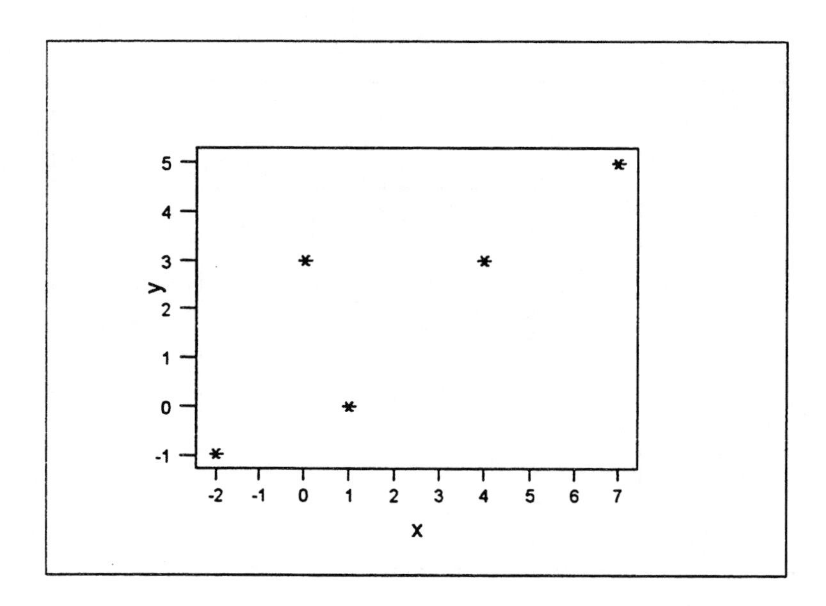

Correlation of x and y = 0.837

3.6.

a.

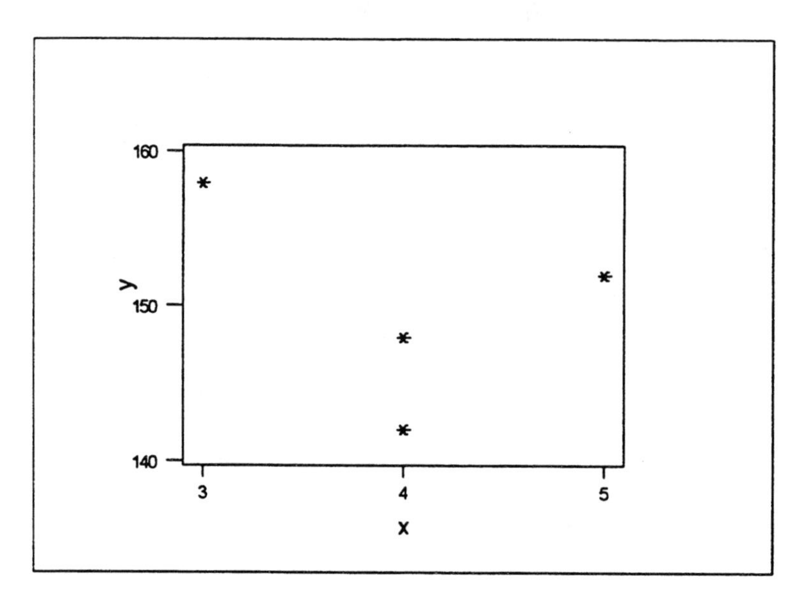

Correlation of x and y: r = -0.363

b. Correlation of x and y/100: r = -0.363

The result is expected since the correlation does not change with this linear transformation.

3.7.

a.

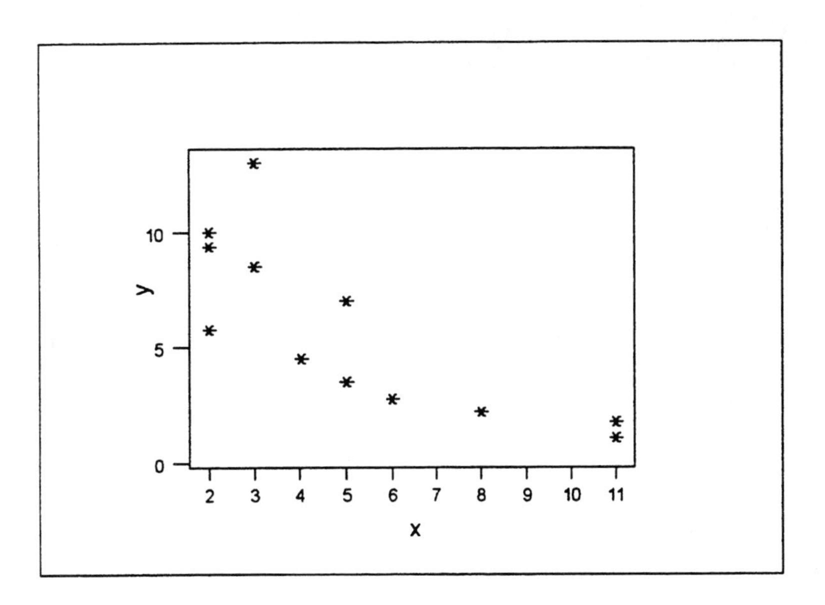

My guess at the correlation coefficient is -0.8

b.

Mean of x = 5.167
Mean of y = 5.800
Standard deviation of x = 3.271
Standard deviation of y = 3.792
Correlation of x and y = r = -0.791

3.8.

x	y	Xtilde	Ytilde	Product/11
2	10.0	-0.96822	1.10750	-0.097481
2	5.8	-0.96822	0.00000	-0.000000
11	1.8	1.78355	-1.05476	-0.171020
8	2.2	0.86630	-0.94928	-0.074760
6	2.8	0.25479	-0.79107	-0.018324
5	3.5	-0.05096	-0.60649	0.002810
2	9.4	-0.96822	0.94928	-0.083555
11	1.1	1.78355	-1.23934	-0.200948
4	4.5	-0.35671	-0.34280	0.011116
5	7.0	-0.05096	0.31643	-0.001466
3	8.5	-0.66246	0.71196	-0.042877
3	13.0	-0.66246	1.89857	-0.114339

Sum of (Prod/11) = -0.791 (Same answer as above.)

3.9.

Σx	Σy	Σ(x*y)	Σ(x²)	Σ(y²)
17.2	640	1296.25	38.54	55650

S_{xx} = 1.56
S_{yy} = 4450
S_{xy} = -79.75
r = -0.957

3.10. In order to show that linear transformations of x and y do not change the value of r if b and d have the same signs, first note that $Mean(\tilde{x}) = \frac{\sum \tilde{x}}{n} = \frac{\sum (a+bx)}{n}$ and

$$Mean(\tilde{y}) = \frac{\sum \tilde{y}}{n} = \frac{\sum (c+dy)}{n}$$

Also,

$$S_{\tilde{x}\tilde{y}} = \sum\left(a+bx-\sum\frac{(a+bx)}{n}\right)\left(c+dy-\sum\frac{c+dy}{n}\right)$$

$$=\sum\left(a+bx-\frac{na}{n}-b\sum\frac{b}{x}\right)\left(c+dy-\frac{nc}{n}-d\sum\frac{y}{n}\right)$$

$$=bd\sum(x-\bar{x})(y-\bar{y})$$

Similarly,

$$S_{\tilde{x}\tilde{x}} = \sum\left(a+bx-\sum\frac{a+bx}{n}\right)^2$$

$$=\sum\left(a+bx-\frac{na}{n}-b\sum\frac{x}{n}\right)^2$$

$$=b^2\sum(x-\bar{x})^2$$

Also,

$$S_{\tilde{y}\tilde{y}} = \sum\left(c+dy-\sum\frac{c+dy}{n}\right)^2$$

$$=\sum\left(c+dy-\frac{cn}{n}-d\sum\frac{y}{n}\right)^2$$

$$=d^2\sum(y-\bar{y})^2$$

Using the above equations,

$$r = \frac{S_{\bar{x}\bar{y}}}{\sqrt{S_{\bar{x}\bar{x}}}\sqrt{S_{\bar{y}\bar{y}}}} = \frac{bd\sum(x-\bar{x})(y-\bar{y})}{\sqrt{b^2d^2\sum(x-\bar{x})^2(y-\bar{y})^2}}$$

$$= \frac{bdS_{xy}}{|b||d|\sqrt{S_{xx}}\sqrt{S_{yy}}} = \frac{S_{xy}}{\sqrt{S_{xx}}\sqrt{S_{yy}}}$$

If b and d have opposite signs, r will be the negative of its original value.

3.11. The sign of the correlation coefficient is negative when there is an inverse relationship. The key to prosperity is not to reduce the number of central bankers per capita in order to increase the GDP growth rate. The lurking variable may be the financial ability of the nation to create large central banks which would mean a higher GNP and a smaller number of those banks.

3.12. Storks obviously do not bring babies. The lurking variable might be related to the positive growth rate of both the storks and the population.

3.13.

a.

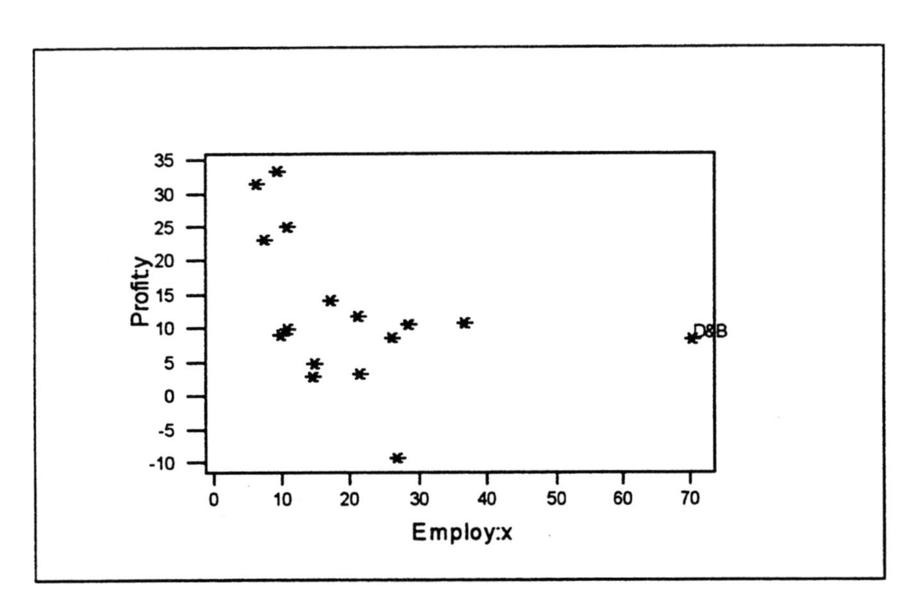

Correlation of Employ:x and Profit:y = -0.387

b. Correlation of Employ:x and Profit:y without the D&B data = -0.561
This correlation coefficient without the Dunn & Bradstreet data is closer to -1 as compared to the r value in part 'a'.

c.

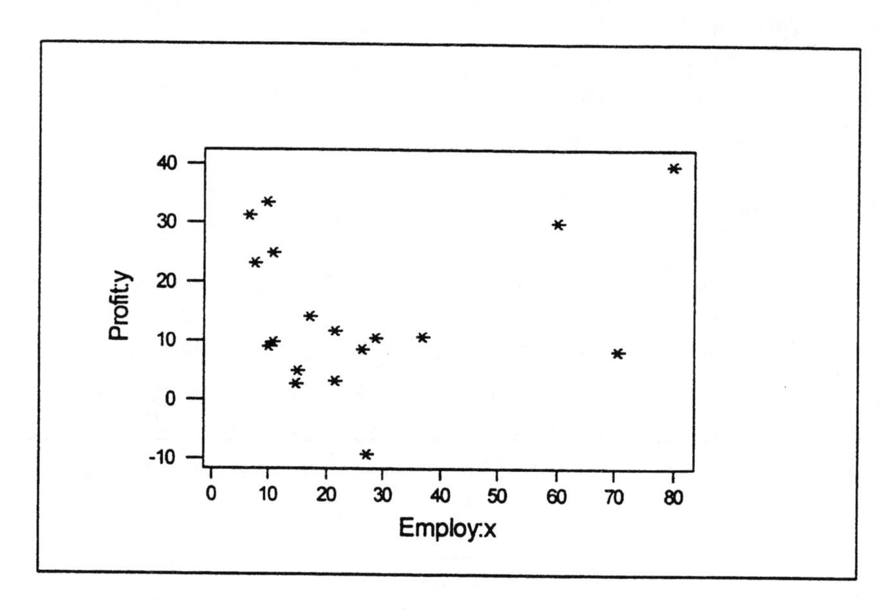

Correlation of Employ:x and Profit:y (includes two new points) = 0.219

The correlation coefficient has changed from negative to positive. The addition of two outlying points can have a drastic effect on the correlation coefficient.

3.14.

a.

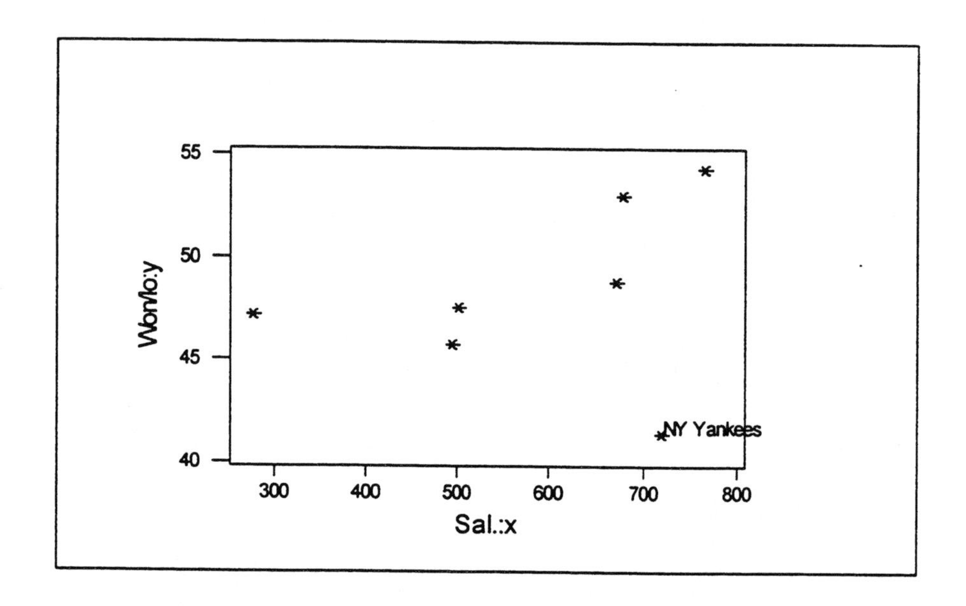

Correlation of Sal.:x and Won/lo:y = 0.281

There is a modest positive correlation between the salary and the won/lost percent.

b.

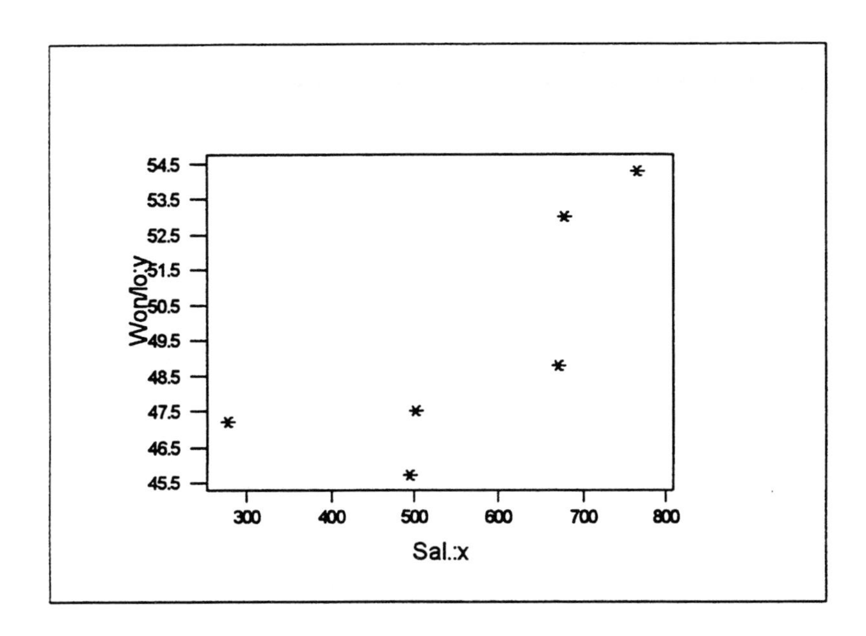

Correlation of Sal.:x and Won/lo:y = 0.768
Removing the NY Yankees increased the correlation coefficient by a large extent.

c. The data without the NY Yankees seems to indicate that larger salaries are associated with more games won so perhaps a championship team can be bought by hiring better players at higher salaries. The data in part 'a' does not indicate that buying teams is as likely a possibility. Causality can be established by showing consistency, responsiveness, and mechanism. The teams paying a higher salary must precede a future higher won/lost percentage.

3.15.

r value	Diagram
(a.) r = -.7	(c)
(b.) r = .9	(d)
(c.) r = .5	(a)
(d.) r = 0	(b)

3.16.

Data Set 1

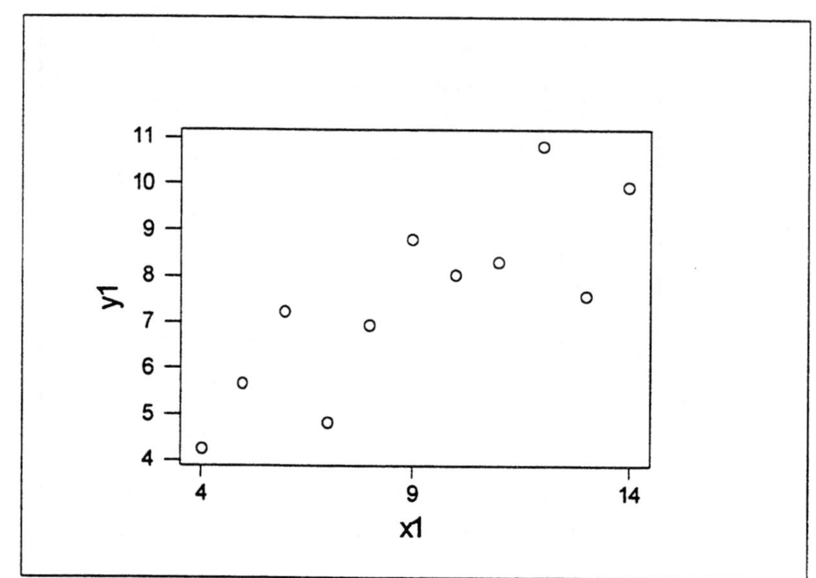

Correlation of x1 and y1 = 0.816

Data Set 2

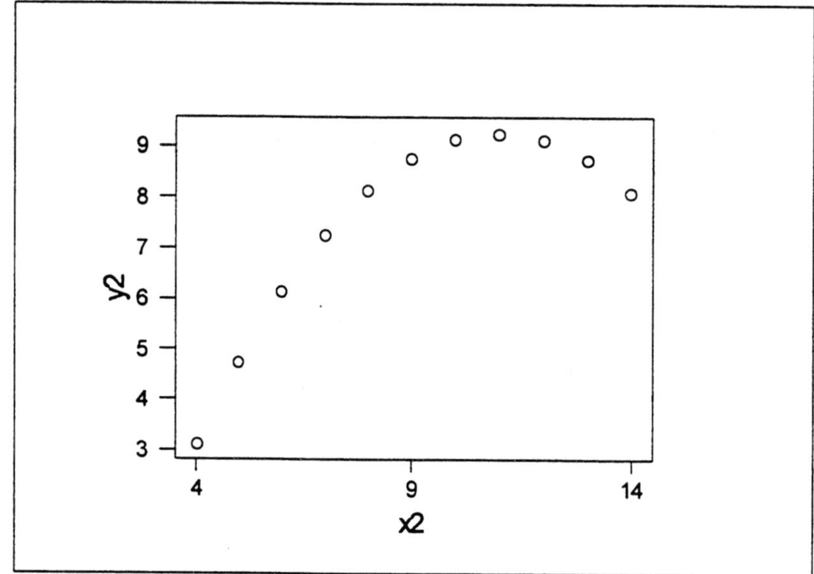

Correlation of x2 and y2 = 0.816

Data Set 3

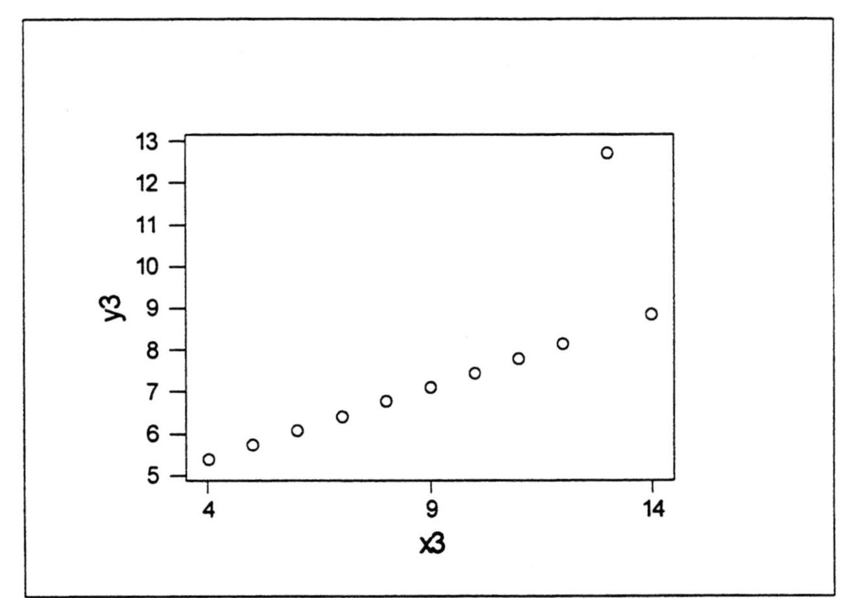

Correlation of x3 and y3 = 0.816

Data Set 4

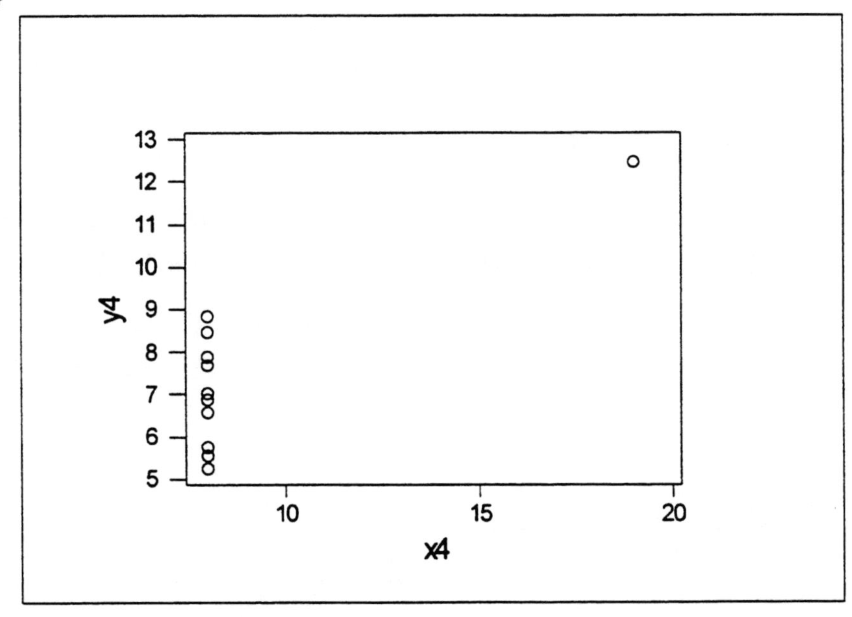

Correlation of x4 and y4 = 0.816

Even though the four diagrams look different, their correlation coefficients are all the same.

3.17.

a.

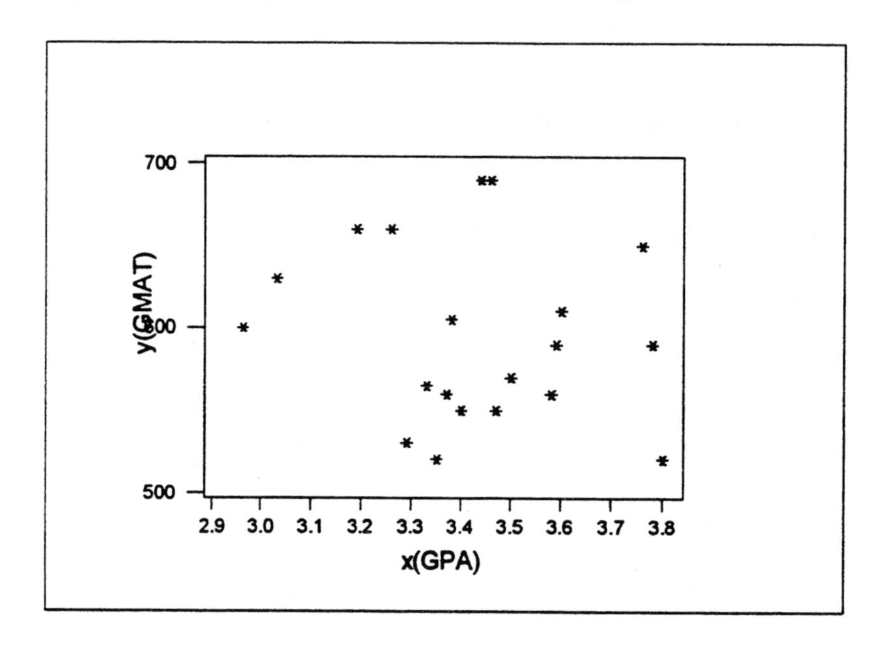

Correlation of x(GPA) and y(GMAT) = -0.158

b. There should be a strong positive correlation between GPA and GMAT for students accepted to the graduate program, but the negative correlation in the admitted students leaves doubt about the students who were not admitted.

c. Variables that could be plotted against GPA and/or GMAT scores include work experience in the business world, interpersonal skills profile score, and writing ability (essay) score.

3.18.

a.

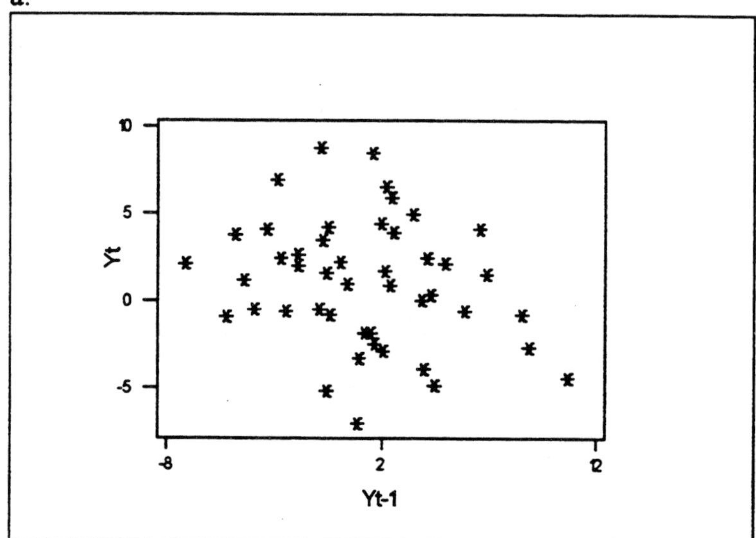

b. The lag 1 correlation coefficient is -0.230.

These observations have a slightly negative lag 1 correlation of -0.230. The linear relation between y_t and y_{t-1} is not strong.

3.19.

a. The r_1 and r_{12} values for Example 3.6 are 0.77 and 0.88; these are the same values as those shown in this problem.

b. The lag 1 correlation coefficient is 0.773, farther from zero than the -0.230 value obtained in 3.18b.

3.20.

a.

Mean of DoorOp1 = 0.17
Mean of DoorOp2 = 0.08
Mean of DoorOp3 = 0.15
Mean of DoorOp4 = 0.16
Mean of DoorOp5 = 0.18
Mean of DoorOp6 = 0.16
Mean of DoorOp7 = 0.25
Mean of DoorCl1 = 0.12
Mean of DoorCl2 = 0.07
Mean of DoorCl3 = 0.07
Mean of DoorCl4 = 0.11
Mean of DoorCl5 = 0.13
Mean of DoorCl6 = 0.16
Mean of DoorCl7 = 0.26

b.

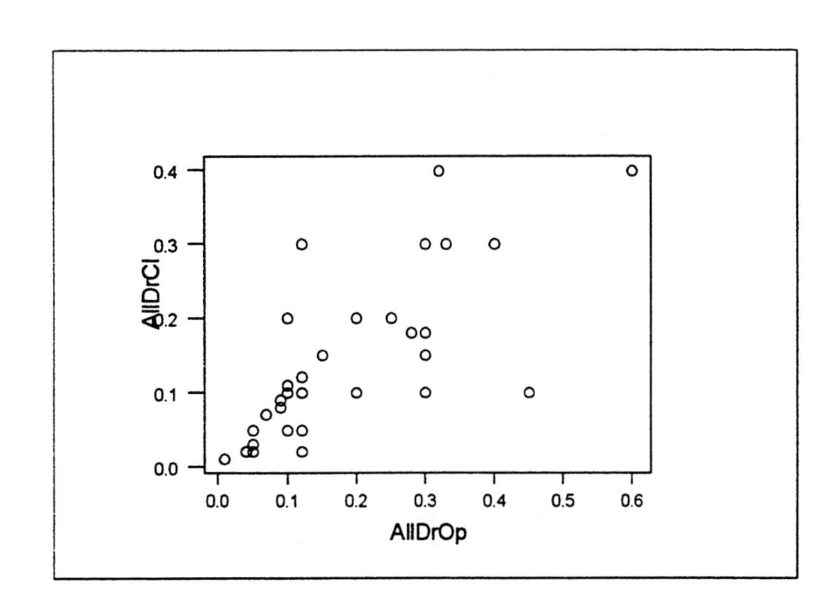

c.

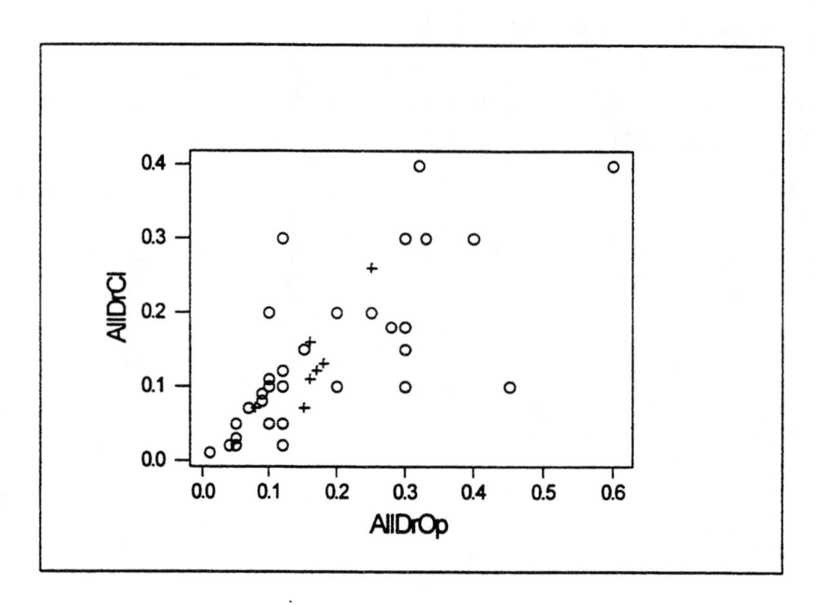

A guess at the correlation coefficient for original data: 0.7
A guess at the correlation coefficient for the mean data: 0.85
(Answers will vary.)

d.
Calculated Correlation of AllDrOp and AllDrCl = 0.747
Calculated Correlation of MeanOp and MeanCl = 0.868

e.
Correlations calculated from means or aggregates can be misleading when applied to individual cases since the effect of outliers is minimized when means or aggregates are used. The spread in scatter in both directions is reduced.

3.21.

a. Graph of $y = 3 + 7x$. Slope = 7; y-intercept = 3.

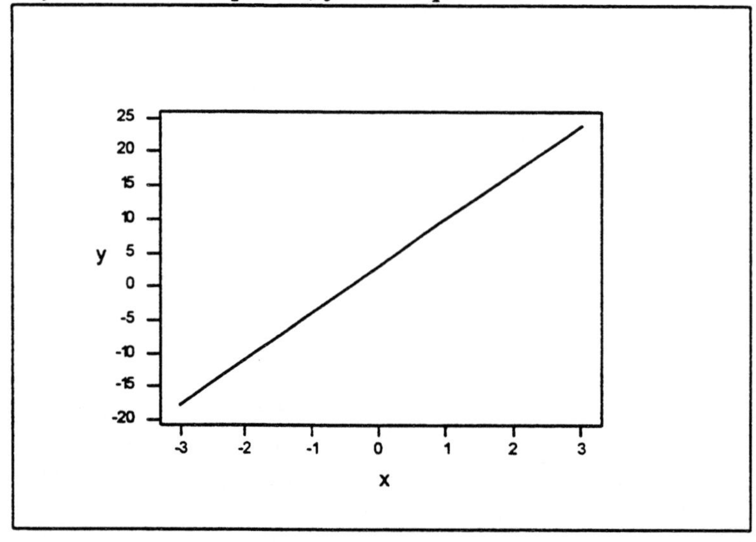

b. Graph of $y = -4 + 2x$. Slope = 2; y-intercept = -4.

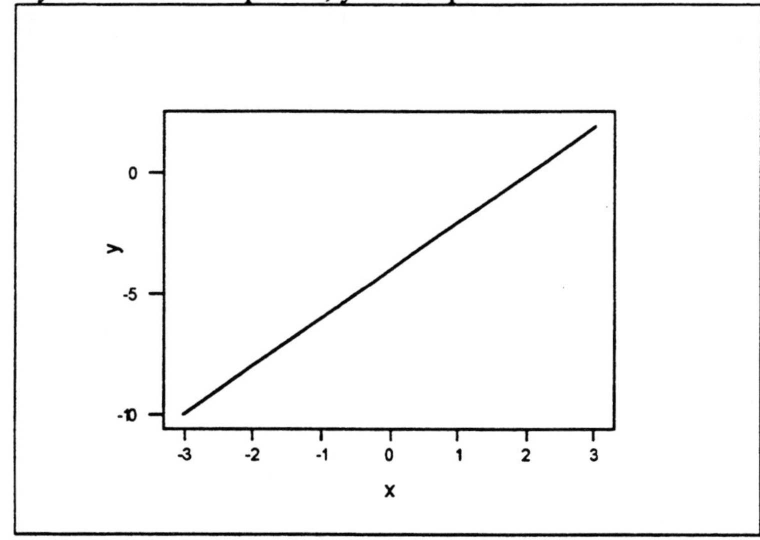

c. Graph of $y = 7 - x$. Slope = -1; y-intercept = 7.

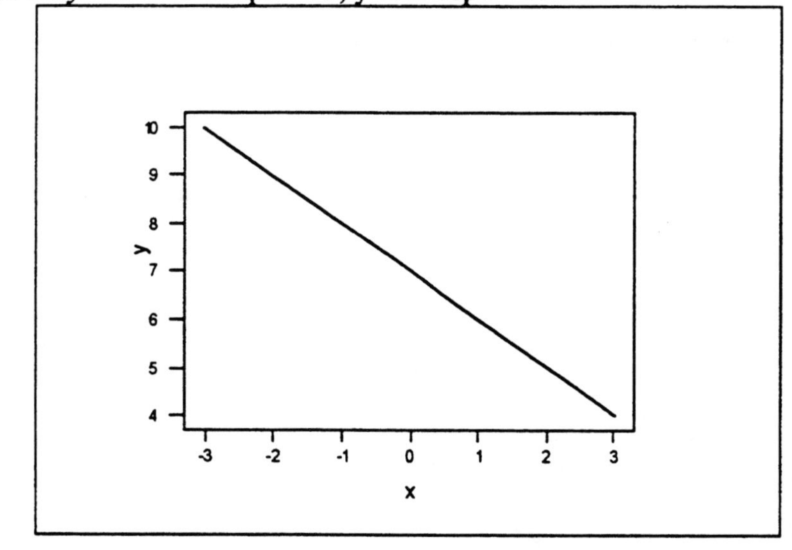

3.22.

a. Graph of $y = -1 - 8x$:
Slope = -8
y-intercept = -1

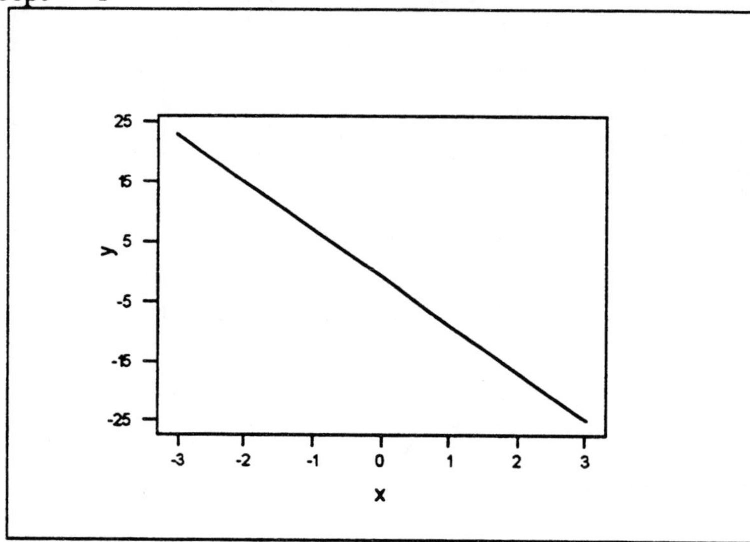

b. Graph of $y = 4x$:
Slope = 4
y-intercept = 0

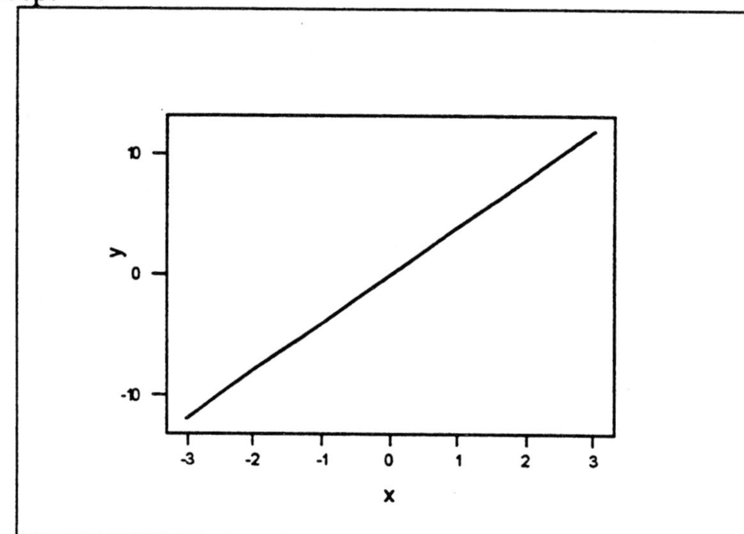

c. Graph of $y = -10x$:
Slope = -10
y-intercept = 0

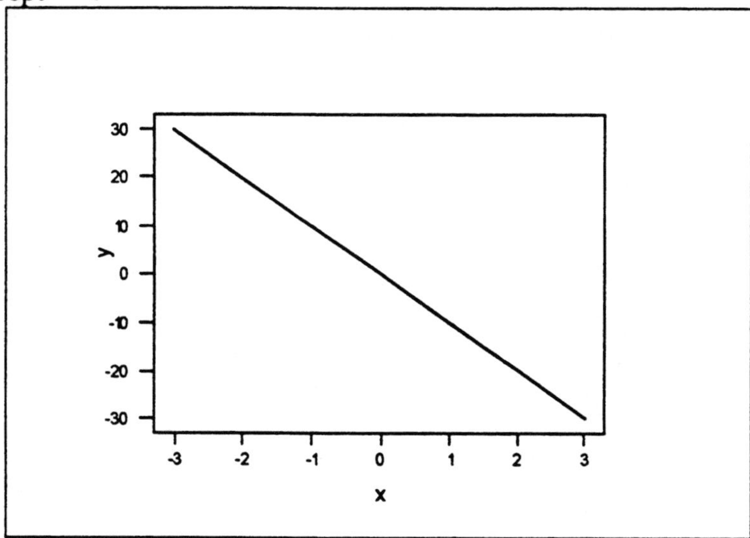

3.23.

a.

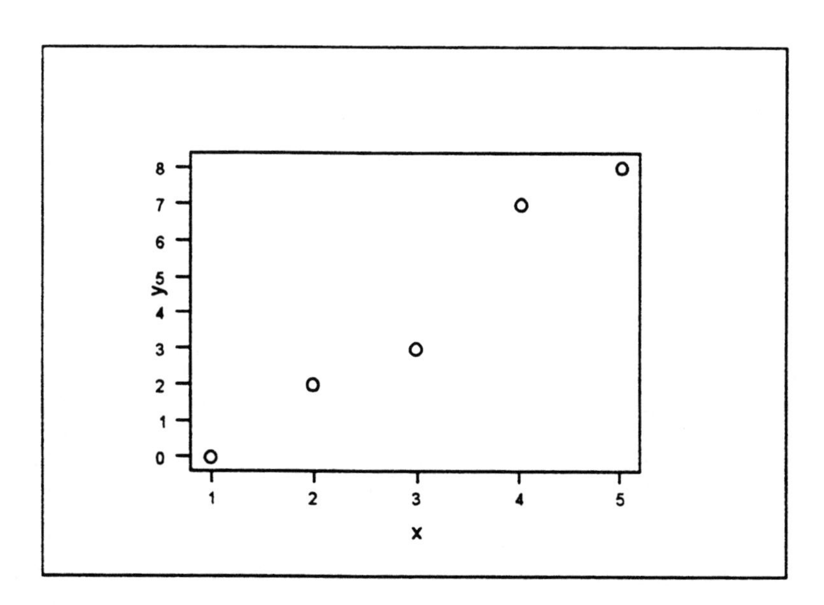

b. The slope is positive.

c.

x	y	$x-\bar{x}$	$y-\bar{y}$	$(x-\bar{x})^2$	$(x-\bar{x})(y-\bar{y})$
1	0	−2	−4	4	8
2	2	−1	−2	1	2
3	3	0	−1	0	0
4	7	1	3	1	3
5	8	2	4	4	8

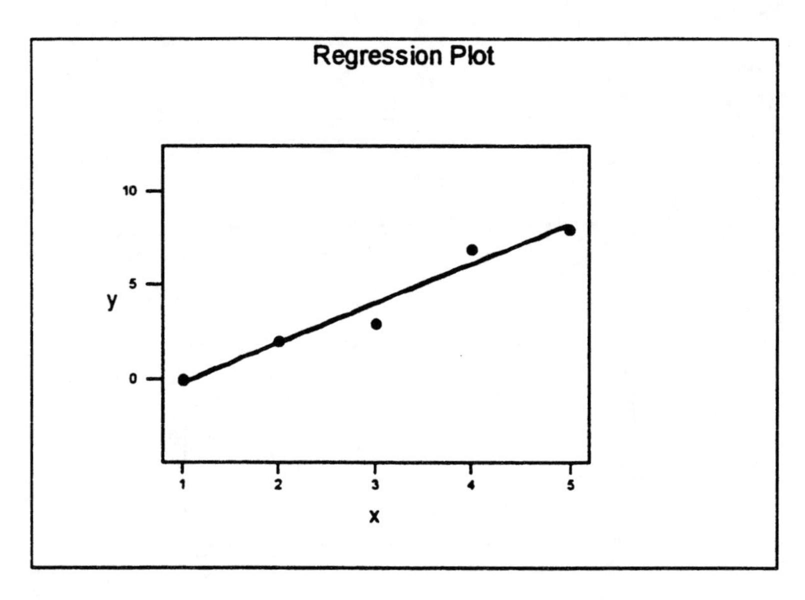

$y = -2.3 + 2.1\,x$

d. When $x = 4$, $y = -2.3 + 2.1 * 4 = 6.1$.

3.24.

a.

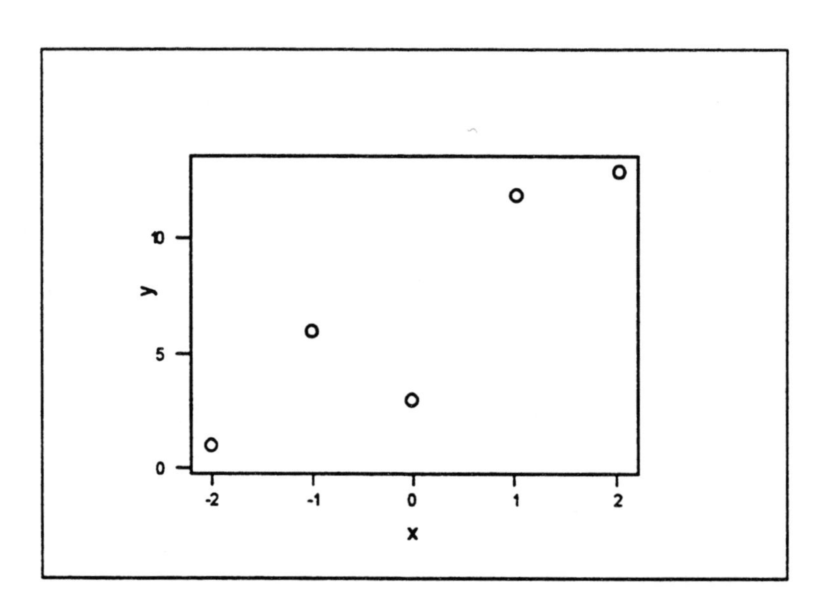

b. Slope is positive.

c.

x	y	$x-\bar{x}$	$y-\bar{y}$	$(x-\bar{x})^2$	$(x-\bar{x})(y-\bar{y})$
−2	1	−2	−4	4	8
−1	6	−1	−2	1	2
0	3	0	−1	0	0
1	12	1	3	1	3
2	13	2	4	4	8

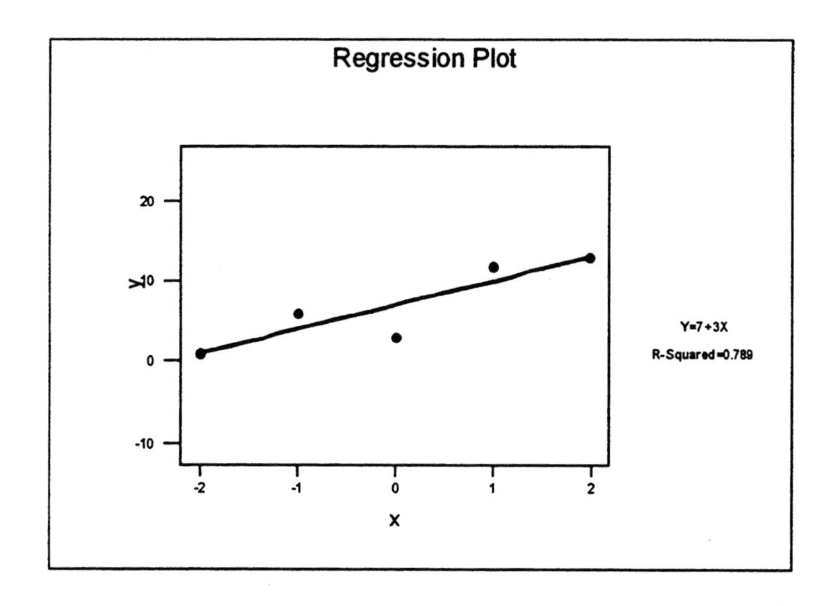

$y = 7 + 3x$

d. When $x = 1$, $y = 7 + 3 * 1 = 10$.

3.25.

a.

Residuals:

0.2
0.1
-1.0
0.9
-0.2

b. Standard deviation of the residuals: 0.689

3.26.

a.

Residuals:

0
2
-4
2
0

b. Standard deviation of the residuals: 2.450

3.27.

a. Least squares regression equation: y=162-3x
Slope = -3
y-intercept = 162

b.

Correlation coefficient from Exercise 3.6: r = -0.363

x	y	$x-\bar{x}$	$y-\bar{y}$	$(x-\bar{x})^2$	$(y-\bar{y})^2$	$(x-\bar{x})(y-\bar{y})$
4	142	0	-8	0	64	0
5	152	1	2	1	4	2
4	148	0	-2	0	4	0
3	158	-1	8	1	64	-8
Totals:				2	132	
Square root:				1.44	11.489	

Slope = b = (11.489/1.44)(-0.363) = -3

3.28.

a.

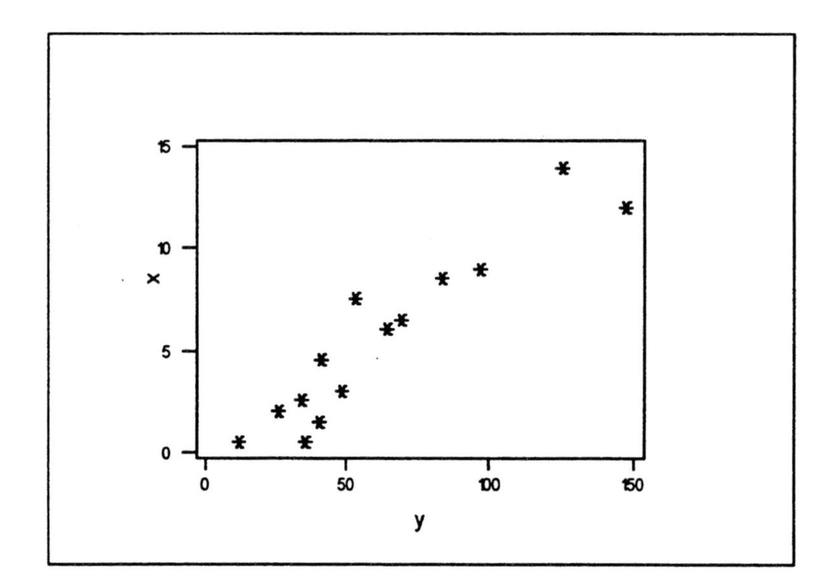

b.

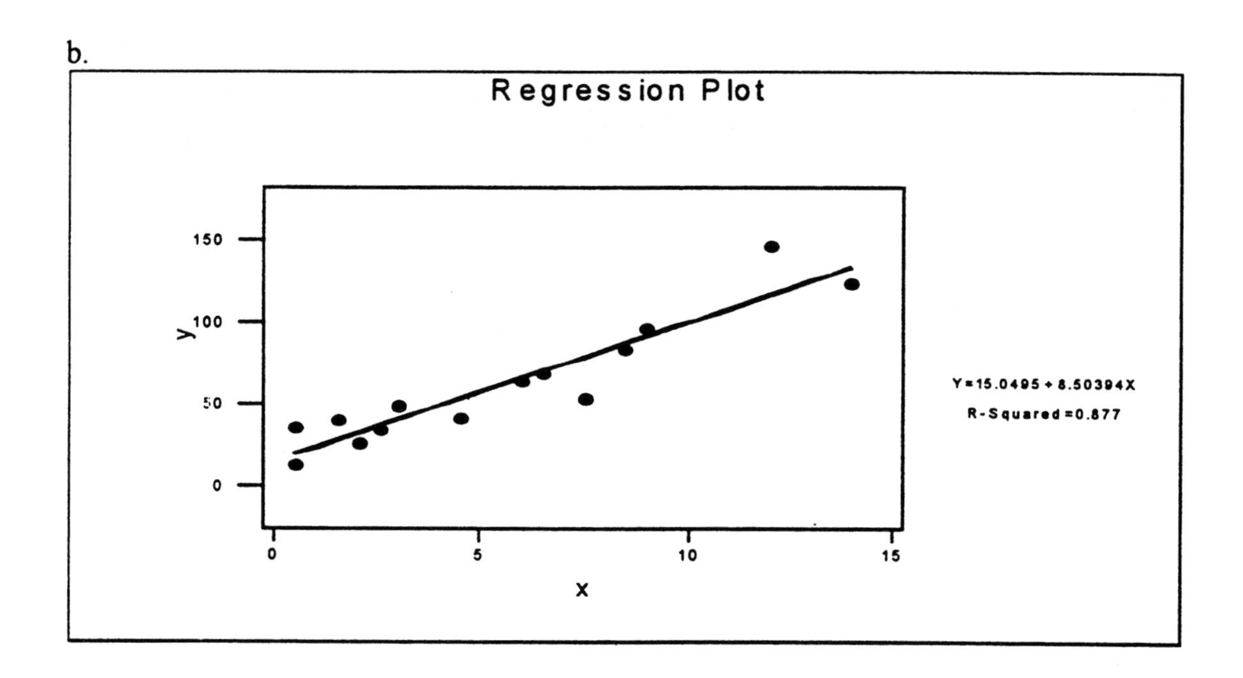

$y = 15.050 + 8.504x$

c.

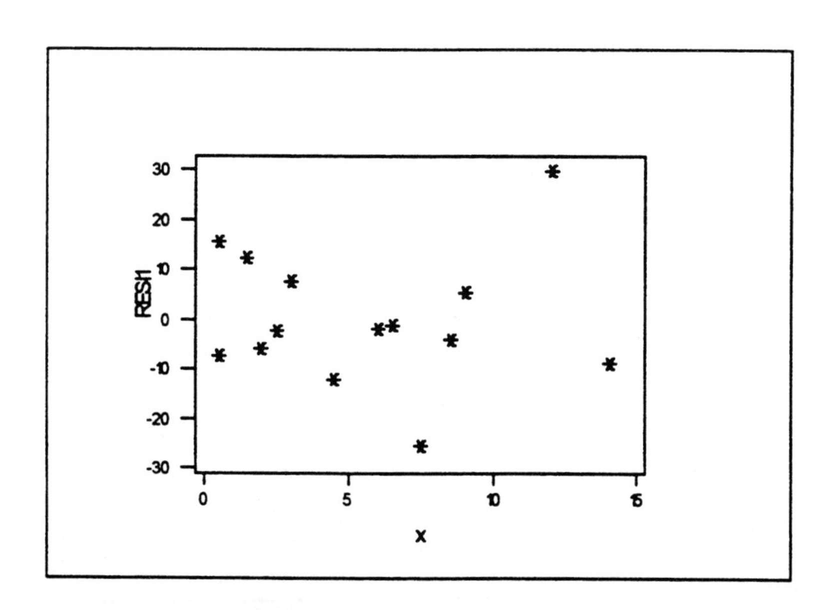

The predicted values are within plus or minus 30 units of the actual values which is a fairly adequate description of the data.

d. Earnings for R&D of 5.5: $y = 15.0495 + 8.50394(5.5) = 61.821$.

3.29.

a.

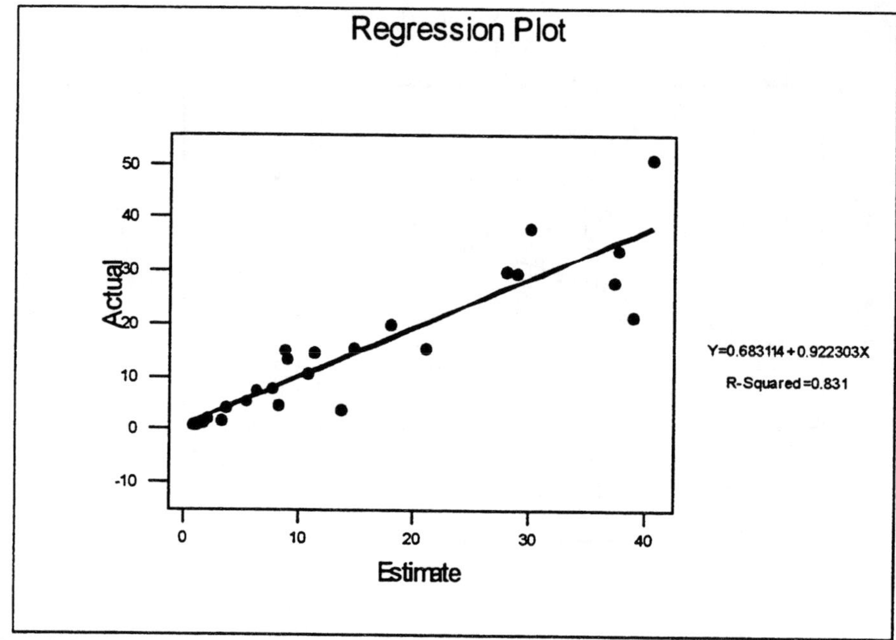

$y = 0.683 + 0.922x$

b.

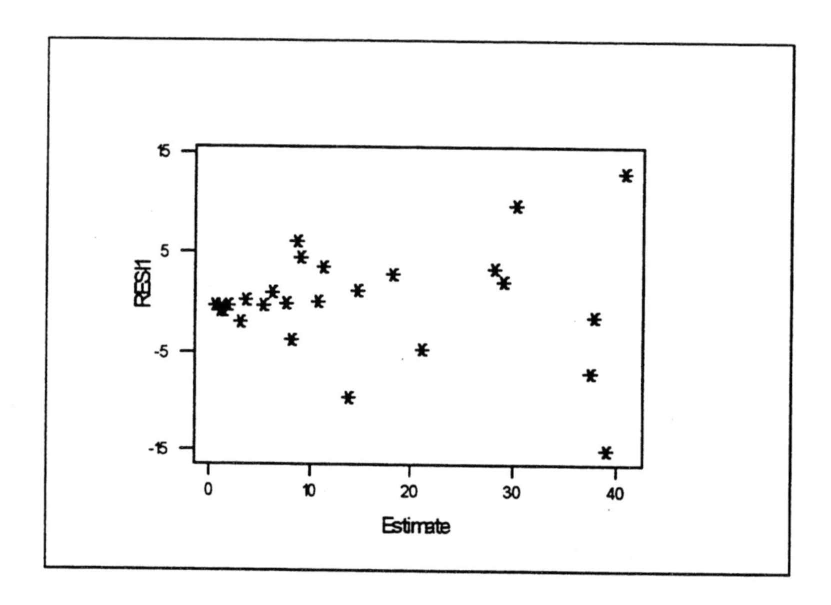

Residuals are closer to zero for smaller estimates; therefore, estimated costs are better predictors of actual costs for smaller projects.

3.30.

a.

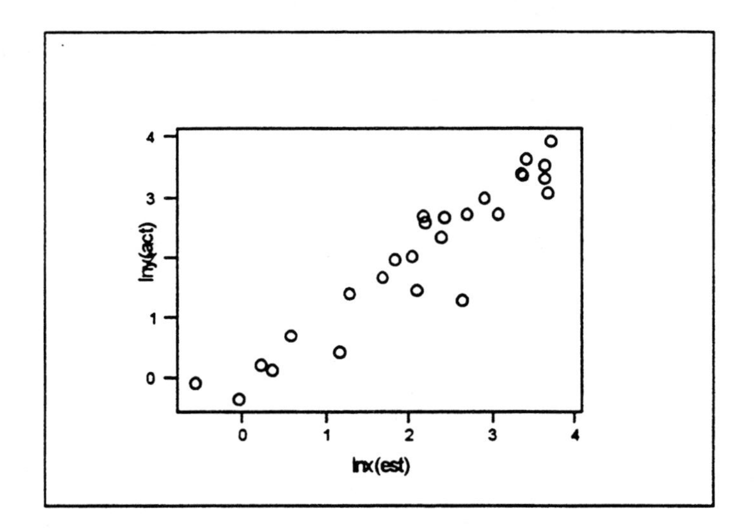

b.

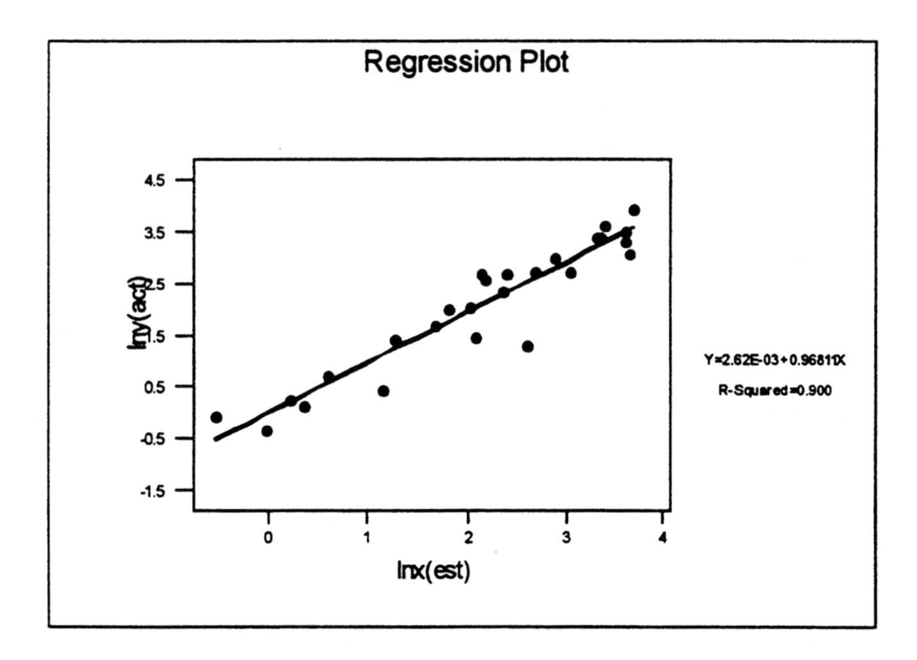

lny(act) = 0.003 + 0.968 lnx(est)

c.

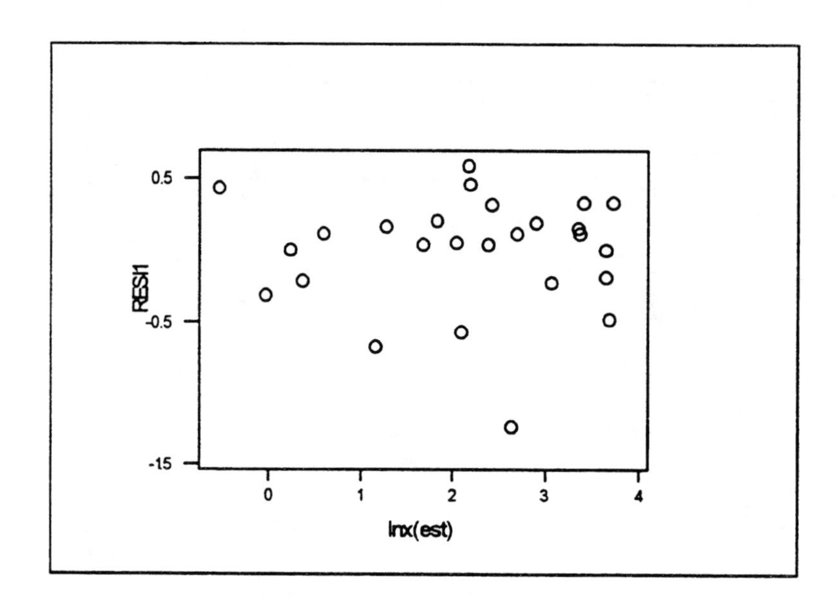

The deviations are not the same as those in 3.29b throughout the fitting region.

d.

Actual cost with an estimated cost of 19.5 million dollars:
ln(19.5) = 2.970
lny(act) = 0.003 + 0.968 * 2.970 = 2.878
antilog(2.878) = 17.779 million dollars

3.31.

a.

Regression Analysis

```
The regression equation is
Actual = 0.948 Estimate

Predictor        Coef        Stdev    t-ratio        p
Noconstant
Estimate      0.94804      0.05512      17.20    0.000

s = 5.601
Analysis of Variance

SOURCE         DF          SS          MS          F        p
Regression      1      9280.1      9280.1     295.79    0.000
Error          25       784.4        31.4
Total          26     10064.4

Unusual Observations
Obs. Estimate      Actual       Fit  Stdev.Fit   Residual     St.Resid
  8      40.6       51.28     38.52       2.24      12.77       2.49RX
  9      37.8       34.10     35.84       2.08      -1.74      -0.33 X
 17      37.4       27.97     35.46       2.06      -7.48      -1.44 X
 26      38.9       21.57     36.91       2.15     -15.34      -2.97RX

R denotes an obs. with a large st. resid.
```

```
X denotes an obs. whose X value gives it large influence.
```

b.

Residual plots for the least squares line from Ex.3.29a: $y = 0.683 + 0.922x$

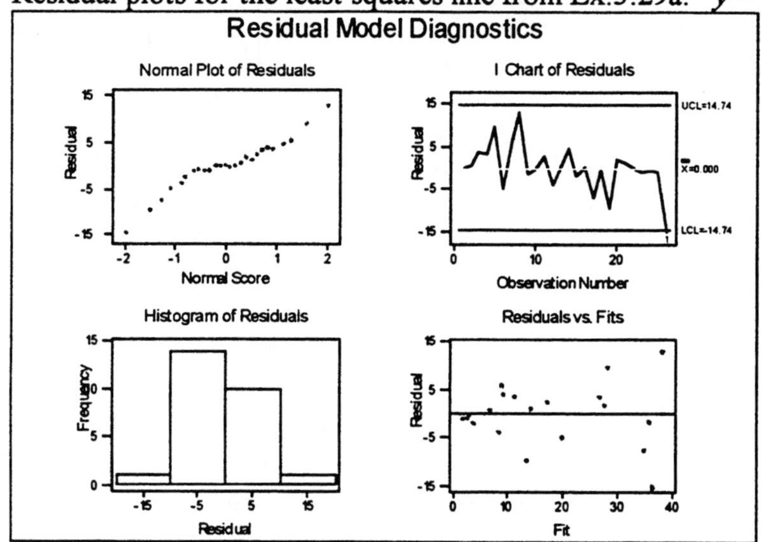

Residual plots for the regression line through the origin from 3.31a.:
Actual = 0.948 Estimate

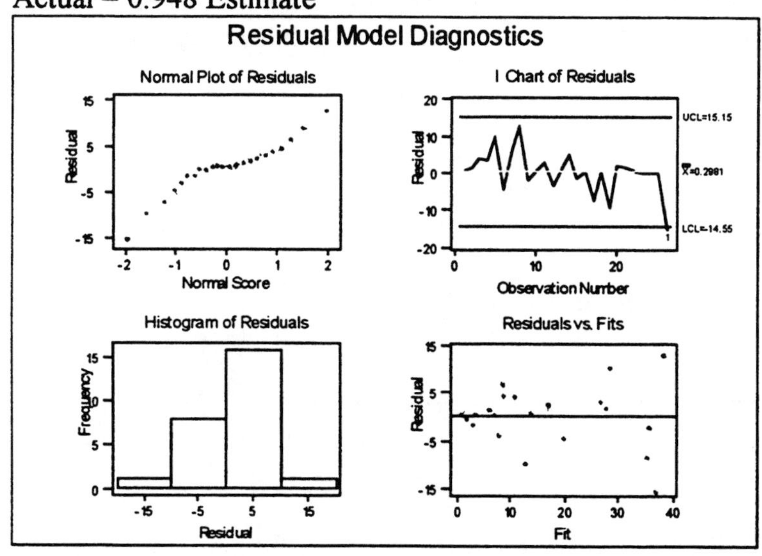

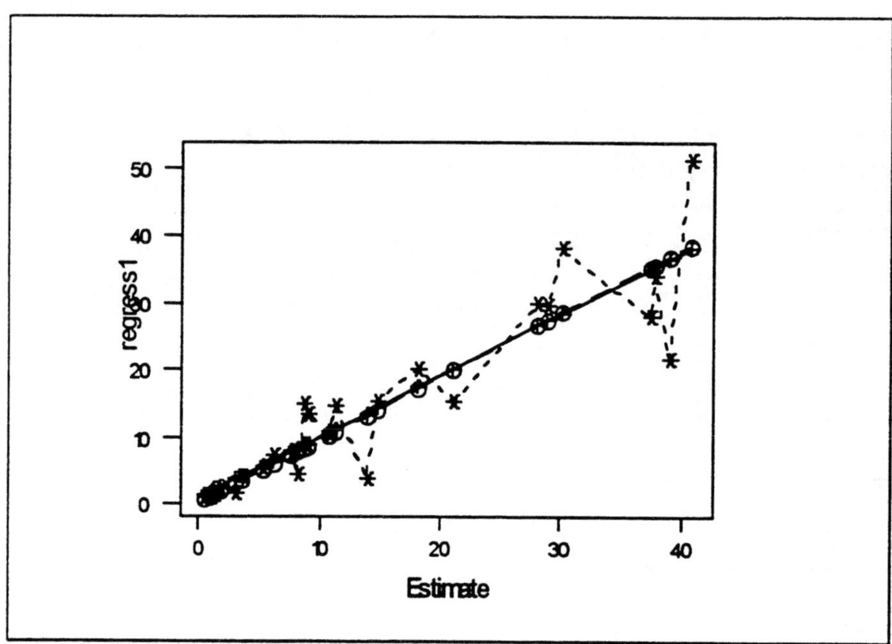

Astericks: actual ordered pairs
Plus : From Ex.3.29a
Triangle: From the least squares line through the origin.
The least squares line from Ex.3.29a is the best fit since it minimizes the distances between the estimated value and the predicted value. The two plots are extremely close, however, as shown on this graph.

c. Sum of residuals: 7.7518

3.32.

a.

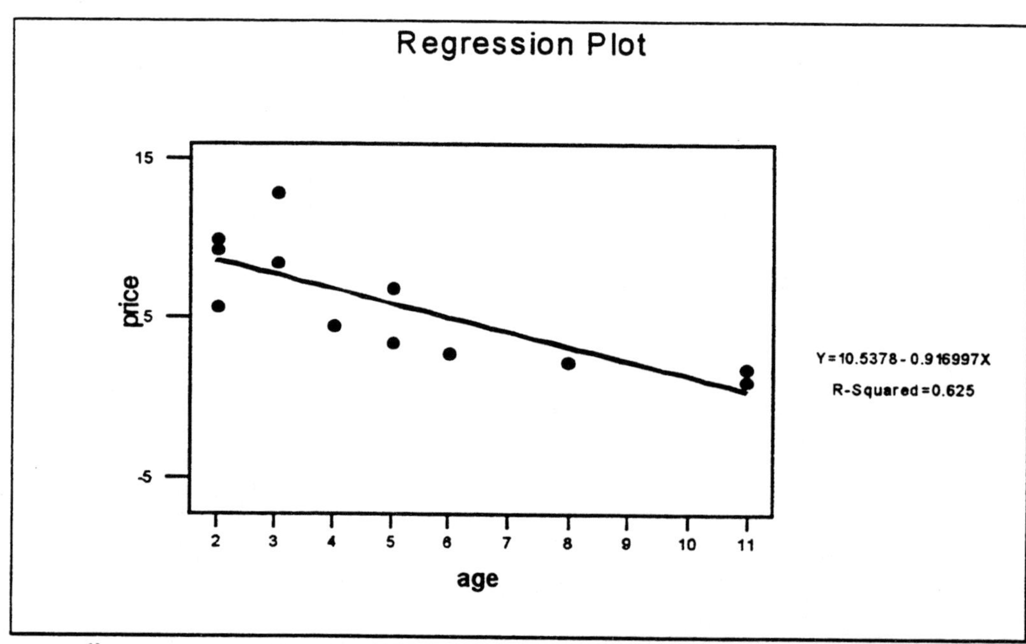

Least squares line: $y = 10.538 - 0.917x$

b. From Ex3.7b: r = -0.791
b = (158.2/117.7)^0.5*(-0.791) = -0.917

c.

```
RESIDUALS
 1.29618
-2.90382
 1.34915
-1.00184
-2.23584
-2.45283
 0.69618
 0.64915
-2.36983
 1.04717
 0.71317
 5.21317
```

Sum of residuals: 0

d. Price of a 7 year old car = 10.5378 - 0.916997*7 = 4.119

3.33.

a. The risk-free rate = 0.045 = 4.5%.
United States treasury bills have a risk free rate.

b.

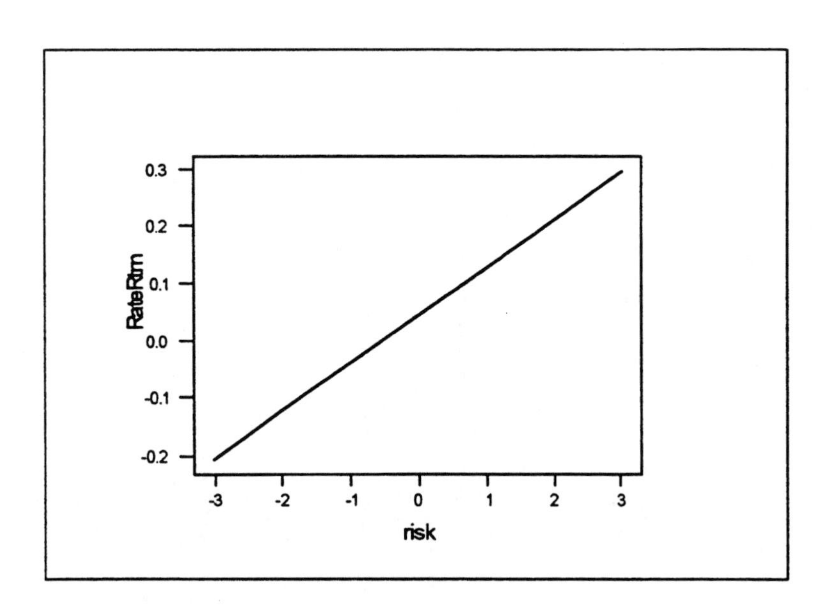

If the systematic risk = 0.70, the required rate of return is y = 0.045 + 0.084 * 0.7 = 0.1038.

3.34.

a.

Least squares line: $y = 60.711 + 0.414x$

b. The percent of market value at which properties are assessed varies since the y-intercept of the line is not zero. Only for properties with a fair market value of 325.53 will the assessed value be 60%.

c.

Parcel	Assessed Value	Market Value	RESIDUALS
1	68.2	87.4	-1.51137
2	74.6	88.0	-3.55778
3	64.6	87.2	-0.22276
4	80.2	94.0	0.12661
5	76.0	94.2	2.06332
6	78.0	93.6	0.63631
7	76.0	88.4	-3.73668
8	77.0	92.2	-0.35019
9	75.2	90.4	-1.40588
10	72.4	90.4	-0.24808
11	80.0	91.4	-0.90208
12	76.4	91.4	-0.90208
13	70.2	89.6	-0.13837
14	75.8	91.8	-0.25398
15	79.2	94.8	1.34012
16	74.0	88.4	-2.90968
17	72.8	93.6	2.78652
18	80.4	92.8	-1.15609
19	74.2	90.6	-0.79238
20	80.0	91.6	-2.19069

21	81.6	92.8	-1.65229
22	75.6	89.0	-2.97128
23	79.4	91.8	-1.74258
24	82.2	98.4	3.69961
25	67.0	89.8	1.38483
26	72.0	97.2	6.71732
27	73.6	95.2	4.05572
28	71.4	88.8	-1.43457
29	81.0	97.4	3.19581
30	80.6	95.4	1.36121

Unusual Observations

Obs.	Assessed	Market	Fit	Stdev.Fit	Residual	St.Resid
3	64.6	87.200	87.423	1.199	-0.223	-0.10 X
26	72.0	97.200	90.483	0.578	6.717	2.83R

R denotes an obs. with a large st. resid.
X denotes an obs. whose X value gives it large influence.

3.35.

a.

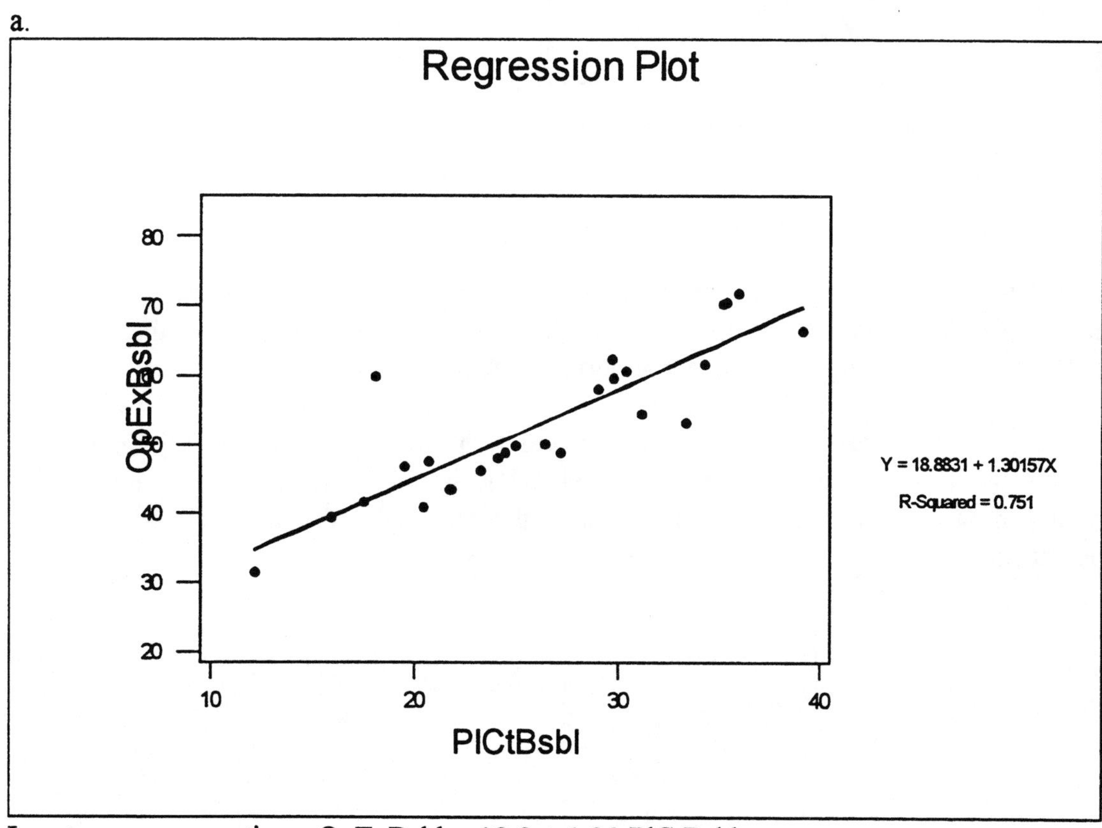

Least squares equation: OpExBsbl = 18.9 + 1.30 PlCtBsbl

b. For a baseball team with a player cost of 15 million dollars, operating expense = 18.8831 + 1.30157*15 = 38.407.

Player cost is a very large percent of the operating expense. The proportion of the operating expense is slightly smaller (31.34%) for teams with smaller player cost as compared to teams with a larger player cost (56.38%).

c. Residual standard deviation = s = 5.273

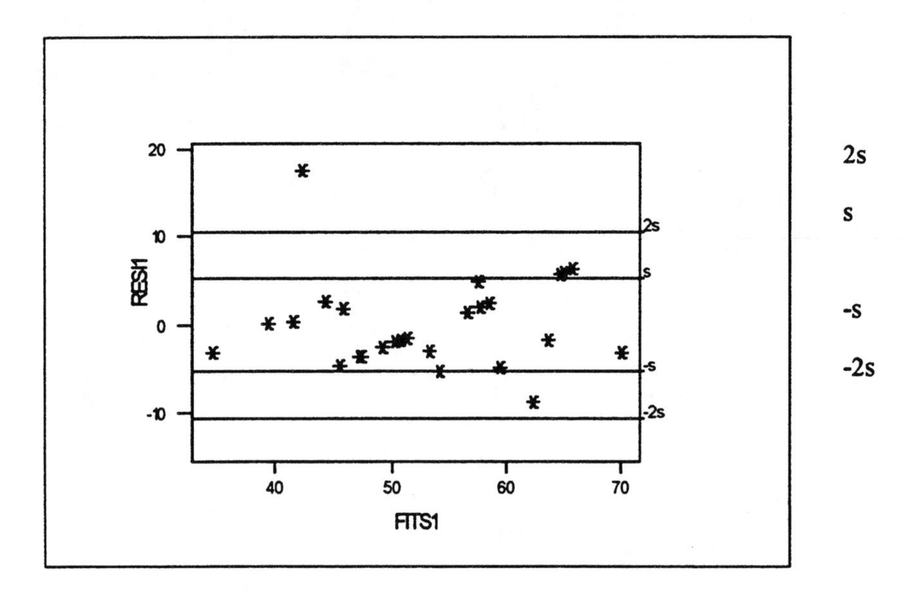

The residuals are generally within a horizontal band centered at zero. This is important since any deviation from a horizontal band would indicate a need for adjusting the original model, which was a straight line, to accommodate the displayed pattern.

d. The empirical rule states that about 68% and 95% of the data will be within one and two standard deviations of zero respectively. The actual numbers are 77% and 96%. There are more observations within one standard deviation of zero than predicted by the empirical rule.

e.

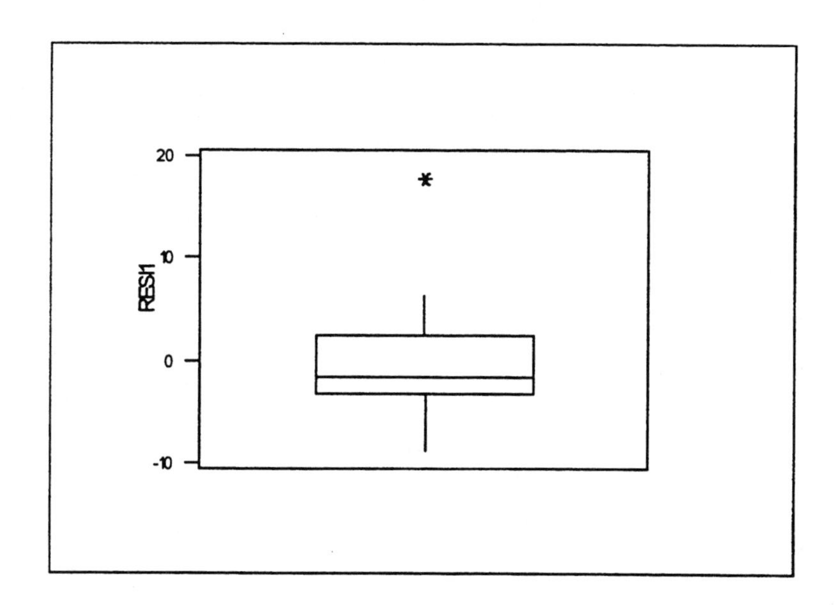

3.36.

a.

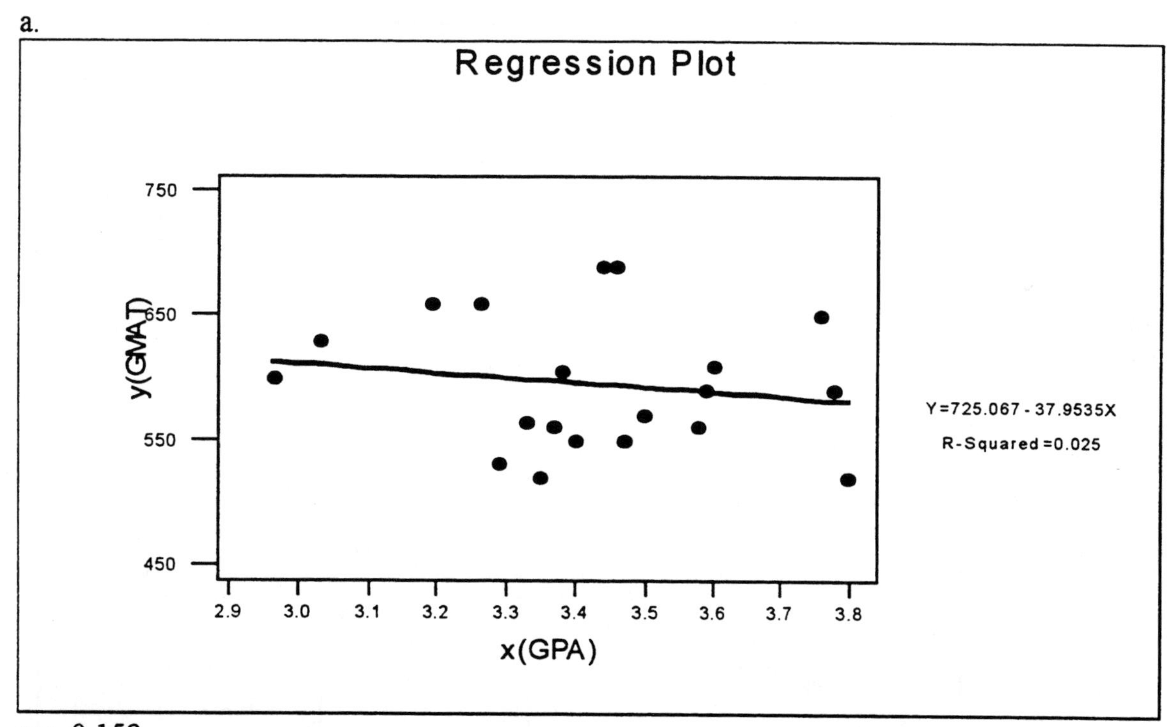

$r = -0.158$

Least squares line: $y = 725.067 - 37.954x$

GPA is not a strong predictor of GMAT scores since the correlation coefficient is close to zero.

b. An undergraduate with a GPA of 3.5 would be predicted to have a GMAT score of $y = 725.067 - 37.9535 * 3.5 = 592.230$.

c. $b = (233.77339 / 0.9813358) * (- 0.158)) = -37.639$

3.37.

a. and b.

First data set:

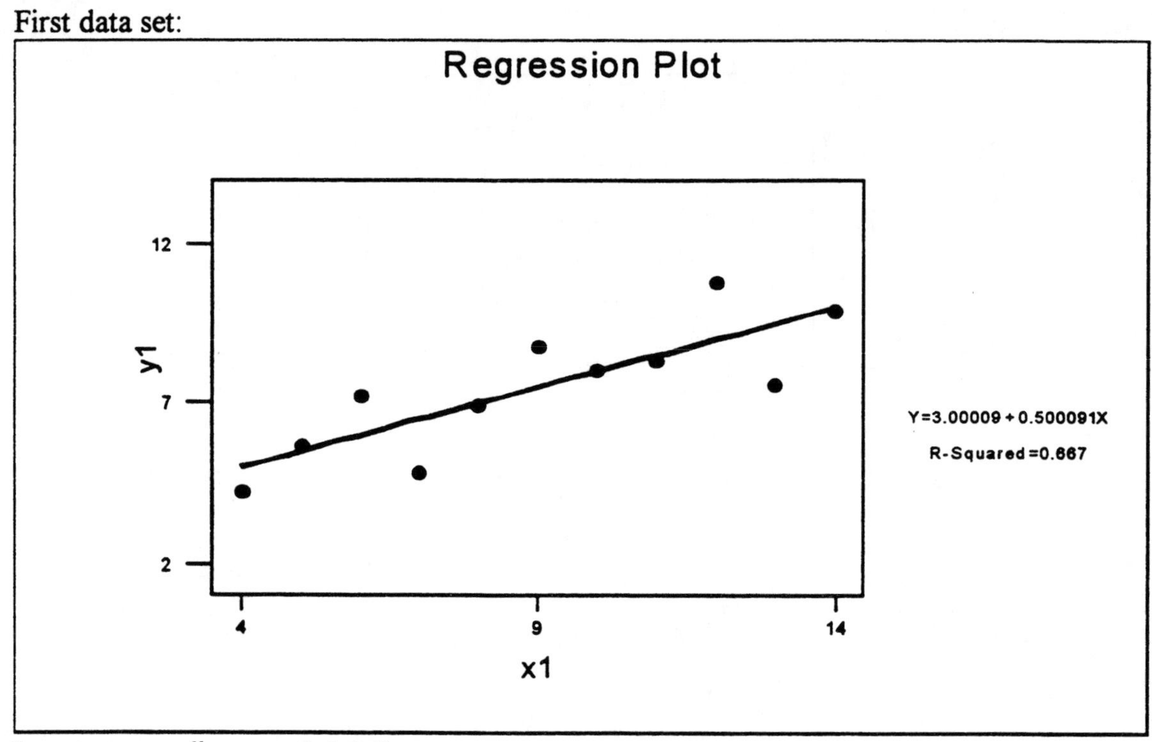

Least squares line: $y = 3 + 0.5\,x$

Second data set:

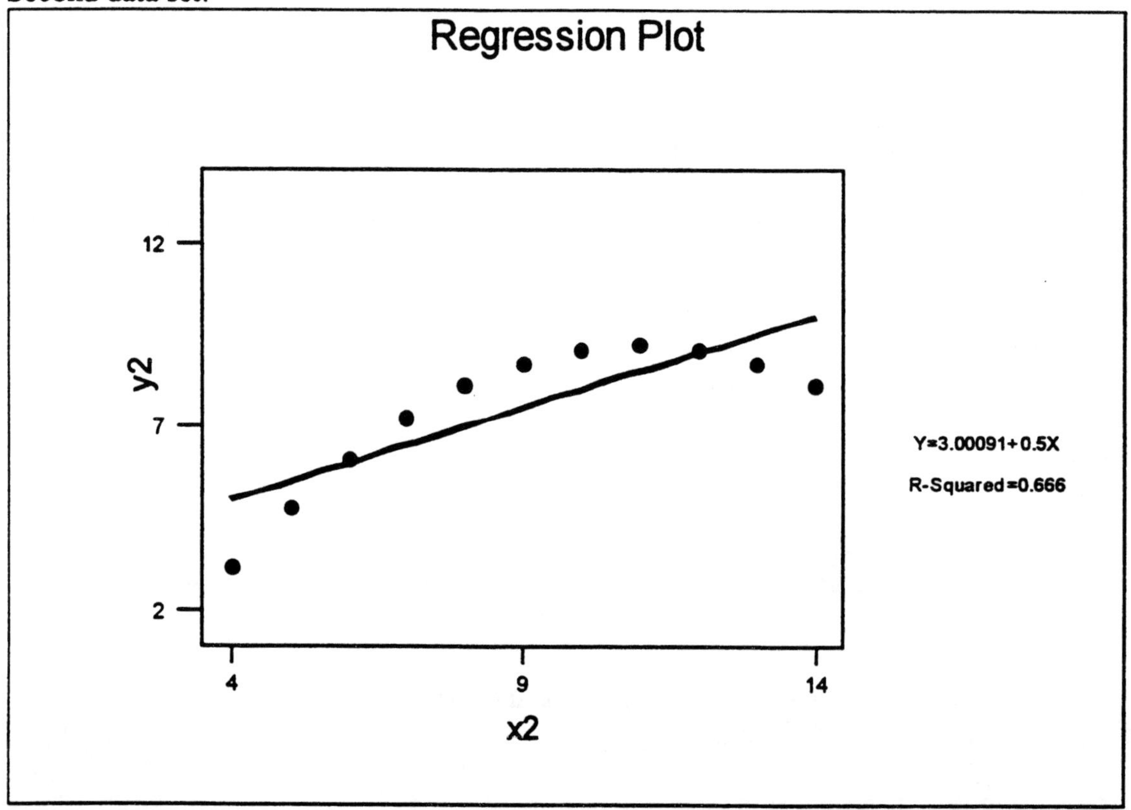

Least squares line: $y = 3 + 0.5\,x$

Third data set:

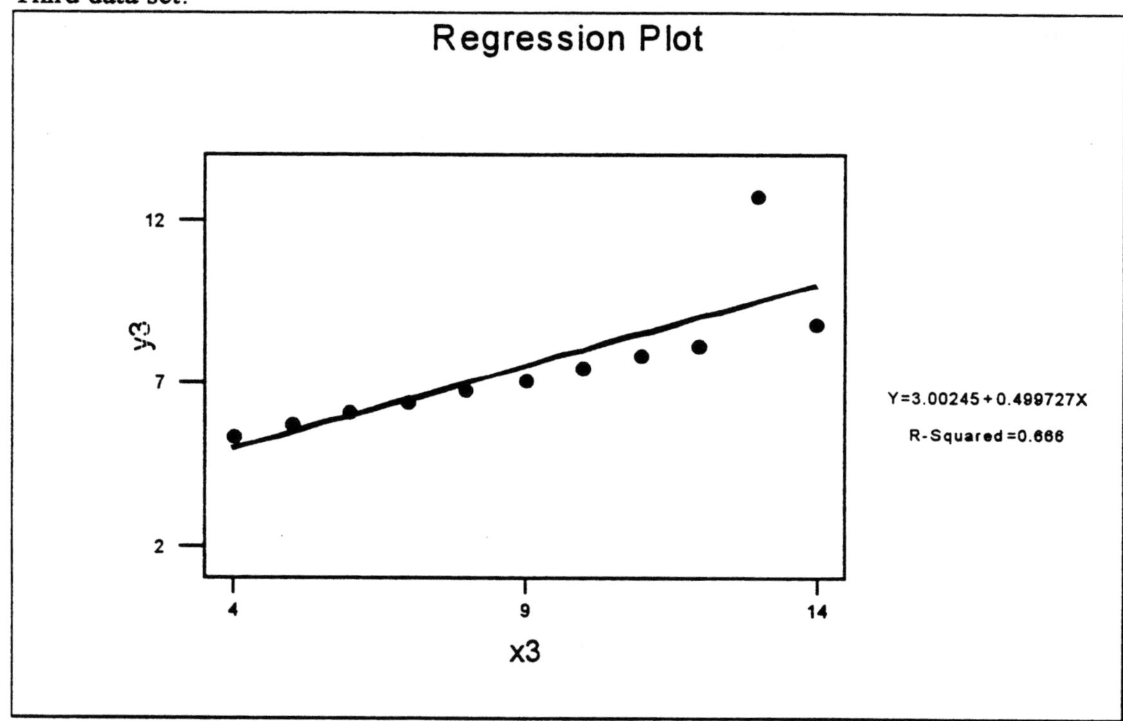

Least squares line: $y = 3 + 0.5\,x$

Fourth data set:

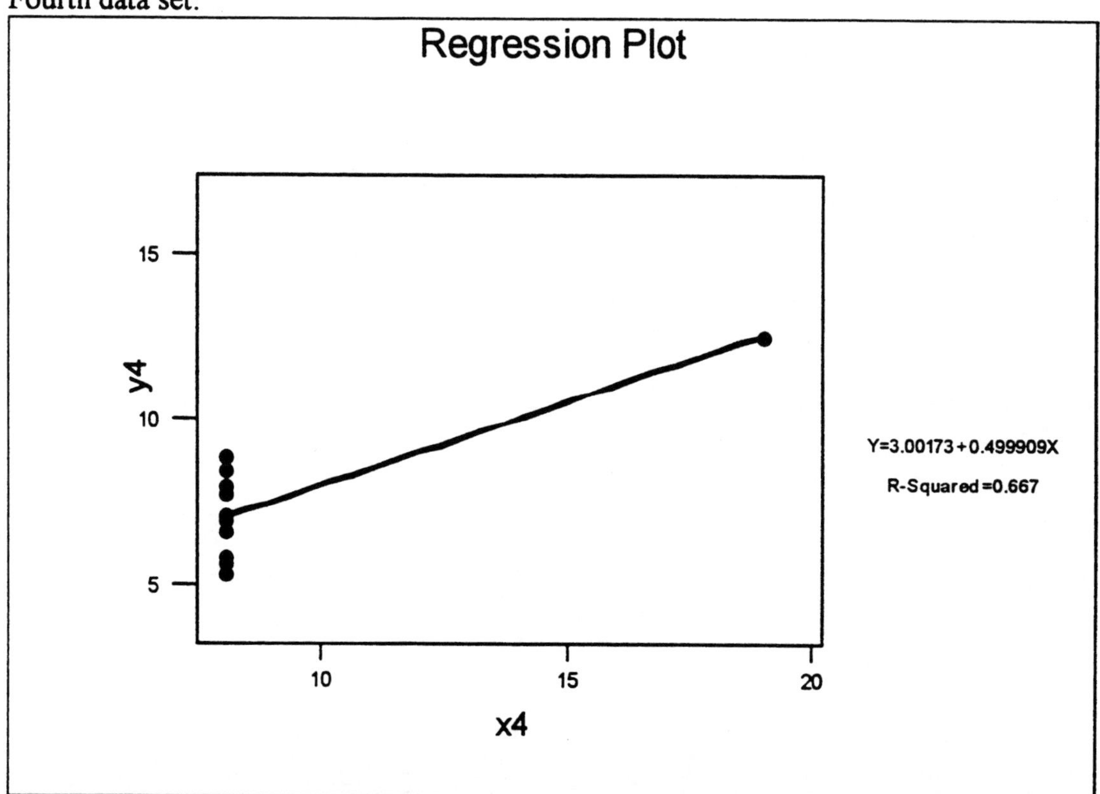

Least squares line: $y = 3 + 0.5\,x$

The first data set seems to be best represented by the least squares line for the range of x values considered.

c.

First data set:

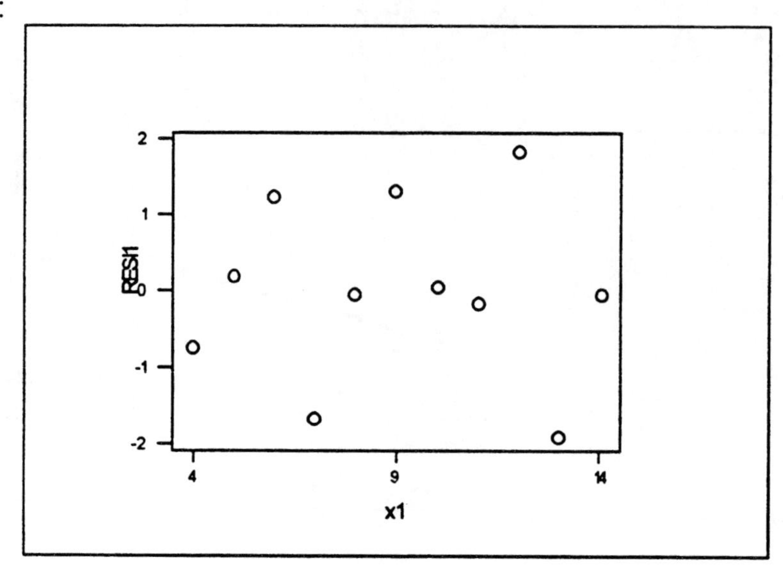

Second data set:

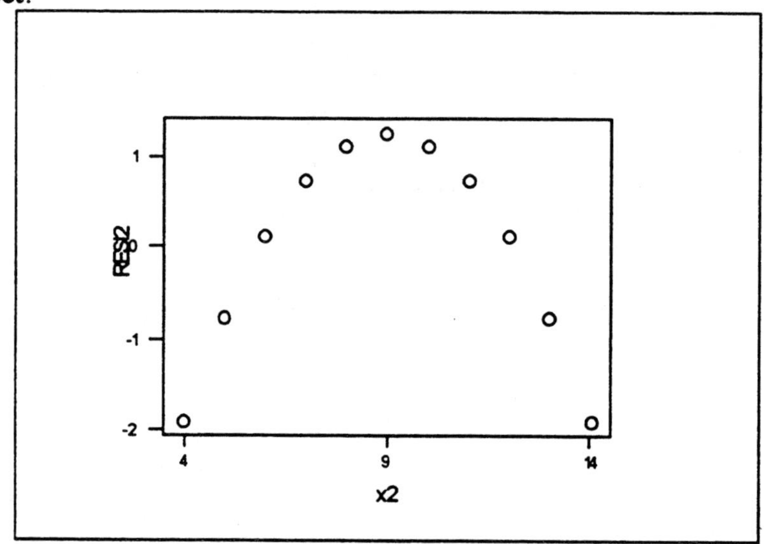

Third data set:

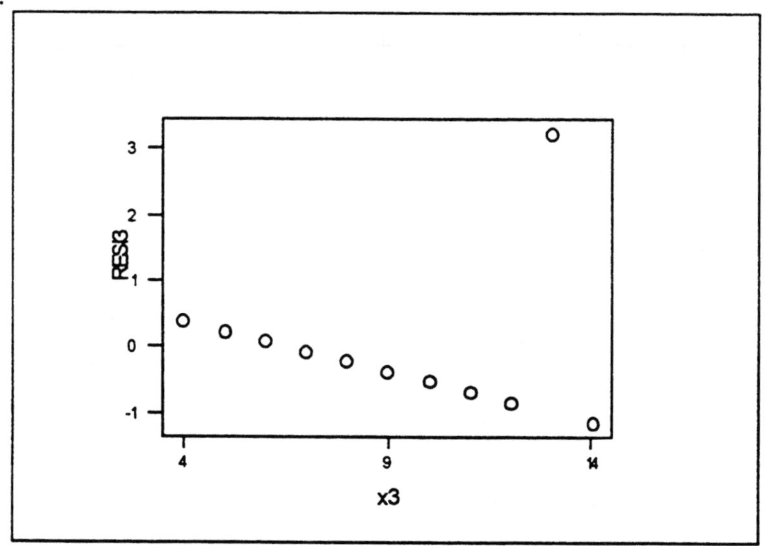

Fourth data set:

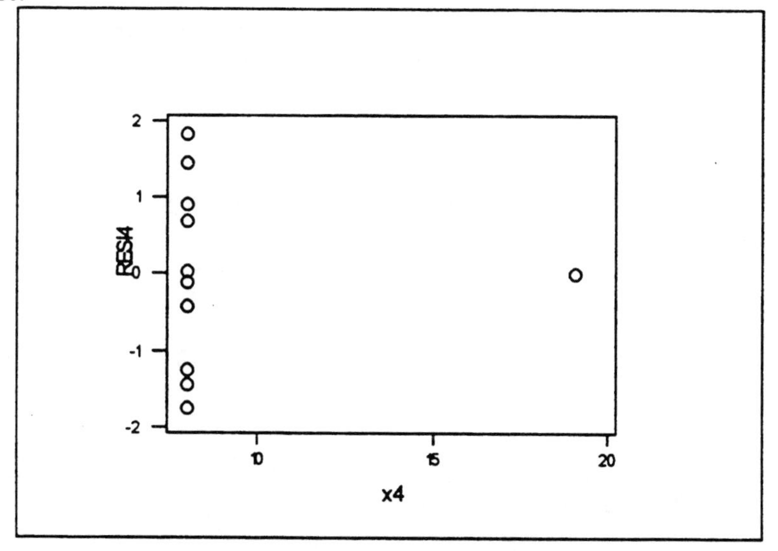

The first data set still appears to be best represented by the least squares line just as it was in part b. The residual plots for the last three data sets indicate an inadequate model (second data set) or potential outliers or unrepresentative observations. Keep in mind these data sets are artificial and were created to show that the same least squares line can represent quite different (x,y) configurations. That is why it is important to plot the data and examine the residuals.

d. The influential observation in data set 4 is the point (19, 12.50) since it is the only point without 8 as the first coordinate. It determines the slope of the fitted line.

3.38.

a.

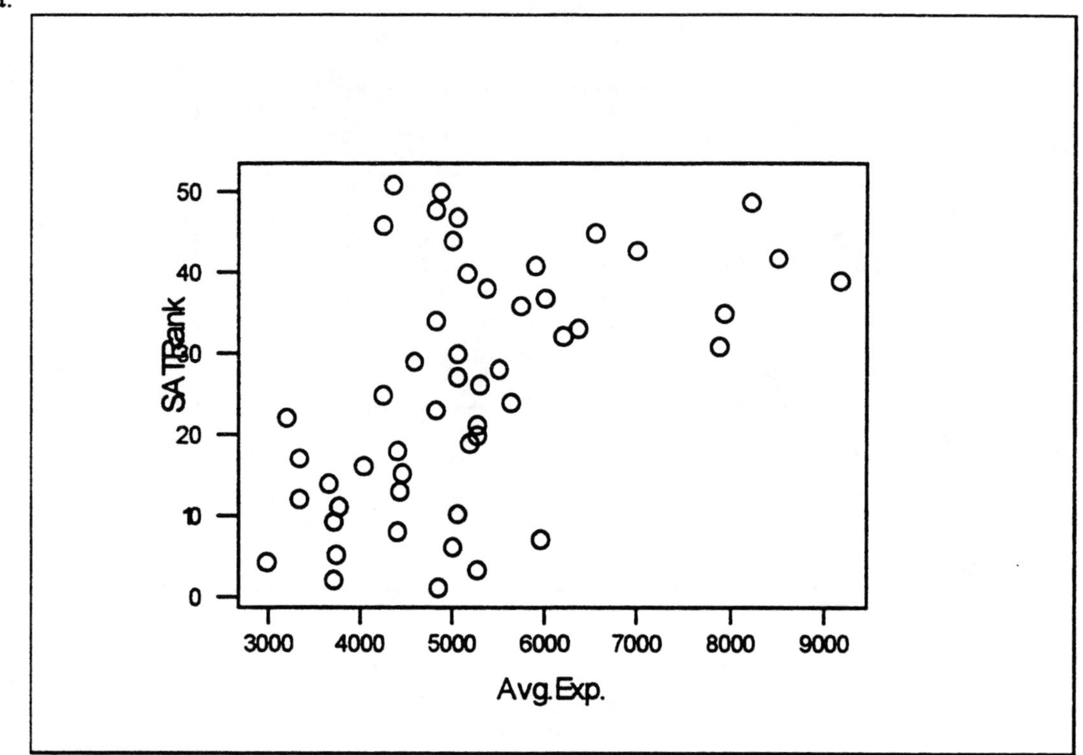

This plot suggests that higher expenditures result in a slightly higher performance on the math section of the SAT.

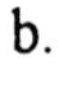

b.

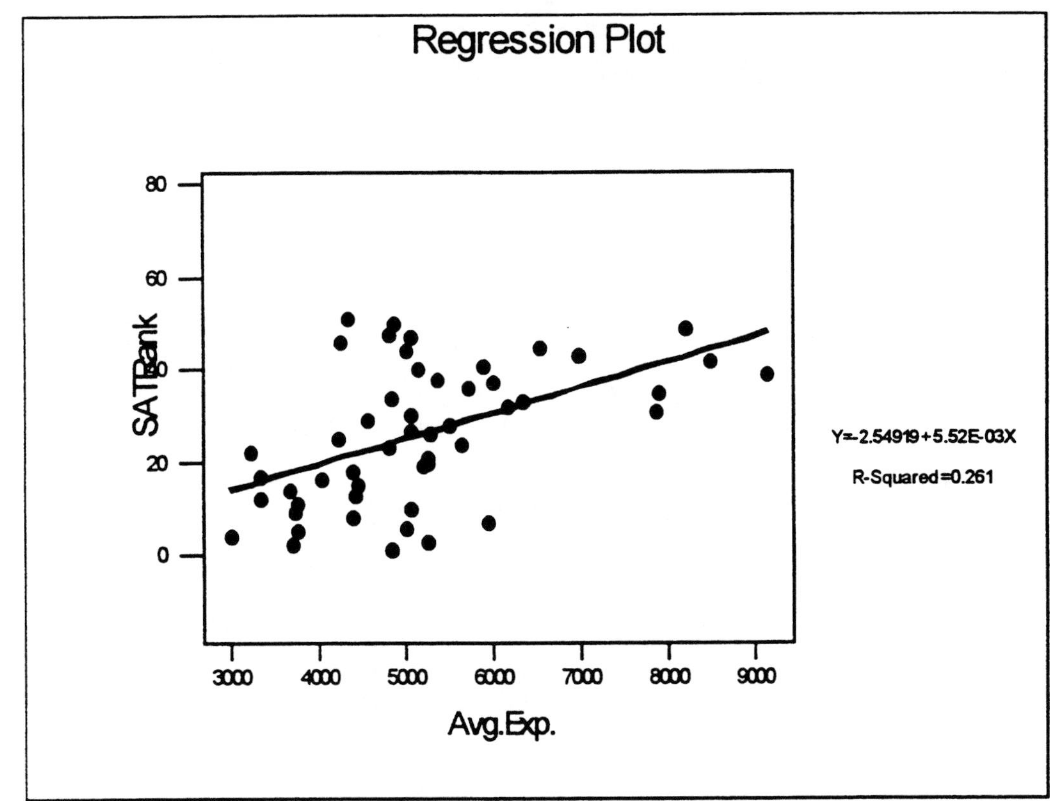

Least squares line: $y = -2.549 + 0.006x$

For an average expenditure of $7000, the SAT rank is predicted to be:
-2.54919 + 0.00525* 7000 = 34.2

c. For expenditures of $7000 or higher, the SAT rank prediction appears to be closer, but there is a wide variation. The positive association between expenditure and rank may be misleading. If there is a very small percentage of high school seniors taking the SAT, then probably only the students who are college bound are taking the test, resulting in higher scores for that state.

3.39.

a.

Regression Analysis

```
The regression equation is
Yt = 1.27 - 0.217 Yt-1

46 cases used 2 cases contain missing values

Predictor         Coef       Stdev    t-ratio       p
Constant        1.2670      0.5491       2.31   0.026
Yt-1           -0.2167      0.1375      -1.58   0.122

s = 3.564       R-sq = 5.3%      R-sq(adj) = 3.2%
```

```
Analysis of Variance

SOURCE          DF          SS          MS          F          p
Regression       1       31.53       31.53       2.48      0.122
Error           44      559.03       12.71
Total           45      590.56

Unusual Observations
Obs.      Yt-1        Yt        Fit   Stdev.Fit   Residual   St.Resid
 20        1.5     8.470      0.933       0.528      7.537      2.14R
 26        0.9    -7.130      1.083       0.527     -8.213     -2.33R
 30       -0.9     8.800      1.460       0.596      7.340      2.09R
 48       10.6    -4.490     -1.025       1.399     -3.465     -1.06 X

R denotes an obs. with a large st. resid.
X denotes an obs. whose X value gives it large influence.
```

b. Residuals plotted over time:

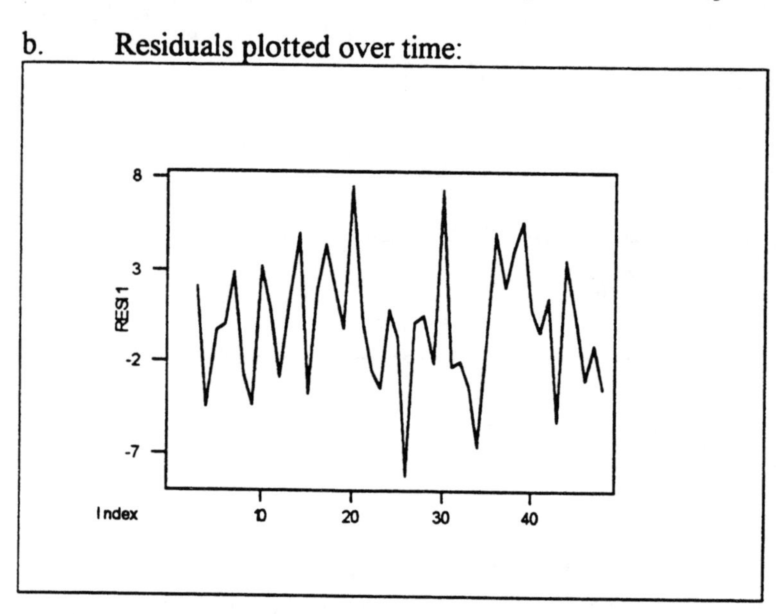

The lag 1 autocorrelation coefficient is 0.028.

There is no obvious pattern in the residuals that appears over time. The residuals appear to vary randomly.

c. and d.

Month	Actual Rate of Return	Forecast Rates of Return	
		Least Squares Equation	Random Walk Model
Jan-91	4.07	0.73835	2.45
Mar-91	2.20	-0.14267	6.51
Jun-91	-4.91	0.44757	3.79
Sep-91	-1.93	0.84685	1.95

Month	Error in Forecast Rate of Return	
	Least Squares Equation	Random Walk Model
Jan-91	3.33165	1.6200
Mar-91	2.34267	-4.3100
Jun-91	-5.35757	-8.7000
Sep-91	-2.77685	-3.8800
Total of absolute values:	13.80874	18.5100

The least squares equation gives the slightly better forecast.

3.40.

a.

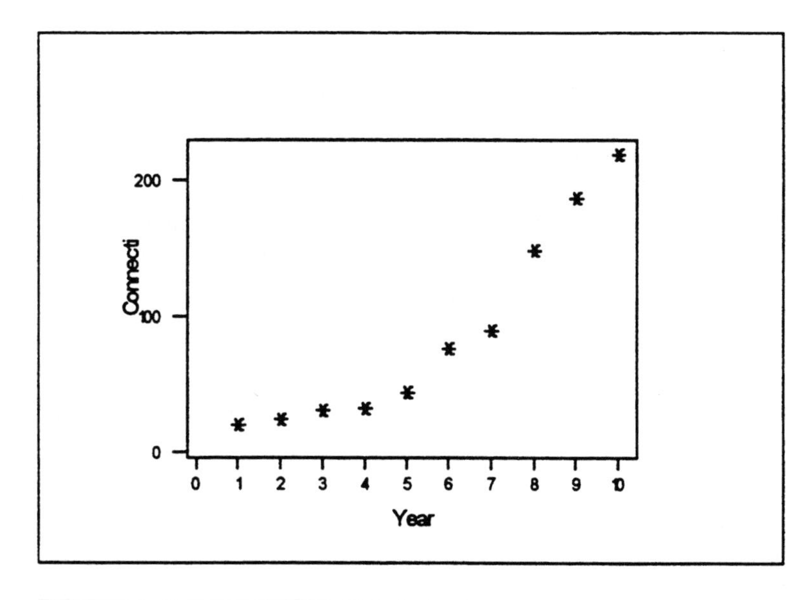

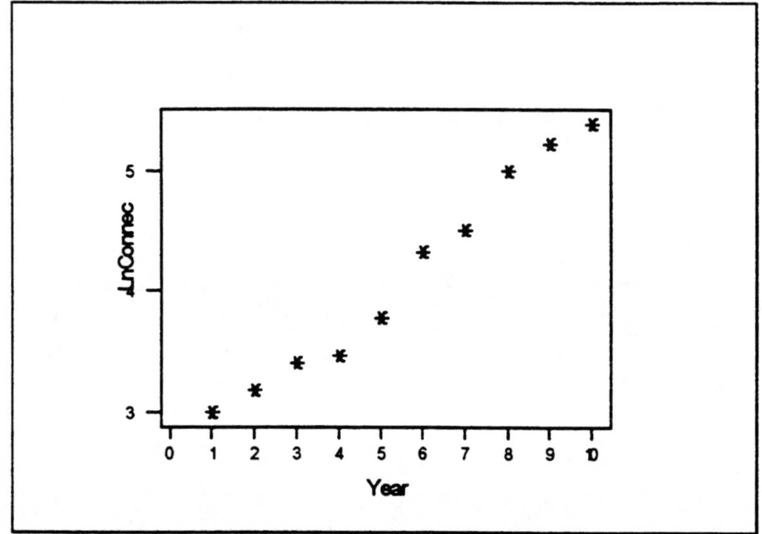

The first graph curves upward and the second graph is fairly straight. The log transformation “straightens out” the scatterplot.

b.

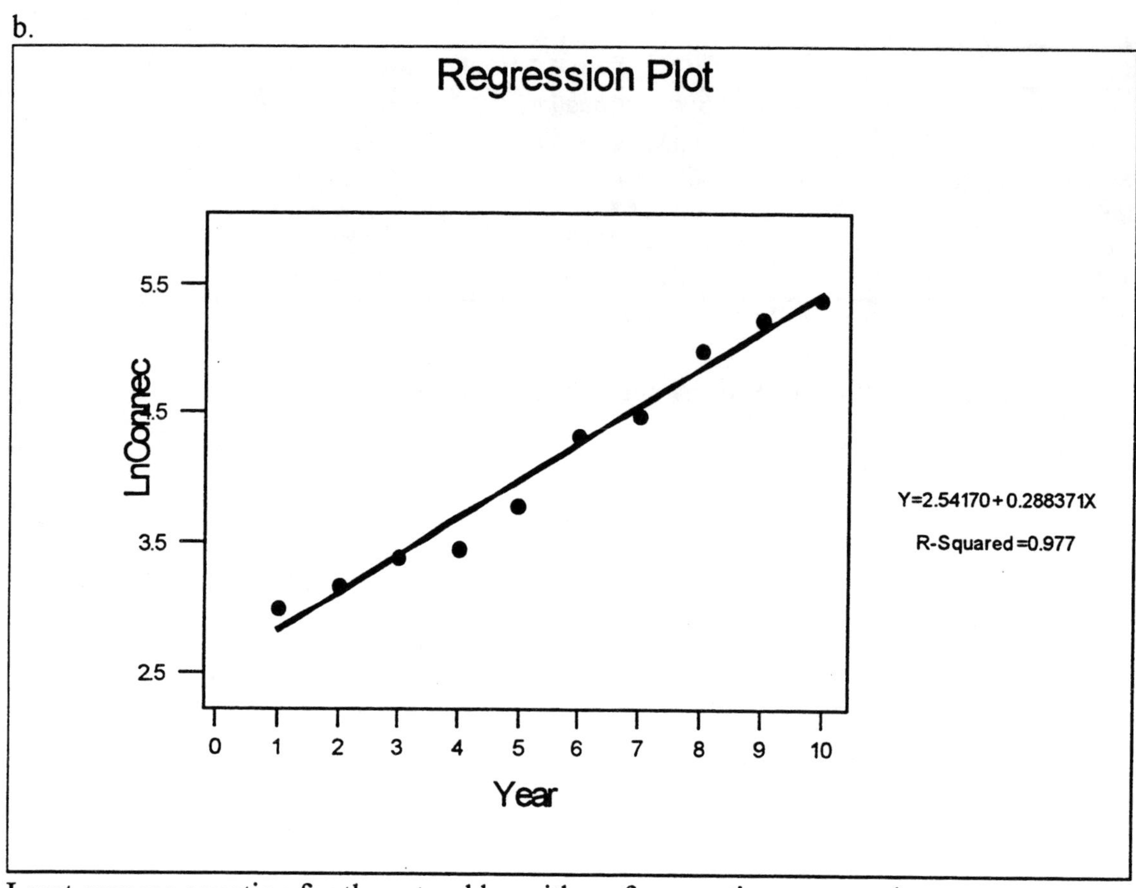

Least squares equation for the natural logarithm of connections versus the year.:
$y = 2.542 + 0.288x$

c. Annual growth rate = 100(antilog(slope coefficient) -1)% =
100(antilog(0.2884)-1)% = 100(1.3343-1)% = 33.4%

d. For year 11, y = 2.54170 + 0.28837*11 = 5.71377
Antilog(5.71377) = 303
It is too far into the future to confidently predict the number of connections for year 20.

3.41.

a.

	Type of Crime				
Status	Burglary	Drugs	Larceny	Other	Total
Discharged	735 (24.53%)	632 (21.09%)	411 (13.72%)	525 (17.52%)	2303 (76.8%)
Received	203 (6.78%)	131 (4.37%)	61 (2.04%)	298 (9.95%)	693 (23.13%)
Total:	938 (31.31%)	763 (25.47%)	472 (15.75%)	823 (27.47%)	2996 (100%)

b. Relative frequencies for discharged inmates:

	Type of Crime				
Status	Burglary	Drugs	Larceny	Other	Total
Discharged	735 (31.91%)	632 (27.44%)	411 (17.85%)	525 (22.80%)	2303 (100%)

Relative frequencies for received inmates:

	Type of Crime				
Status	Burglary	Drugs	Larceny	Other	Total
Received	203 (29.29%)	131 (18.90%)	61 (8.80%)	298 (43.00%)	693 (100%)

c. The largest percentage of criminal types discharged is burglary, followed by drugs. The largest percentage received is other, followed by burglary.

3.42.

a.

	Age of Criminal				
Status	17-26 years old	27-36 years old	37-46 years old	47 years or older	Total
Discharged	833	998	362	110	2303
Received	271	235	141	46	693
Total:	1104	1233	503	156	2996

b. Relative frequencies for discharged criminals:

	Age of Criminal				
Status	17-26 years old	27-36 years old	37-46 years old	47 years or older	Total
Discharged	833 (36.17%)	998 (43.33%)	362 (15.72%)	110 (4.78%)	2303 (100%)

Relative frequencies for received criminals:

	Age of Criminal				
Status	17-26 years old	27-36 years old	37-46 years old	47 years or older	Total
Received	271 (39.11%)	235 (33.91%)	141 (20.35%)	46 (6.64%)	693 (100%)

c. The largest percent of age discharged is 27-36 year old, followed by 17-26 year old. The largest percent of age received is 17-26 year old, followed by 27-36 year old.

3.43. a. and b.

	Volume		
Problem	Low	High	Total
No	180 (45%)	120 (30%)	300 (75%)
Yes	60 (15%)	40 (10%)	100 (25%)
Total	240 (60%)	160 (40%)	400 (100%)

c. Customers who experience problems do proportionately as much low (high) volume business as those who are not experiencing problems. That is, there appears to be no relationship between problems and volume.

3.44.

a.

	Cars Owned		
Income	2 or fewer	More than 2	Total:
< $30,000	140	160	300
$30,000 or more	110	90	200
Total:	250	250	500

b. Lower income households are more likely to have more than 2 cars (160 as compared to 140). This is a counterintuitive result.

c.

	Cars Owned		
Household size	2 or fewer	More than 2	Total:
4 or fewer	225	150	375
More than 4	25	100	125
Total:	250	250	500

d. Smaller households (4 or fewer) are more likely to have 2 or fewer cars. Larger households are more likely to have more than 2 cars.

3.45. Hospital I had 88/2200 = 4% of surgery patients who died.
Hospital II had 21/700 = 3% of surgery patients who died. From this information, Hospital II is preferred as a location for surgery.

3.46. a.
Patients in good condition who died after surgery in Hospital I = 14/700 = 2%.
Patients in good condition who died after surgery in Hospital II = 15/600 = 2.5%.
Patients in poor condition who died after surgery in Hospital I = 74/1500 = 4.93%.
Patients in poor condition who died after surgery in Hospital II = 6/100 = 6%.

b. Hospital I is preferred with this additional information. This answer differs from that in problem 3.45. (Lurking variables changed the relationship between these two categorical variables, an example of Simpson's paradox.)

c. Overall attrition rate for hospital I as a weighted average:

86/2200 = (700/2200) (14/700) + (1500/2200) (74/1500)

0.039 = (0.318) (0.020) + (0.6818) (0.0493)

Of the 2200 surgery patients at Hospital I, 700 were in good condition. Of the 700 who were in good condition, 14 died. Of the 2200 patients at Hospital I, 1500 were in poor condition. Of the 1500 patients who were in poor condition, 74 died. Of the 2200 patients at hospital I, 88 died.

Overall attrition rate for hospital II as a weighted average:

21/700 = (600/700) (15/600) + (100/700) (6/100)

0.03 = (0.857) (0.025) + (0.1429) (0.06)

Of the 700 patients at Hospital II, 600 were in good condition. Of those 600, 15 died. Of the 700 patients at hospital II, 100 were in poor condition. Of the 100 who were in poor condition, 6 died. Of the 700 patients at hospital II, 21 died.

3.47.

a. There does not appear to be any association between promotional expenditure and sales.

Cell frequencies rearranged so that there is a positive association between promotional expenditure and sales: (Answers may vary.)

	Sales		
Promotional expenditure	Below median	Above median	Total
Below median	150	50	200
Above median	50	150	200
Total	200	200	400

Cell frequencies rearranged so that there is a negative association between promotional expenditure and sales: (Answers may vary.)

	Sales		
Promotional expenditure	Below median	Above median	Total
Below median	50	150	200
Above median	150	50	200
Total	200	200	400

Marginal totals cannot change since there still must be half of the expenditures and sales below the median and half above the median by the definition of median.

b.

For stores with promotional expenses above the median:

	Sales		
Price	Below median	Above median	Total
Below median	30	70	100
Above median	60	40	100
Total	90	110	200

The largest group of sales/price combination is for the sales above the median and price below the median, i.e. low price and high promotions create high sales.

c.

For stores with promotional expenses below the median:

	Sales		
Price	Below median	Above median	Total
Below median	45	55	100
Above median	65	35	100
Total	110	90	200

The largest group is for high price and low sales, followed by low price and high sales. Therefore, without as much promotion, low sales are associated with high price and high sales are associated with low price.

d. Promotion increases sales from 55 to 70 when the price is low. The manager should increase promotion and lower price to increase sales.

3.48.

Month	1961	1962	1963	1964	1965	1966	1967	1968	1969	1970	1971	1972	1973	1974	1975	Average
January	56.3	55.3	53.3	53.4	52.9	53.4	52.8	52.6	53.6	53.5	52.1	52.3	53.3	54.8	55.8	53.69
February	55.7	54.9	52.8	53.0	52.6	52.7	52.8	52.1	53.4	53.0	51.5	51.5	53.1	54.2	54.7	53.20
March	55.8	54.9	53.0	53.0	52.8	53.0	53.2	52.4	53.5	53.2	51.5	51.7	53.5	54.6	55.0	53.41
April	56.3	54.9	53.4	53.2	53.0	52.9	55.3	51.6	53.3	52.5	52.4	51.5	53.5	54.3	55.6	53.58
May	57.2	54.6	54.3	54.2	53.6	55.4	55.8	52.7	53.9	53.4	53.3	52.2	53.9	54.8	56.4	54.38
June	59.1	57.7	58.2	58.0	56.1	58.7	58.2	57.3	52.7	56.5	55.5	57.1	57.1	58.1	60.6	57.39
July	71.5	68.2	67.4	67.5	66.1	67.9	65.3	65.1	61.0	65.3	64.2	63.6	64.7	68.1	70.8	66.45
August	72.2	70.6	71.0	70.1	69.8	70.0	67.9	71.5	69.9	70.7	69.6	68.8	69.4	73.3	76.4	70.75
Septembr	72.7	71.0	69.8	68.2	69.3	68.7	68.3	69.9	70.4	66.9	69.3	68.9	70.3	75.5	74.8	70.27
October	61.5	60.0	59.4	56.6	61.2	59.3	61.7	61.9	59.4	58.2	58.5	60.1	62.6	66.4	62.2	60.60
November	57.4	56.0	55.6	54.9	57.5	56.4	56.4	57.3	56.3	55.3	55.3	55.6	57.9	60.5		56.60
December	56.9	54.4	54.6	54.0	54.9	54.5	53.9	55.1	54.3	53.4	53.6	53.9	55.8	57.7		54.79
Average	61.0	59.4	58.6	58.0	58.3	58.6	58.5	58.3	57.6	57.7	57.2	57.3	58.8	61.0	62.2	58.83

There is very little variation from year to year. The variation comes in August and September when employment is higher and December through May when employment is lower each year.

3.49.

Month	*1988*	*1989*	*1991*	*1991*	*1992*	*Average*
Jan	701	923	863	1255	994	947.2
Feb	681	862	1072	1166	847	925.6
Mar	620	881	1440	1037	719	939.4
Apr	618	903	1638	914	778	970.2
May	543	948	1669	993	866	1003.8
Jun	551	952	1640	1008	804	991.0
Jul	848	915	1699	1094		1139.0
Aug	1114	1096	1714	1454		1344.5
Sep	1079	965	1571	1334		1237.3
Oct	1143	1067	1583	1184		1244.3
Nov	1040	1049	1481	1180		1187.5
Dec	1018	948	1418	1033		883.4
Averages	995	1125	1648	1304	1167	1559.7

There are higher inventory levels in July through November and lower levels in December through April.

3.50.

	Age		
Model	<=20 Years	>20 Years	Total
B07	90	123	213.0
	3.09%	3.09%	3.09%
B27	1214	435	1649.0
	41.68%	14.93%	56.61%
B37	1042	9	1051.0
	35.77%	0.31%	36.08%
Total:	2346	567	2913
	80.54%	19.46%	100.00%

B27 and B37 both less than 20 years of age have the largest problem with skincracking rivots.

3.51.

a.

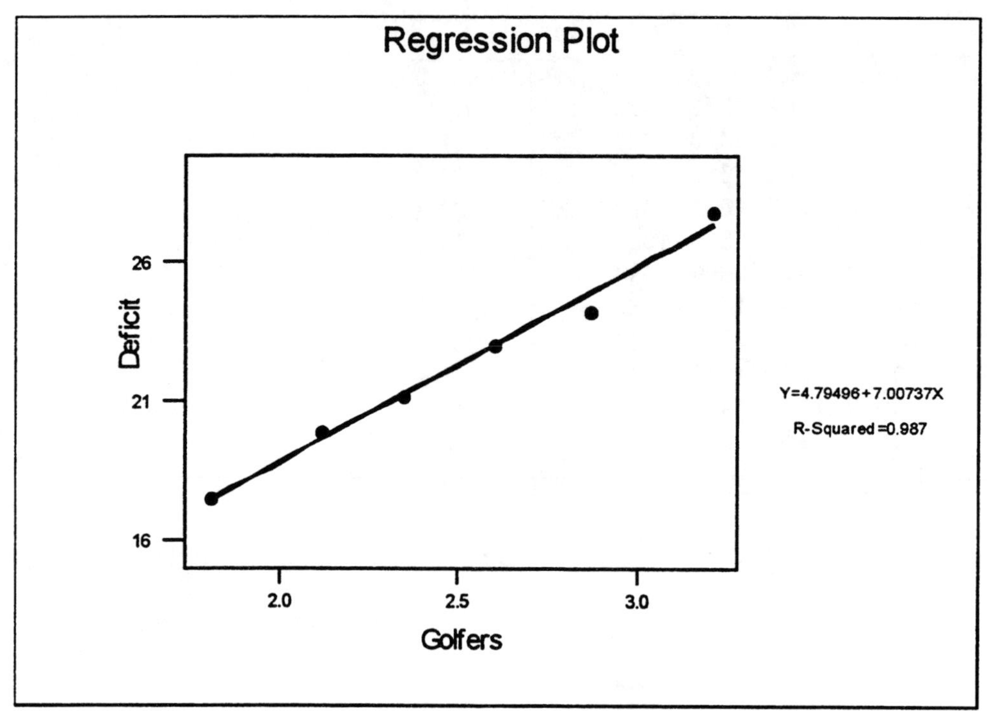

Correlation coefficient = r = 0.993.

b. The correlation coefficient remains r = 0.993.

c. The correlation changes to -0.993.

3.52.

a.

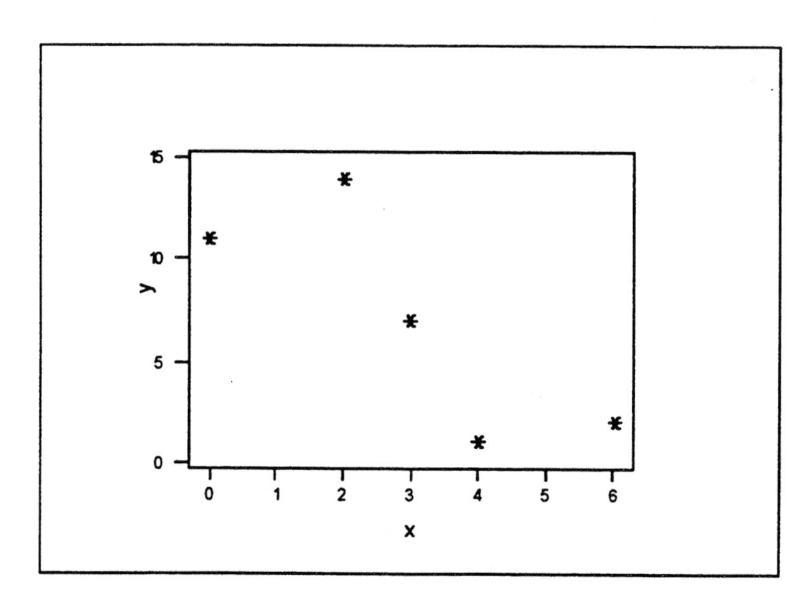

b. The correlation coefficient and the slope are both negative.

c.

x	y	x-xBar	y-yBar	x-xBarSq	Product
0	11	-3	4	9	-12
2	14	-1	7	1	-7
6	2	3	-5	9	-15
3	7	0	0	0	0
4	1	1	-6	1	-6

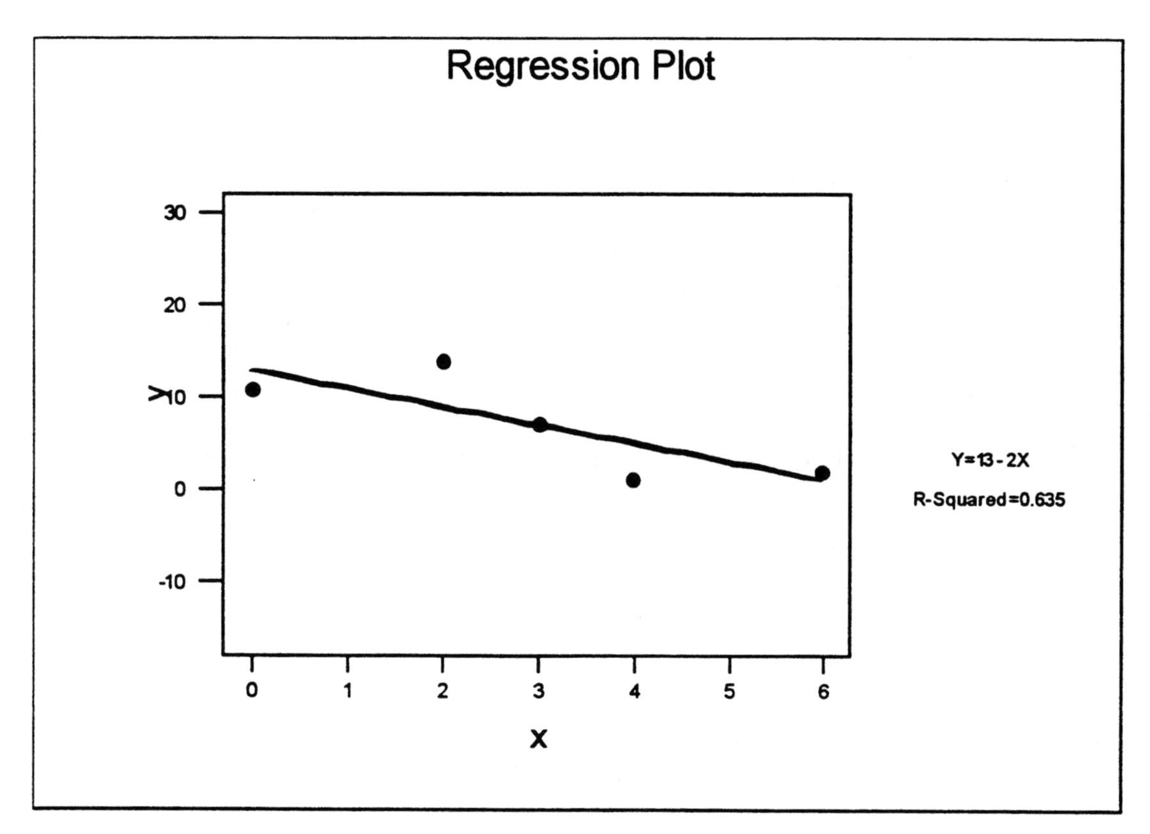

Least squares line: $y = 13\text{-}2x$

d. When x = 5, y = 13 - 2 * 5 = 3.

3.53. The correlation coefficient is r = -0.797.

3.54.

a. The residuals are: -2, 5, 1, 0, and -4.

b. The standard deviation of the residuals is 3.391.

3.55.

a.

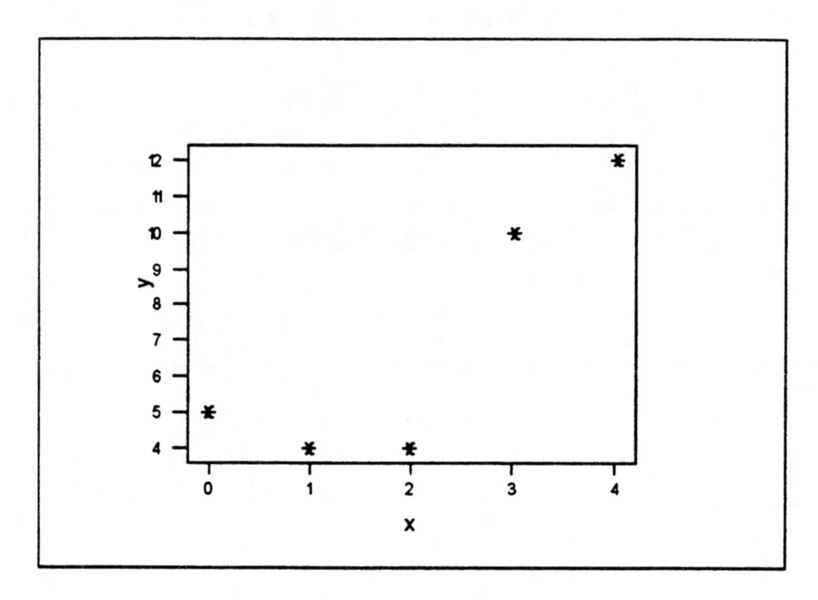

b. The sign of the correlation coefficient and the slope of the least squares line are both positive.

c.

x	y	x-xBar	y-yBar	x-xBarSq	Product
0	5	-2	-2	4	4
1	4	-1	-3	1	3
2	4	0	-3	0	0
3	10	1	3	1	3
4	12	2	5	4	10

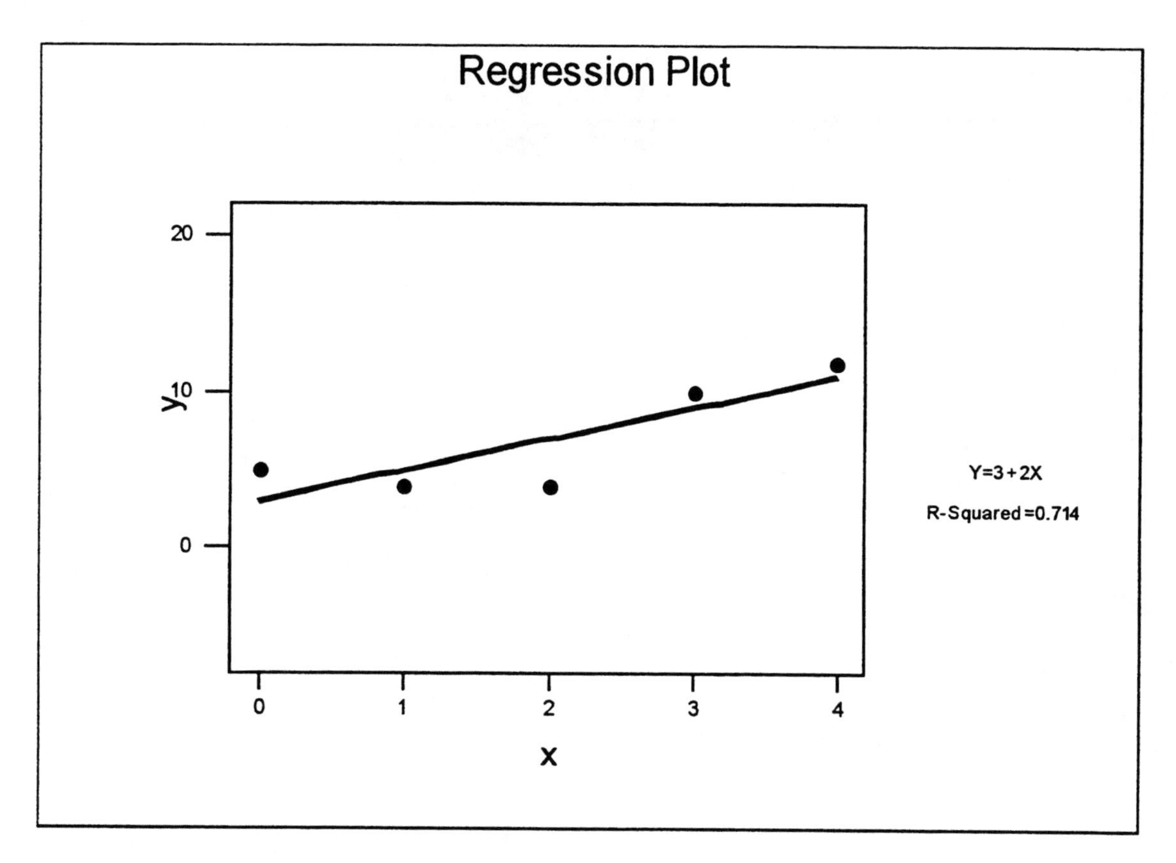

Least squares line: $y = 3 + 2x$

d. When x = 1.5, y = 3 + 2 * 1.5 = 6.

3.56. The correlation coefficient is r = 0.845.

3.57.

a. The residuals are: 2, -1, -3, 1, and 1.
b. The standard deviation of the residuals = s = 2.

3.58.

a. and b.

	Sales		
	< 50 billion	50 billion or more	Total
Japan	7 (35%)	9 (45%)	16 (80%)
U.S.	3 (15%)	1 (5%)	4 (20%)
Total	10 (50%)	10 (50%)	20 (100%)

The largest percent of sales is the bracket in Japan ($50 billion or more); the smallest percent of sales is in the U.S. (50 billion or more.)

3.59.

	Sales		
	< 50 billion	50 billion or more	Total
Japan	7 (28%)	9 (36%)	16 (64%)
U.S.	3 (12%)	1 (4%)	4 (16%)
Other	5 (20%)	0 (0%)	5 (20%)
Total	15 (60%)	10 (40%)	25 (100%)

The largest percent of sales remains that of Japan (50 billion or more). The smallest now is Other (50 billion or more).

3.60. a. and b.

	Assests		
	< 300 billion	300 billion or more	Total
Asia	9 (36%)	8 (32%)	17 (68%)
Europe	6 (24%)	2 (8%)	8 (32%)
Total	15 (60%)	10 (40%)	25 (100%)

The largest category is Asia with less than $300 billion assets . The smallest category is Europe with $300 billion or more assets.

3.61.

a.

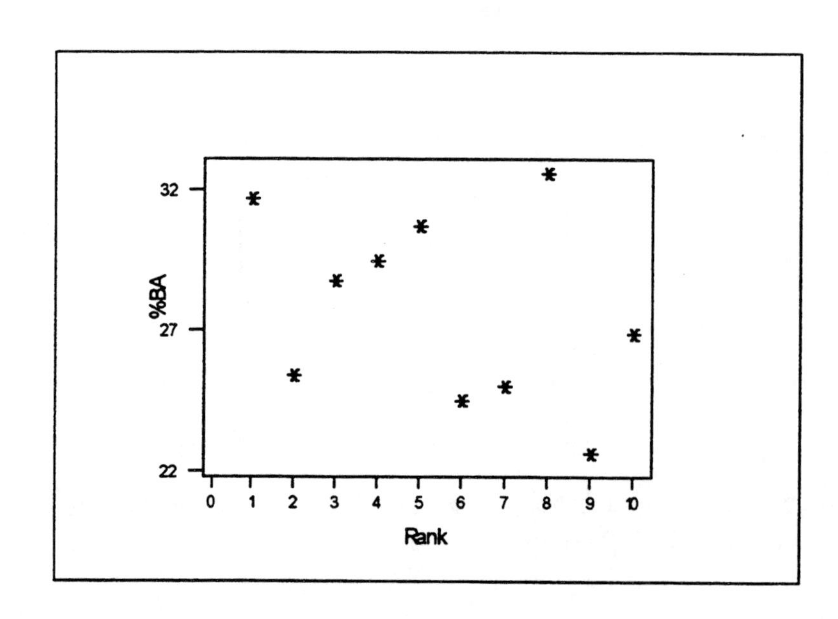

b. The correlation coefficient = -0.345.

3.62. Removing fuzzy dice hanging from rear view mirrors would not improve compliance with the speed limit. The lurking variable is the age of driver who put hanging dice on their rear view mirror. Young drivers (or those who want to be) are more likely to speed.

3.63. The lurking variable is the type of individuals who are likely to drive Corvettes, Camaros, Chargers, and Mustangs or the attitude of those who drive this list of cars. Young drivers are more likely to drive these cars and more likely to have fatal accidents.

3.64.

a. The growth appears to be exponential since the numbers are growing more rapidly in '92 and '93 than they are in earlier years.

b.

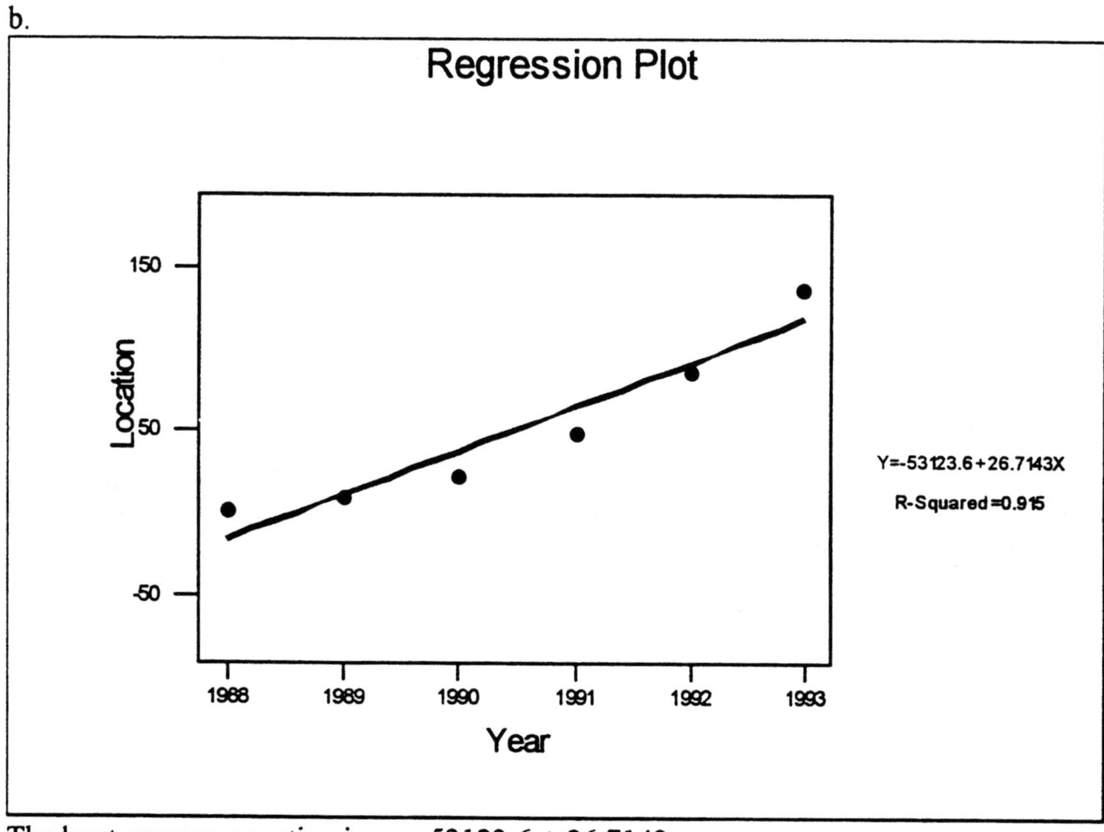

The least squares equation is $y = -53123.6 + 26.7143x$.

c. The predicted number for 1994 based on the least squares equation is y = -53123.6 + 26.7143 * 1994 = 144.7 (almost 145).

3.65.

a. The regression equation is the same as that in Example 3.9.

b. b = Sqrt(422.41 / 718.99) * (0.915) = 0.701

3.66.
The regression equation is $y = 1.89 + 0.002\ x$ from Minitab.

3.67.
a.

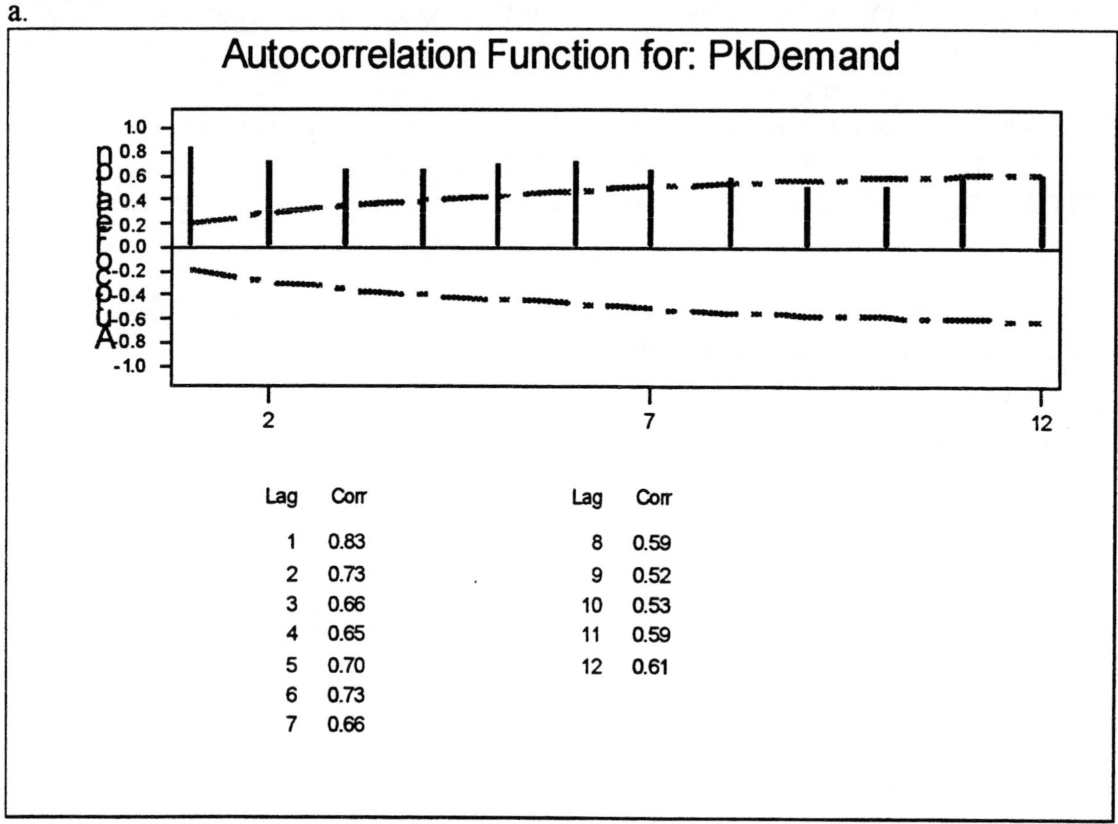

Lag	Corr
1	0.83
2	0.73
3	0.66
4	0.65
5	0.70
6	0.73
7	0.66

Lag	Corr
8	0.59
9	0.52
10	0.53
11	0.59
12	0.61

There is a strong annual seasonal pattern since the autocorrelation coefficients for lags 1-12 are large. The autocorrelation is largest for lag 1 which means that the association between the peak demands in adjacent months is stronger than for the peak demands separated by more than one month.

b.

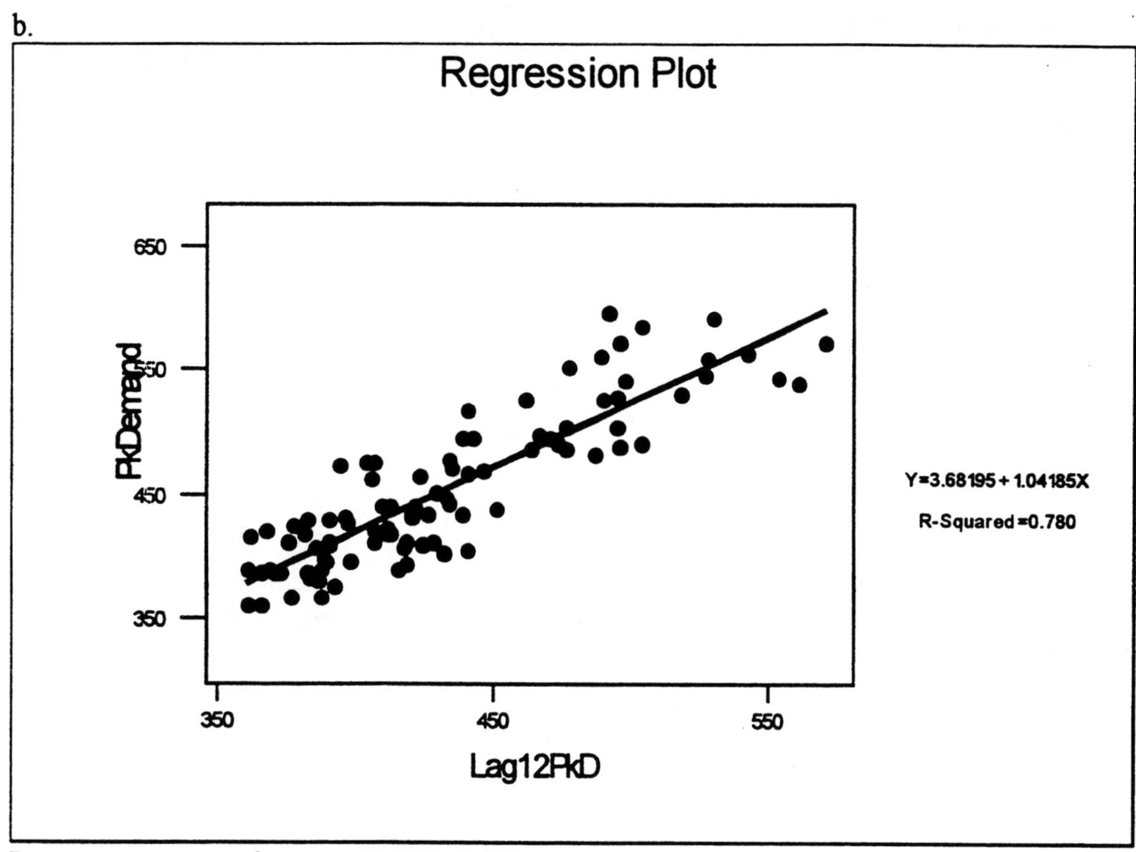

Least squares equation: $y = 3.682 + 1.042x$

c.

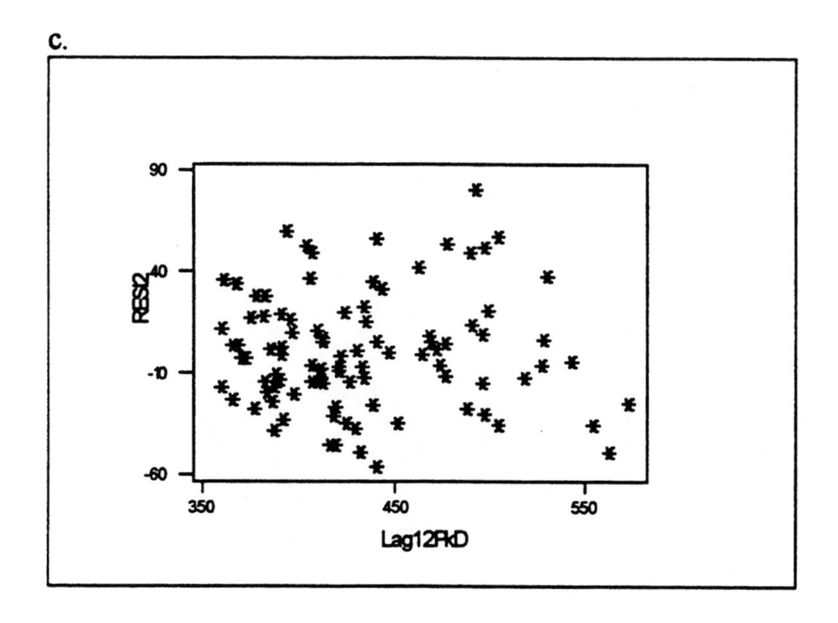

Autocorrelation Function

```
            -1.0 -0.8 -0.6 -0.4 -0.2  0.0  0.2  0.4  0.6  0.8  1.0
              +----+----+----+----+----+----+----+----+----+----+
  1    0.121                          XXXX
  2    0.141                          XXXXX
  3    0.124                          XXXX
  4    0.062                          XXX
  5   -0.040                         XX
  6    0.075                          XXX
  7    0.021                          XX
  8   -0.017                          X
  9   -0.027                         XX
 10   -0.071                        XXX
 11   -0.039                         XX
 12   -0.362                  XXXXXXXXXX
```

There appears to be some correlation among the residuals. The model could be improved by including the peak demand from one or two months ago in predicting the next month's demand.

d. Next May's peak demand is predicted to be $y = 3.68195 + 1.04185 * 500$ = 524.607.

3.68.

a.

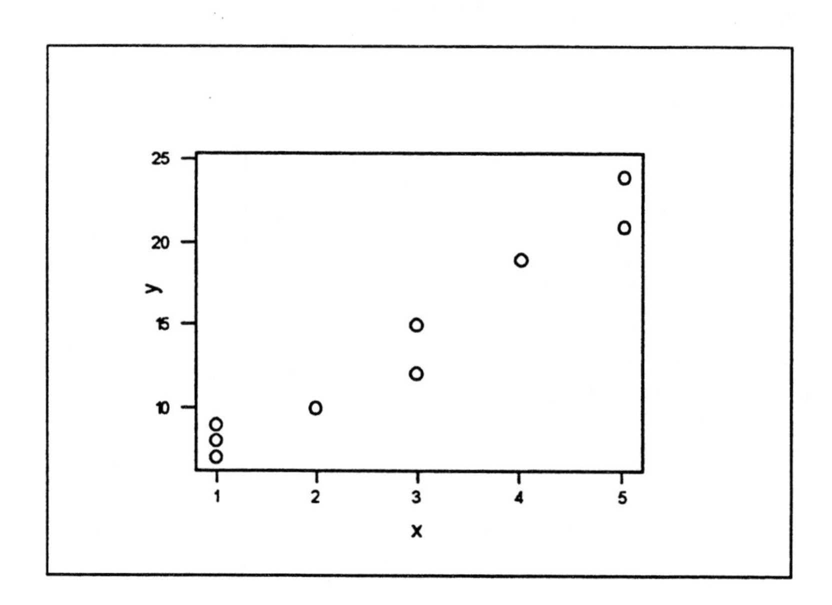

b.

x	y	x-xBar	y-yBar	x-xBarSq	y-yBarSq	Product
1	9	-1.77778	-4.8889	3.16049	23.901	8.6914
1	7	-1.77778	-6.8889	3.16049	47.457	12.2469
1	8	-1.77778	-5.8889	3.16049	34.679	10.4691
2	10	-0.77778	-3.8889	0.60494	15.123	3.0247
3	15	0.22222	1.1111	0.04938	1.235	0.2469
3	12	0.22222	-1.8889	0.04938	3.568	-0.4198
4	19	1.22222	5.1111	1.49383	26.123	6.2469
5	24	2.22222	10.1111	4.93827	102.235	22.4691
5	21	2.22222	7.1111	4.93827	50.568	15.8025

Mean('x') = 2.77778
Mean('y') = 13.8889
S_{xx} = 21.5556
S_{yy} = 304.889
S_{xy} = 78.7778

c.

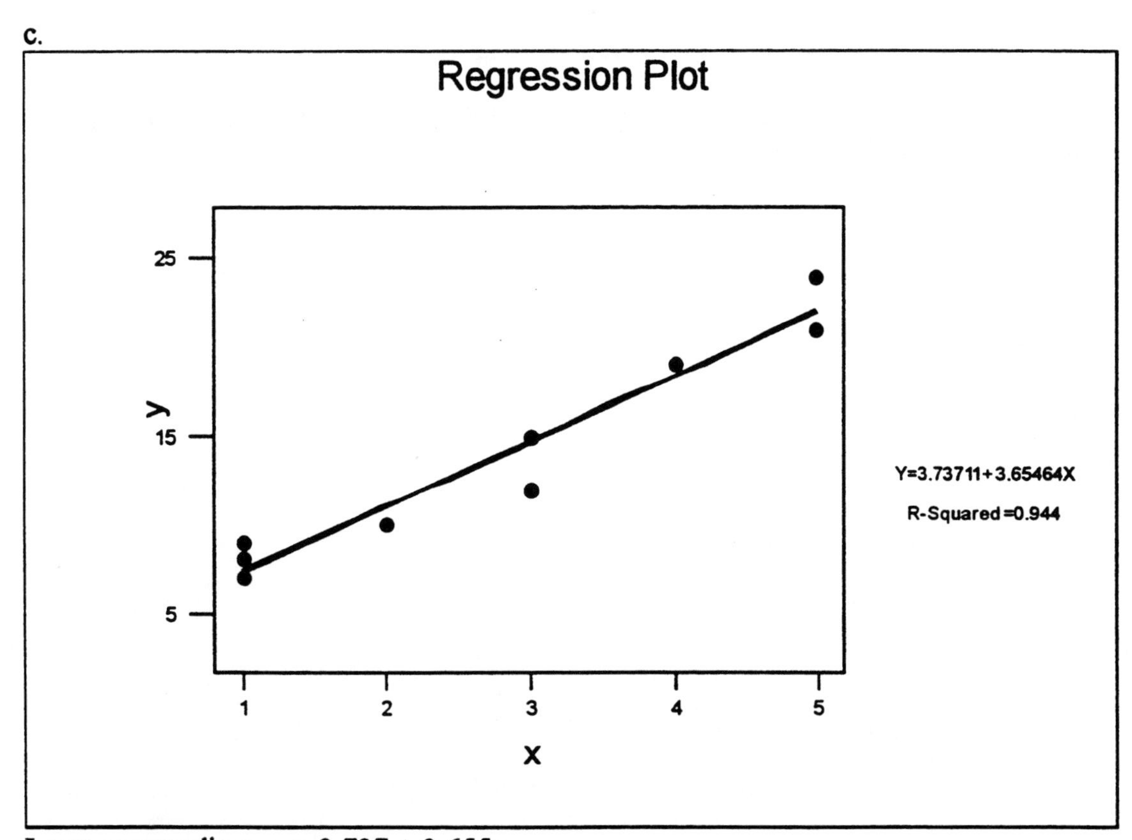

Least squares line: $y = 3.737 + 3.655x$

d. When x = 3, y = 3.73711 + 3.65464*3 = 14.701.

3.69.

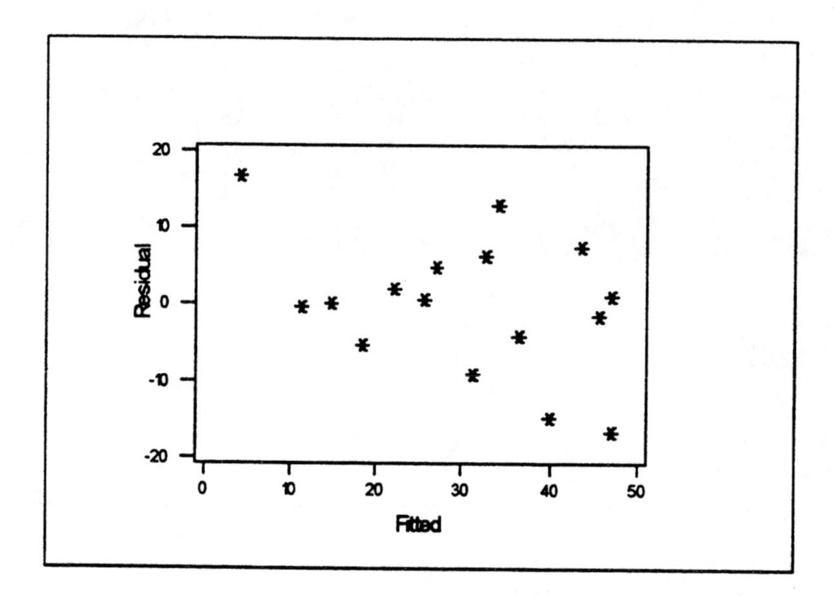

The fitted line does appear to adequately represent the data since the residuals are random and within a horizontal band about zero.

4.1. I agree with the statement. Data collected directly by one person are primary data to that person. When the data are used by another person, it becomes secondary data to the second person.

4.2.

a. Financial information in the Business Week 1000 is secondary data.
b. Customer perceptions are primary data.
c. Annual salaries and bonuses of the highest paid CEO are secondary data.
d. Unemployment rate of the county is secondary data.
e. National choice of TV channels is secondary data.
f. Trucking firm data is primary data.

4.3. Steps in the data collection process:
The social club will collect data with the purpose of determining the reasons why past members have discontinued their memberships and why current members are disenchanted, if at all, with the organization. Previous and current member rosters will be gathered. A questionnaire will be designed and mailed to a predetermined portion of the group. The number of questionnaires mailed will be determined by the funds available and the number of members on the rosters. Personnel will be trained to record and organize data collected. The data will be analyzed and the results interpreted with the original objectives in mind.

4.4. Data collection process:
The highway welcome centers will collect data with the purpose of evaluating their influence on traveler's behavior. Individuals who make use of the welcome center will fill out a questionnaire on their destination, length of stay, reasons for visiting the state and welcome center, state policies on littering, speeding, and so forth. Alternatively, the personnel at the center will ask the questions and record the answers. A predetermined percentage of the travelers will be questioned. Personnel will be trained to record and organize the data collected. The data will then be analyzed and the results interpreted with the original objectives in mind.

4.5. Factors that influence the choice of a method for collecting data include objectives of the study, cost, structure of the population, type of information sought, administrative facilities and personnel available to carry out the plan.

4.6. Factors which might be responsible for lower scores by one ethnic group on the GMAT test include socio-economic background, classroom experiences, and out of classroom environment.

4.7. Additional questions on the GMAT exam could be asked to determine the income level of the student's family, work experience, and other experiences which could effect performance on the exam.

4.8.

a. Television viewers are a nonprobability sample.
b. Every twelfth invoice is an example of a probability sample.
c. The engines chosen represent a probability sample.

4.9.

a. The top 20 plates are a non-probability sample.
b. Material cut from the end of the roll is a non-probability sample.
c. Call length samples are a probability sample.

4.10.

a. The group of viewers is a non-probability sample since the group is selected from interested viewers instead of from the general population.
b. The invoices chosen represent a systematic random sample.
c. The automobile engines chosen represent a simple random sample.

4.11.

a. The plates chosen are all from the tops of the boxes, therefore this is non-probability sample.
b. The material chosen is a non-probability sample since the pieces are all from the ends of the rolls.
c. The calls chosen are a systematic random sample.

4.12. Answers will vary.

4.13. These two assertions imply that public opinion polls should not be used in politics. Politicians, however, should be aware of the ideas of their constituents in order to use their opinions in forming new laws. Polls perhaps should be banned prior to elections so that individuals who might otherwise be swayed to vote for the winning candidate will instead make their own decision.

4.14.

a. The frame consists of all past and current employees at the plant. A stratified random sampling method may be used. Segment the employees by job title.

b. The frame consists of all diskettes produced by the company. A systematic random sampling method may be used. Choose a random disk from the first, say 100 disks, then choose every 100th disk thereafter.

c. The frame consists of all students at the large university. A simple random sample may be selected. Alternatively, a stratified sampling method may be used, segmenting the student population by, for example, classification, then choosing random samples from each segment.

d. The frame consists of all hospitals in the country. A single random sample may be selected. Alternatively, a stratified sampling method may be used, segmenting the hospitals by, for example, size of hospital, size of city, and part of the country, then choosing random samples from each segment.

4.15.

For the manufacturing plant:

a. The actual frame may have to consist of only current employees since previous employees may be difficult if not impossible to contact.

b. Personal verification, questionnaire, or a telephone call could all be used. A questionnaire could be used first, followed by random personal verification, and telephone calls to those who did not respond to the initial questionnaire.

c. The advantage of using a questionnaire is its ease as compared to personal contact. The disadvantage of using a questionnaire is that some employees may not respond.

For the software company:

a. The frame should be relatively easy to construct since all disks produced during a certain period of time are in the control of the company.

b. The sample must be conducted by loading and running the software program through most or all of its features to prove that it is not defective.

c. The disadvantage in using this sampling method is that it may not be possible to use all features of the software due to time constraints.

For the large university:

a. The problem with constructing the frame is connected with having incorrect addresses for some students.

b. Questionnaires should be mailed to each student who is randomly selected. A follow-up mailing could be used.

c. The advantage of a questionnaire is the convenience and cost as compared to personal questioning. The disadvantage of this method is related to incorrect addresses and students who do not respond to the questionnaire.

For the pharmaceutical company:

a. The difficulty in construction of the frame is finding out the names and addresses of all hospitals.

b. Questionnaires are to be mailed to the selected hospitals.

c. The advantage of using a questionnaire is the convenience and cost relative to on-site interviews. The disadvantage is the possible lack of response by some units.

4.16.

a. The frame consists of all drop sites where the newspaper is deposited for distribution. A simple random sample could be selected. Alternatively, a stratified random sampling method may be used for choosing sampling units, segmenting the drop sites by, for example, geographical area.

b. The frame consists of all individuals with access to the county's computer system. A simple random sampling method may be used for choosing sampling units.

c. The frame consists of all unemployment offices in the U.S. A simple random sampling method may be used to choose sampling units. The number of unemployed at the chosen offices is recorded.

d. The frame consists of the judge's voter constituency. A simple random sample may be chosen from the constituency list.

4.17.
For the community newspaper:
a. The frame should be easy to construct since the newspaper has a list of all drop sites.
b. Direct measure is used at each drop site, comparing the number of newspapers deposited each day with the number remaining the next day.
c. The advantage of using this method of sampling is that the employee of the newspaper does not have to make an extra trip to go to each location. The disadvantage comes if some of the newspapers are thrown away instead of being picked up by a reader.

For the computer system:
a. The frame should be easy to construct; the computer system can print out a list of users. This list can be compared to the list of county enplanes with a computer program since the employees' names are also in the computer's memory for payroll purposes.
b. The disadvantage is that a computer program may have to be written to achieve the comparison. The advantage is that accuracy should be 100% if all of the information is in the computer and correct.

For the unemployed workers:
a. The construction of the frame is easy since the U.S. Bureau of Labor Statistics has a list of all unemployment offices.
b. The selection of offices could be done using a random number table. A mail questionnaire can be sent to each office.
c. The advantage of a mail questionnaire are the convenience and cost. The disadvantage comes from nonresponse or those offices who may not respond to the questionnaire.

For the judge:
a. The frame could be constructed by using a list of all registered voters. This should be readily available.
b. A table of random numbers can be used to select individuals for the sample. A mail questionnaire is sent to all voters in the sample.
c. The advantages of this sampling method are convenience and cost. The disadvantage is that non-registered members of the community are not included. Other lists would have to be merged with the voter registration list to include all community

members, such as vehicle registration, driver's license registration, and/or tax rolls. Also, mail surveys always have the possibility of a large nonresponse.

4.18.

a. Only households with someone home during 8 to 5 are included in this survey. Those households where all family members are gone during this time are not included.

b. Only listeners of the radio station who are interested in taking the time to call are included.

c. It is possible that locations around the perimeter of Texas would not be as likely to be included.

d. This sample does not include customers who are more likely to shop during morning hours (perhaps retired people) or customers who shop at night (perhaps those who work during the day.)

4.19. (Answers will vary.) One solution: choose 10 random numbers from a random number table. leaving out numbers greater than 32 and repeats:

04 23 17 13 09 30 21 11 20 02

giving these values:

60
61
70
64
60
65
64
68
64
67

The sample mean is 64.3.

4.20. (Answers will vary.) Random numbers (leave out numbers greater than 588 and duplicates):

476
5
491
467
360
128
495
212
197
216
467
310
285
2
107
252

412
256
526
233
74
266
468
172
549
50
332
442
166
2
372
567
513
222
273
204
209
434
1
493
9
370
311
167
504
340
516
248
199
108
492
76
73
386
500
132
519
459
398
203
61
531

Sample:

17
9
21
26

22
11
45
33
14
27
24
27
24
34
29
44
27
31
20
12
30
21
37
45
39
23
41
48
39
17
30
33
31
33
39
62
51
41
56
23
31
46
30
42
48
16
79
54
81
63
72
53
71
85
54

55
28
28
52
74

b. The sample mean is 38.3.

4.21. (Answers will vary.)

a.

Small sample	Medium sample	Large sample
34	33	56
37	44	45
15	25	48
21	24	28
19	48	54
11	18	77
26	65	74
14	54	57
39	39	95
26	66	35
Means: 24.2	41.6	56.9

b. Overall sample mean = (215/588) * 24.2 + (258/588) * 41.6 + (115/588) * 56.0
=38.054

c. Means:

Small	Medium	Large
27.702	36.693	58.904

The population mean is 37.092 compared to 38.054 for the overall sample mean and 38.3 for the unstratified sample mean.

4.22.

a. The sample proportion of faulty bills is 48/500 = 0.096.
b. This is enumerative study since the auditor needs to the current situation, rather than predict what will happen in the future.

4.23.

a. The statement can compare the mean age of users to the mean age of nonusers.
b. The image is a qualitative description, not related to a numerical measurement, so the company cannot legitimately make a statement involving the phrase “twice as positive.”

4.24.

a. Mean response of first sample = (1*10 + 2*10 + 3*70 + 4*60 + 5*20 + 6*30)/200 = 3.8. The mean response makes sense since the categories are arranged in order from least to most likely to buy.

b. Mean response of second sample = (1*120 + 2*40 + 3*10 + 4*10 + 5*10 + 6*10)/200 = 1.9.

c. The mean response for the second sample is half of that for the first sample. This does not mean that the first sample is twice as likely to buy a car since these numerical measures cannot be compared to one another. It is only true that the second sample is less likely to buy overall than the first sample.

4.25.

a. What portion of your college expenses are you paying?

b. What portion of your college expenses are you paying?
Check one of the following: ______ More than half
______ Less than half

c. What portion of your college expenses are you paying?
Check one of the following: ______ None
______ Between 1% and 25%
______ Between 25% and 50%
______ Between 50% and 75%
______ Between 75% and 99%
______ 100%

4.26.

a. The responses to question "4.25.a" above yield a ratio scale. Quantitative data is obtained.

b. The responses to question "4.25.b" above yield a nominal scale. Qualitative data is obtained.

c. The responses to question "4.25.c" above yield a interval scale. Qualitative data is obtained.

4.27. (Answers will vary.)

a. A double-barreled question: Are department stores concerned about their customers and the quality of goods that are sold?

b. A leading question: Don't you think finance companies try to cheat the consumer?

c. A one-sided question: How important is a college education to you?
______ extremely important
______ very important
______ important
______ not important

4.28. (Answers will vary.)

a. Are department stores concerned about their customers?
Are department stores concerned about the quality of goods sold?

b. How do finance companies treat the consumer?

______ fairly
______ unfairly

c. How important is a college education to you?

______ very important
______ important
______ unknown
______ not important
______ very unimportant

4.29.

a. This is a leading question since the reader may be more likely to agree.
Better wording: How has deregulation affected the consumers?

_____ Benefited consumers
_____ Hindered consumers

b. The respondent may not know how to answer this question. Is a yes or no answer required?
Better wording: Give your opinion of the proposed 5% increase in the city tax rate.

_____ Favorable
_____ Indifferent
_____ Unfavorable

c. This is a leading question.
Better wording: How does the quality of products today compare with the quality of products 10 years ago?

4.30. (Answers may vary.)

a. _____ 1 year
_____ 2 years
_____ 3 years
_____ 4 years

b. _____ Television
_____ Radio
_____ Newspapers

c. _____ A lot
_____ A little

4.31. (Answers will vary.)

4.32.

a. (Tom, Erik), (Chris, Andy), (Sue, Grace)
(Tom, Chris), (Erik, Andy), (Sue, Grace)
(Tom, Andy), (Erik, Chris), (Sue, Grace)

b. Two pairs are available for the experiment if they are paired by gender.

4.33.
(Al, Bob), (Carol, Dennis, Ellen)
(Al, Carol), (Bob, Dennis, Ellen)
(Al, Dennis), (Bob, Carol, Ellen)
(Al, Ellen), (Bob, Carol, Dennis)
(Bob, Carol), (Al, Dennis, Ellen)
(Bob, Dennis), (Al, Carol, Ellen)
(Bob, Ellen), (Al, Carol, Dennis)
(Carol, Dennis), (Al, Bob, Ellen)
(Carol, Ellen), (Al, Bob, Dennis)
(Dennis, Ellen), (Al, Bob, Carol)

4.34.
a. These are independent samples. The experimental units are the 40 donors. The treatment is the solicitation for donations. The response is the donation amount.

b. These are matched pairs. The experimental units are the 10 husbands and wives. The treatment is the question as to how many children. The response is the answer to the question of how many children.
c. These are independent samples. The experimental units are the 100 checks. The treatment is the question as to why checks are not cashed. The response is the reason why the checks are not cashed.

4.35.
a. The experimental units are the 27 trees. The explanatory variables are the amount of insect spray, the weather, and the resistance of the tree itself. The response variable is the amount of fruit yield from each tree.
b. The experimental units are 100 people. The explanatory variables are the quality of the training, the ability of each individual to do work without making careless errors, and the number of vouchers created. The response variable is the number of errors on each voucher.
c. The experimental units are 22 workers. The explanatory variables are the quality of the work design and the difficulty level of assembling the piece of equipment. The response variable is the time it takes in minutes to assemble a standard piece of equipment.

4.36.
a. This is not an experiment since there are no experimental units established and there is not a treatment or a response. This is just an observation on the part of the MTV executive.
b. This is not an experiment since there are no experimental units established and there is not a treatment or a response. This is just an observation on the part of the sales manager.
c. This is not an experiment since there are no experimental units established and there is not a treatment or a response. This is just an observation on the part of the manufacturer.

4.37.

a. The explanatory variable is whether the area was wired for cable. The response variable is the number of Duran Duran albums purchased

b. The explanatory variable is the number of calls per customer. The response variable is sales.

c. The explanatory variable is the payment terms. The response variable is the "days until payment" for each distributor.

4.38.

a. Potential lurking variables are the socio-economic status and age of the residents in each area of the city.

b. The potential lurking variable is sales ability of each of the sales employees.

c. The potential lurking variable is the other financial aspects of each company including the type of product sold by the company and whether the company is getting into financial trouble.

4.39. A matched pair test for Pepsi vs. Coke would allow each individual to taste each of the drinks blindfolded and choose their favorite. The order of the drinks for each individual would be selected at random, perhaps by a flip of a coin. This matched pair test would control for any variation in taste preferences that might occur between two people.

4.40. Each subject will look at each advertisement and rate it on a scale of 1 (worst) to 5 (best) as to the more preferable brand image that is portrayed. The means for the advertisements are compared.

4.41. (Answers will vary.)

4.42. The respondents tend to be long-time residents, educated, have low incomes, but they are younger than 40 instead of over 40.

a. The proportion of respondents who are over 40 is 10 + 17 + 19 = 46%

b. The proportion of respondents who have some post high school education is 12 + 18 + 19 + 23 = 72%

c. The proportion of respondents who have lived in the area more than 5 years is 8 + 66 = 74%.

d. The proportion of respondents who have an annual income of $30,000 or less is 17 + 13 + 11 + 11 + 8 = 60%.

4.43.

Age group	Excellent or good	Other	% Exc or good
Under 18	0	1	0.00%
18-21	23	8	74.19%
22-30	129	31	80.63%
31-40	114	28	80.28%
41-50	56	8	87.50%
51plus	224	15	93.72%
No reply	39	4	90.70%

Therefore these older customers tend to have a more favorable image of the bank's services relative to those customers in other age groups.

4.44.

Which is your category of employment?

_____ University
_____ Downtown
_____ State government

What is your degree of support for the rapid transit system?

_____ Support
_____ Do not support
_____ No opinion

4.45. (Answers will vary.)

a. Nominal scale: Brand of toothpaste used
b. Ordinal scale: Highest college degree earned
c. Interval scale: Heat index
d. Ratio scale: The amount of your weekly grocery bill

4.46.

a. What is your current income? Check one of these intervals:

_____ Less than $20,000
_____ Between $20,000 and $40,000
_____ Between $40,000 and $60,000
_____ Greater than $60,000

What is the highest level of education you completed?

_____ Junior high school
_____ High school
_____ Junior college
_____ Bachelor's degree
_____ Master's degree
_____ Doctoral degree

b. For each of the following features, rate its importance to you in selecting a prefabricated home:

		No importance	Little importance	Very important
1.	Cost	________	____________	____________
2.	Floor plans	________	____________	____________
3.	Availability	________	____________	____________
4.	Financing	________	____________	____________
5.	Sales staff	________	____________	____________

c. Each of the questions in part "b" is structured and uses an ordinal scale.

4.47. Sell the same clothes at two prices in two stores that are alike in size and customer base. Compare the total profit for the sold clothing items from the line sold at each store.

4.48. Randomly divide the students into two groups of 30 students each. One group is taught with the traditional method. The other is taught by the new method. Test the students before the instruction, then again after the instruction. Compare the scores of each group.

4.49. After 10,000 miles of driving, record the tread depth remaining on each rear tire for each of the 20 cars. Compare the mean depths remaining for each tire brand.

4.50. Twenty-five homes are to be appraised by each appraiser. The difference between appraised values is recorded and compared.

4.51. (Answers will vary.)
a. An example of independent samples: Current annual profit per square foot for department stores of a large chain will be used to determine whether profit is higher in small towns or large cities. Ten stores in small cities are randomly selected and 20 stores in large cities are randomly selected for the comparison. Average profit per store is calculated for each group.
b. An example of a paired sample: A chain of fast food restaurants is interested in the relative popularity of two menu items: a deluxe hamburger and a chicken breast sandwich. The numbers of hamburgers and chicken sandwiches sold in a week at each of ten randomly selected restaurants of the chain are recorded. The mean difference between the number of hamburgers sold and the number of chicken sandwiches sold is computed.

6.1.

a. Discrete
b. Continuous
c. Discrete
d. Continuous
e. Discrete
f. Discrete

6.2.

a. Continuous
b. Continuous
c. Discrete
d. Discrete
e. Continuous

6.3.

a.

Elementary Outcomes	X = Difference
2, 4	2
2, 6	4
2, 7	5
2, 8	6
4, 6	2
4, 7	3
4, 8	4
6, 7	1
6, 8	2
7, 8	1

b.

X	P(X)
1	2
2	3
3	1
4	2
5	1
6	1

6.4.

a. All ratings that A can receive: (1, 1), (1, 2), (1,3), (2,2), (2, 3), (3, 3)
b. All distinct values of X: 2, 3, 4, 5, 6

6.5.

a. All ratings that A can receive: (1,1,1,1), (1,1,1,2), (1,1,2,1), (1,1,2,2), (1,2,1,1), (1,2,1,2), (1,2,2,1), (1,2,2,2), (2,1,1,1), (2,1,1,2), (2,1,2,1), (2,1,2,2), (2,2,1,1), (2,2,1,2), (2,2,2,1), (2,2,2,2)

b. All distinct values of X: 4, 5, 6, 7, 8

6.6.

Let H = higher and L = lower.
All possible outcomes and the number of days, X, that the night shift has a higher production rate:
HHH (3), HHL (2), HLH (2), HLL (1), LHH (2), LHL (1), LLH (1), LLL (0)

6.7.

a. S = {CCC, CCB, CBB, CBC, BCC, BCB, BBC, BBB}

b. Number of switches: CCC (0), CCB (1), CBB (1), CBC (2), BCC (1), BCB (2), BBC (1), BBB (0)

6.8.

a. Ann may chose B or C. Barbara may chose A or C. Carol may chose A or B.
{BAA, BAB, BCA, BCB, CAA, CAB, CCA, CAB}

b. The possible values of X are 0, 1, or 2.

6.9.

x	f(x)
0	0.29 + 0.08 + 0.11 + 0.13 = 0.61
2	0.08 + 0.15 = 0.23
4	0.16

6.10.

a.

Pairs of integers	X
1, 2	3
1, 6	7
1, 7	8
1, 9	10
2, 6	8
2, 7	9
2, 9	11
6, 7	13
6, 9	15
7, 9	16

b. Distinct values of X: 3, 7, 8, 9, 10, 11, 13, 15, 16

c.

Sum	Times occurred
3	1
7	1
8	2
9	1
10	1
11	1
13	1
15	1
16	1

x	f(x)
3	0.1
7	0.1
8	0.2
9	0.1
10	0.1
11	0.1
13	0.1
15	0.1
16	0.1

6.11.

a. $P(x \leq 3) = 0.05 + 0.10 + 0.15 + 0.20 = 0.50$

b. $P(x \geq 5) = 0.15 + 0.10 = 0.25$

c. $P(2 \leq x \leq 5) = 0.15 + 0.20 + 0.25 + 0.15 = 0.75$

6.12.

a. This is a legitimate probability distribution.

b. This is not a legitimate probability distribution since the probabilities do not sum to 1.

c. This is a legitimate probability distribution.

d. This is not a legitimate probability distribution since there cannot be a negative number as a probability.

6.13.

a.

x	f(x)
3	0.1
4	0.2
5	0.3
6	0.4

This is a legitimate probability distribution since the probabilities sum to 1.

b.

x	f(x)
1	-0.5
2	0.0
3	0.5
4	1.0

This is not a legitimate probability distribution since there cannot be a negative number as a probability.

c.

x	f(x)
-2	0.0
-1	0.1
0	0.2
1	0.3
2	0.4

This is a legitimate probability distribution.

d.

x	f(x)
2	0.75
3	0.375
4	0.1875
5	0.09375

This is not a legitimate probability distribution since the probabilities do not sum to 1.

6.14.

a. $P(X=2) = 0.4$

b. $P(X \text{ is odd}) = P(X=1) + P(X=3) = 0.267 + 0.267 = 0.533$

6.15.

From 6.7: CCC (0), CCB (1), CBB (1), CBC (2), BCC (1), BCB (2), BBC (1), BBB (0)

x = # of switches	f(x)
0	0.25
1	0.50
2	0.25

6.16.

x	f(x)
0	0.250
1	0.625
2	0.125

6.17.

Squares are 0, 1, and 4.

x	f(x)
0	0.2
1	0.4
4	0.4

6.18.

x	f(x)
2	0.400
3	0.533
4	0.067

6.19.

a. $X = 0, 1, 2,$ or 3 questions can be answered correctly.

b.

x	f(x)
0	0.222
1	0.444
2	0.278
3	0.056

c. $P(X \geq 1) = 0.444 + 0.278 + .056 = 0.778$

d.

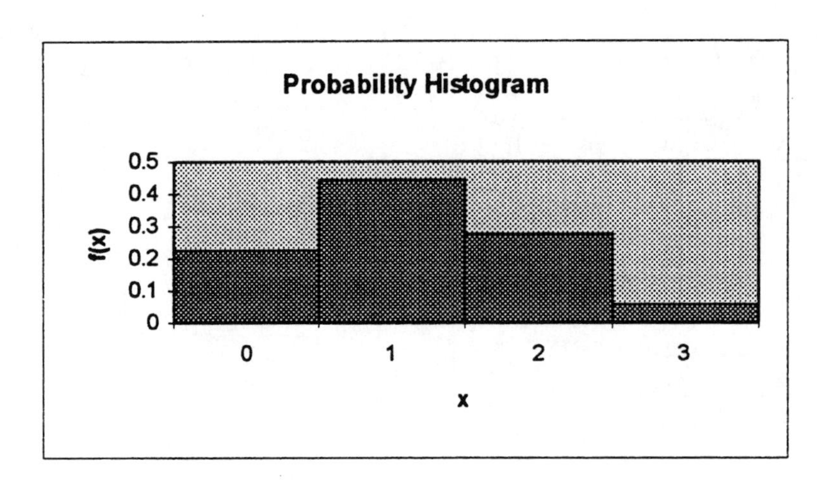

6.20.

x	f(x)
1	0.2
2	0.1
3	0.2
4	0.1
5	0.3
6	0.1

6.21.

x	f(x)
1	0.25
2	0.28
3	0.30
4	0.17

6.22. (Answers will vary.)

a. In an investment of a certain type, 36 percent of the time the investment doubles in 5 years, 42 percent of the time it quadruples in 5 years, and 22 percent of the time it increases by a power of 6 in 5 years.

b. In a certain gambling game, there are 11 possible outcomes. In 3 of them, you lose 2 dollars. In 4 of the outcomes, you break even. In 2 outcomes, you make 4 dollars. In 2 of the outcomes, you make 5 dollars.

6.23. a.

x	f(x)
0	70/200 = 0.35
1	84/200 = 0.42
2	36/200 = 0.18
3	10/200 = 0.05

b. $P(X \geq 2) = 0.18 + 0.05 = 0.23$

6.24.

x	f(x)
1	106/203 = 0.522
2	72/203 = 0.355
3	25/203 = 0.123
4	0/203 = 0

6.25.

a. $P(X \leq 3) = 0.12 + 0.25 + 0.43 + 0.12 = 1 - 0.08 = 0.92$

b. $P(X \geq 2) = 1 - 0.12 - 0.25 = 0.63$

c. $P(1 \leq X \leq 3) = 1 - 0.12 - 0.08 = 0.80$

6.26.

x	f(x)
0	0.029
1	0.343
2	0.514
3	0.114

6.27.

a. $P(X \geq 3) = 0.25 + 0.15 + 0.05 = 0.45$

b. $P(X < 5) = 1 - 0.05 = 0.95$

c. The capacity must increase to a minimum of 4 customers.

6.28. a.

x	f(x)
0	2/381 = 0.005
1	82/381 = 0.215
2	161/381 = 0.423
3	89/381 = 0.234
4	47/381 = 0.123

b. This is an approximation since it is based on a sample of past history which could change in the future.

c.

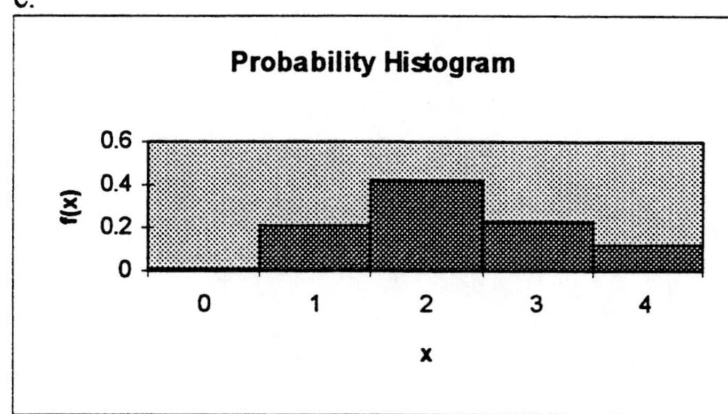

6.29.

a.

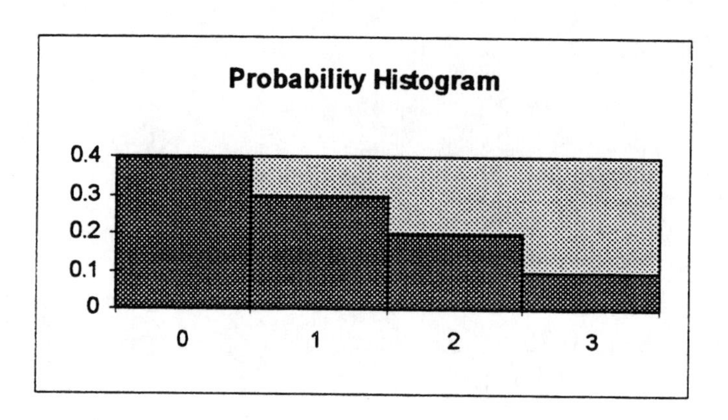

b.

x	f(x)	xf(x)	$x^2f(x)$
0	0.4	0.0	0.0
1	0.3	0.3	0.3
2	0.2	0.4	0.8
3	0.1	0.3	0.9
Total	1.0	1.0	2.0

$E(X) = 0 * 0.4 + 1 * 0.3 + 2 * 0.2 + 3 * 0.1 = 1$

$\sigma^2 = 2 - 1*1 = 1$

$\sigma = 1$

6.30.

x	f(x)	xf(x)	$x^2f(x)$
0	0.3	0.0	0.0
1	0.5	0.5	0.5
2	0.1	0.2	0.4
3	0.1	0.3	0.9
Total	1.0	1.0	1.8

$E(X) = 1$

$\sigma^2 = 1.8 - 1*1 = 0.800$

$\sigma = (.8)^{1/2} = 0.894$

6.31.

x	f(x)	xf(x)
$50,000	0.25	$12,500.00
$ (500)	0.75	$ (375.00)
Total	1	$12,125.00

$E(X)$ = $12,125.00

6.32.

a.

x	f(x)	xf(x)
$50,000	0.000001	$ 0.05
$ 5,000	0.000004	$ 0.02
$ 100	0.000200	$ 0.02
$ 20	0.002000	$ 0.04
$ 0	0.977795	$ 0.00
Total	1.000000	$ 0.13

b. $E(X)$ = $ 0.13

6.33.

x	f(x)	xf(x)	$x^2f(x)$
0	0.2401	0.0000	0.0000
1	0.4116	0.4116	0.4116
2	0.2646	0.5292	1.0584
3	0.0756	0.2268	0.6804
4	0.0081	0.0324	0.1296
Total	1.0000	1.2000	2.28

$E(X) = 1.200$

$\sigma^2 = 2.28 - 1.200 * 1.200 = 0.840$

$\sigma = 0.917$

6.34.

x	f(x)	xf(x)
-800	0.05	-40
0	0.95	0
Total	1.00	-40

Expected payment from insurance company = $E(X) = 40$.

x	f(x)	xf(x)
-750	0.05	-37.5
50	0.95	47.5
Total	1.00	10.0

Expected profit per policy = $E(X)$ = $10.

6.35.

a. P(win A and win B) = 0.5 * 0.65 = 0.325
P(win A and lose B) = 0.5 * 0.35 = 0.175
P(lose A and win B) = 0.5 * 0.65 = 0.325
P(lose A and lose B) = 0.5 * 0.35 = 0.175

b.

Event	x	f(x)
win A, win B	195,000	0.325
win A, lose B	75,000	0.175
lose A, win B	120,000	0.325
lose A, lose B	0	0.175
Total		1.000

c.

Event	x	f(x)	xf(x)
win A, win B	193,000	0.325	62725
win A, lose B	73,000	0.175	12775
lose A, win B	118,000	0.325	38350
lose A, lose B	-2000	0.175	-350
Total		1.000	113,500

Expected net profit = $113,500.

6.36.

a. and b.

Possible outcomes:	x	f(x) = probabilities
win A, win B	195,000	0.4
win A, lose B	75,000	0.1
lose A, win B	120,000	0.2
lose A, lose B	0	0.3
Total		1.0

c.

Event	x	f(x)	xf(x)
win A, win B	$ 193,000	0.4	$ 77,200
win A, lose B	$ 73,000	0.1	$ 7,300
lose A, win B	$ 118,000	0.2	$ 23,600
lose A, lose B	$ (2,000)	0.3	$ (600)
Total		1.0	$ 107,500

Expected net profit = $107,500.

6.37.

x	f(x)	xf(x)
0	0.307	0.000
1	0.286	0.286
2	0.204	0.408
3	0.114	0.342
4	0.064	0.256
5	0.018	0.090
6	0.007	0.042
Total	1.000	1.424

a. Expected value = 1.420
b. Standard deviation of X = 1.351

6.38.

x	f(x)	xf(x)	$x^2f(x)$
2	0.389610	0.78	1.5584
3	0.259740	0.78	2.3377
4	0.194805	0.78	3.1169
5	0.155844	0.78	3.8961
Total	1.000000	3.12	10.9091

Mean = 3.120
Standard deviation = 1.092

6.39.

a.

x	f(x)	xf(x)	$x^2f(x)$
0	0.04762	0.00	0.0000
1	0.35714	0.36	0.3571
2	0.47619	0.95	1.9048
3	0.11905	0.36	1.0714
Total	1.00000	1.67	3.3333

b.

Mean = 1.667
Variance = 3.333-1.667*1.667=0.556
Standard deviation = 0.745

6.40.

a., b., and c.:

x	f(x)	xf(x)	$x^2f(x)$
0	0.05	0.00	0.00
1	0.10	0.10	0.10
2	0.15	0.30	0.60
3	0.20	0.60	1.80
4	0.25	1.00	4.00
5	0.15	0.75	3.75
6	0.10	0.60	3.60
Total	1.00	3.35	13.85

Mean = 3.350

Variance = 13.85 - 3.35 * 3.35 = 2.628

Standard deviation = 1.621

6.41.

a., b., and c.:

x	f(x)	xf(x)	$x^2f(x)$
0	0.05	0.00	0.00
1	0.20	0.20	0.20
2	0.30	0.60	1.20
3	0.25	0.75	2.25
4	0.15	0.60	2.40
5	0.05	0.25	1.25
Total	1.00	2.40	7.30

Mean = 2.400

Variance = 7.30 - 2.4 * 2.4=1.540

Standard deviation = 1.241

6.42.

a.

x	8-2x	y
0	8-2*0	8
1	8-2*1	6
2	8-2*2	4
3	8-2*3	2

b.

x	f(x)	xf(x)	$x^2f(x)$
0	0.1	0.0	0.0
1	0.3	0.3	0.3
2	0.4	0.8	1.6
3	0.2	0.6	1.8
Total	1.0	1.7	3.7

Mean of x = 1.70
Variance of x = 3.7 - 1.7*1.7 =0.81
Standard deviation of x = 0.90

c.

y	f(y)	yf(y)	$y^2f(y)$
2	0.2	0.4	0.8
4	0.4	1.6	6.4
6	0.3	1.8	10.8
8	0.1	0.8	6.4
Total	1.0	4.6	24.4

Mean = 4.60
Variance = 24.4 - 4.6*4.6 =3.24
Standard deviation = 1.80

d. $a = 8$; $b = -2$
Mean for Y = 8 + (-2) * 1.70 = 4.6
Standard deviation for Y = |-2| * 0.9 = 1.8

6.43.

a.

x	f(x)	xf(x)	$x^2f(x)$
0	0.1	0.00	0.00
1.0	0.5	0.50	0.50
1.5	0.2	0.30	0.45
2.0	0.2	0.40	0.80
Total	1.0	1.20	1.75

Mean of x = 1.200
Variation of x= 0.310
Standard deviation of x = 0.557

b.

y	f(y)	yf(y)	y^2 f(y)
-0.100	0.1	-0.01	0.00
0.850	0.5	0.43	0.36
1.325	0.2	0.27	0.35
1.800	0.2	0.36	0.65
Total	1.00	1.04	1.36

Mean =	1.040
Variation =	0.280
Standard deviation =	0.529

6.44.

a.

SSSS (0.0016), SSSN (0.0064), SSNS (0.0064), SSNN (0.0256), SNSS (0.0064), SNSN(0.0256), SNNS (0.0256), SNNN (0.1024), NSSS (0.0064), NSSN (0.0256), NSNS (0.0256), NSNN (0.1024), NNSS (0.0256), NNSN (0.1024), NNNS (0.1024), NNNN (0.4096)

b.

x	f(x)	xf(x)
0	0.4096	0.0000
1	0.4096	0.4096
2	0.1536	0.3072
3	0.0256	0.0768
4	0.0016	0.0064
Total	1.0	0.8000

c. Expected value of x = 0.8

6.45.

a.

y	f(y)	yf(y)
$ 0	0.4096	$ 0
$2,000	0.4096	$ 819.20
$4,000	0.1536	$ 614.40
$6,000	0.0256	$ 153.60
$8,000	0.0016	$ 12.80
Total	1.00	$ 1,600.00

b. Expected value of Y = 1600

6.46. The median is 1 since $P(x \geq 1) = 0.3 + 0.2 + 0.1 = 0.6$ and $P(x \leq 1) = 0.7$, both of which are greater than or equal to 0.5.

6.47.

a. These are not Bernoulli trials since the probability that a customer will buy ketshup varies.

b. These are not Bernoulli trials since they are all from the same geographical area..

c. These are Bernoulli trials.

d. These are not Bernoulli trials since it is probable that one paper not delivered on time means the others in the neighborhood are not delivered on time.

6.48.

a. These are not Bernoulli trials since there are 6 possible outcomes instead of 2.

b. These are Bernoulli trials; $p = 1/6$.

c. These are not Bernoulli trials since there are more than two possible outcomes.

d. These are Bernoulli trials; $p = 12 / 36 = 1/3$.

e. These are Bernoulli trials; p is determined by the long run probability of getting a five. (Assuming that the trials are independent.)

6.49.

a. These are Bernoulli trials; $p = 10 / 25 = 0.4$.

b. These are not Bernoulli trials since the probability changes with each trial.

c. These are not Bernoulli trials since the probability changes with each trial.

6.50.

a. These are Bernoulli trials; $p = 1000 / 2500 = 0.4$.

b. These are not Bernoulli trials since the probability changes with each trial.

c. These are not Bernoulli trials since the probability changes with each trial.

6.51.

a. These are not Bernoulli trials since there is a lack of independence between trials for various customers.

b. These are Bernoulli trials since the trials are independent.

6.52.

a. $P(x \geq 1) = 1 - P(x=0) = 1 - \binom{3}{0} 0.99^0 0.01^3 = 1 - 0.000001 = 0.999999 \approx 1.000$

b. $P(x=2) = \binom{3}{2} 0.99^2 0.01^1 = 0.029$

6.53.

a. $n=4, p=.25;\ P(x=0) = \binom{4}{0} .25^0 .75^4 = 0.316$

b. P(second four are successes | first four are failures) = P(four successes) = 0.0039

c. P(FFFS) = (0.75^3) * (0.25) = 0.105

6.54.

a. $n = 2,\ p = 0.8;\ P(x = 0) = 0.040$

a. $n = 3,\ p = 0.8;\ P(x = 0) = 0.008$

6.55.

a. S = {CC, CNC, CNN, NCN, NCC, NNCC, NNCN, NNNC, NNNN}

b. P(CC) = 0.3333*0.3333 = 0.111
P(CNC) = 0.3333*0.3333*0.6667 = 0.074
P(CNN) = 0.3333*0.6667*0.6667 = 0.148
P(NCN) =0.148
P(NCC) = 0.074
P(NNCC) = 0.6667*0.6667*0.3333*0.3333 = 0.049
P(NNCN) = 0.6667*0.6667*0.6667*0.3333 = 0.099
P(NNNC) = 0.099
P(NNNN) = 0.6667*0.6667*0.6667*0.6667 = 0.198

c.

x	f(x)
0	0.1976
1	0.1481 * 2 + 0.0988 * 2 = 0.4938
2	0.0741 * 2 + 0.1111 + 0.0494=0.3087
Total	1.0

6.56.

a. This is a binomial distribution if the probability that a person answers all questions remains constant for all people; n = 10; p is the probability that a person answers all questions.

b. This is not a binomial distribution.

6.57.

a. This is a binomial distribution with $n = 9$; $p = 1/6$.

b. This is not a binomial distribution.

c. This is a binomial distribution with $n = 3$; $p = 0.6$.

d. This is not a binomial distribution.

6.58.

First trial	Second trial	Third trial	Elementary outcomes	
S	S	S	SSS	p^3
		F	SSF	p^2q^1
	F	S	SFS	p^2q^1
		F	SFF	p^1q^2
F	S	S	FSS	p^2q^1
		F	FSF	p^1q^2
	F	S	FFS	p^1q^2
		F	FFF	q^3

x	f(x)
0	q^3
1	$3 * p^1q^2$
2	$3 * p^2q^1$
3	p^3

6.59.

a. $n = 4, p = 1/3$; $P(x{=}2) = \binom{4}{2} 0.333^2 0.667^2 = 0.296$

b. $n = 6, p = 0.25$; $P(x{=}3) = 0.132$

c. $n = 6, p = 0.75$; $P(x{=}2) = \binom{6}{2} 0.75^2 0.25^4 = 0.033$

6.60.

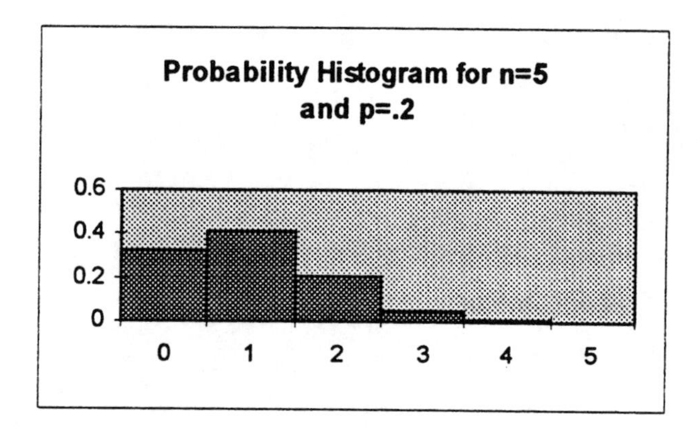

$p = 0.2$; $P(X \geq 4) = 0.007$

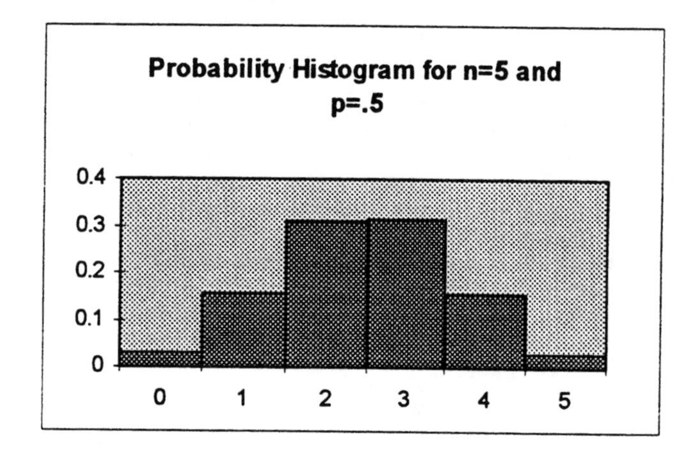

$p = 0.5$; $P(X \geq 4) = 0.187$

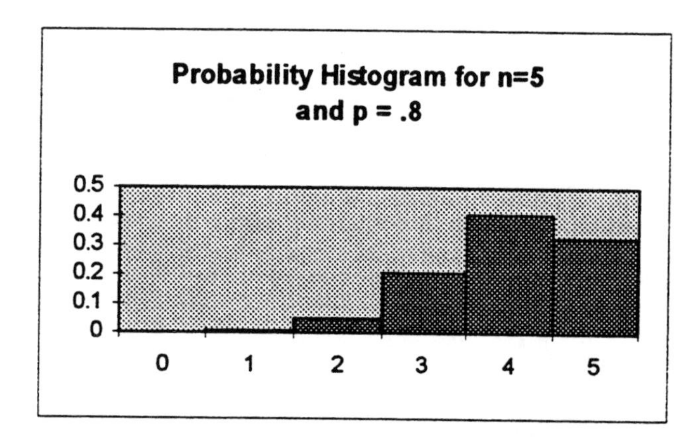

$p = 0.8$; $P(X \geq 4) = 0.737$

6.61.

a. $P(X \leq 2) = 0.765$

b. $P(X \geq 2) = 0.572$

c. $P(X = 2 \text{ or } 4) = 0.312 + 0.049 = 0.385$

6.62. The mode is 2.

6.63. $n = 3; p = 0.46$

a. $P(X = 3) = \binom{3}{3} 0.46^3 0.54^0 = 0.097$

b. $P(X = 1) = \binom{3}{1} 0.46^1 0.54^2 = 0.402$

6.64. $n = 7; p = 0.7$

a. $P(X \leq 4) = 0.353$

b. $P(X \geq 4) = 0.874$

c. $P(X = 4) = 0.227$

6.65. $p = 0.3$

a. $n = 3;\ P(X = 3) = 0.027$

b. $n = 6;\ P(X = 3) = 0.186$

6.66. $n = 15$

a. $p = 0.1;\ P(X \geq 4) = 0.056$

b. $p = 0.3;\ P(X \geq 4) = 0.703$

6.67. $n = 20; p = 0.2$

a. $P(X = 0) = 0.012$

b. $P(X \geq 7) = 0.087$

c. mean = np = 20*0.2 = 4
standard deviation = $npq = (20*0.2*0.8)^{1/2} = 1.789$

6.68.

a. $p = 0.25; n = 4;\ P(X = 0) = 0.316$

b. $p = 0.16/0.52 = 0.30769; n = 4;\ P(X = 0) = \binom{4}{0} 0.3077^0 0.6923^4 = 0.229$

c. p (male from upper management) = 0.16/0.52 = 0.30769
p (female from upper management) = 0.09/0.48 = 0.188
P(X = 0 of 2 females) * P(X = 0 of 2 males) =

$$\binom{2}{0} 0.1875^0 0.8125^2 * \binom{2}{0} 0.30769^0 0.6923^2 = 0.316$$

6.69.

a. $n = 10;\ p = 0.25$
mean = 10*0.25 = 2.5
standard deviation = $(10*0.25*0.75)^{1/2}$ = 1.369

b. $n = 40;\ p = 0.16/0.52 =\ 0.308$
mean = 40 * 0.30769 = 12.308
standard deviation = $(40*0.30769*0.69231)^{1/2}$ = 2.919

c. $n = 40;\ p = 0.39/0.48 = 0.8125$
mean = 40 * 0.8125 = 32.5
standard deviation = $(40*0.8125*0.1875)^{1/2}$ = 2.469

6.70.

a. $n = 24;\ p = 0.4$
mean = 24 * 0.4 = 9.6
standard deviation = $(24 * 0.4 * 0.6)^{1/2}$ =2.4

b. $n = 24; p = 0.6$
mean = 24 * 0.6 = 14.4
standard deviation = $(24 * 0.4 * 0.6)^{1/2}$ =2.4

c. $n = 96; p = 0.4$
mean = 96 * 0.4 = 38.4
standard deviation = $(96 * 0.4 * 0.6)^{1/2}$ = 4.8

6.71.

a.

x	f(x)
0	0.064
1	0.288
2	0.432
3	0.216

b. Mean = $\sum_0^3 x_i f(x_i) = 0*0.064 + 1*0.288 + 2*0.432 + 3*0.216 = 1.8$

Standard deviation =

$$\left(\sqrt{\sum_0^3 x_i^2 f(x_i) - \mu^2}\right) = \sqrt{0*0.064 + 1^2*0.288 + 2^2*0.432 + 3^2*0.216 - 1.8^2} = 0.849$$

c. Mean = *np* = 3*0.6 = 1.8
Standard deviation = *npq* = $(3*0.6*0.4)^{1/2}$ = 0.849

6.72.

a. For y = 1, $f(y) = q^0 p$.
For y = n-1, $f(y) = q^{y-2}p$
For y = n, $f(y) = q^{y-2}p*q = q^{y-1}p$

b. $P(X \leq 3) = P(X = 1) + P(X = 2) + P(X = 3)$
$= 0.5^0 0.5 + 0.5^1 0.5 + 0.5^2 0.5 + 0.5^3 0.5$
$= 0.5 + 0.25 + 0.125 + 0.0625 = 0.938$

6.73.

a. $P(X = 0) = (e^{-3})(3^x)/0! = 0.05$

b. $P(X = 1) = (e^{-3})(3^x)/1! = 0.15$

6.74.

a. $P(X = 0) = (1\text{-}.03)^{100} = 0.048$

b. $P(X = 1) = 100(0.03)(1\text{-}0.03)^{99} = 0.147$

c. $P(X = 0) = 0.05 * 3^0 / 0! = 0.05$
$P(X = 1) = 0.05 * 3^1 / 1! = 0.15$

6.75.

a. Mean = 80* 0.35 = 28
Standard deviation = $(80*0.35*0.65)^{1/2} = 4.266$

b. $\mu = np$
$54 = np$
$\sigma^2 = npq$
$36 = 54q$
$q = 36/54 = 0.667$
$p = 1\text{-}.6667 = 0.333$
$n = 54/p = 54/.3333 = 162$

6.76.

a. Cumulative Distribution Function

Binomial with n = 12 and p = 0.640

x	P(X <= x)
0	0.0000
1	0.0001
2	0.0011
3	0.0070
4	0.0304
5	0.0970
6	0.2352
7	0.4459
8	0.6799
9	0.8648
10	0.9634
11	0.9953
12	1.0000

$P(X \leq 8) = 0.680$

Probability Density Function

Binomial with n = 12 and p = 0.640

x	P(X = x)
0	0.0000
1	0.0001
2	0.0010
3	0.0059
4	0.0234
5	0.0666
6	0.1382
7	0.2106
8	0.2340
9	0.1849
10	0.0986
11	0.0319
12	0.0047

$P(X = 8) = 0.234$

b. **Cumulative Distribution Function**

Binomial with n = 30 and p = 0.420

x	P(X <= x)
2	0.0000
3	0.0001
4	0.0007
5	0.0030
6	0.0099
7	0.0269
8	0.0622
9	0.1249
10	0.2201
11	0.3455
12	0.4893
13	0.6334
14	0.7602
15	0.8581
16	0.9246
17	0.9642
18	0.9849
19	0.9944
20	0.9982
21	0.9995
22	0.9999
23	1.0000

$P(10 \leq X \leq 15) = 0.8581 - 0.2201 = 0.638$

6.77.
Under 25: $P(X \geq 2) = 0.022 + 0.002 = 0.024$
Over 25: $P(X \geq 2) = 0.013 + 0.001 = 0.014$

There are fewer claims for the over 25 group. Older drivers are likely to have fewer claims per policy as shown by the fact that the corresponding numbers in the two tables are lower in the 25 and over group except in the "0 claims" category.

6.78.
Mean = 0 * 0.797 + 1 * 0.179 + 2 * 0.022 + 3 * 0.002 = 0.229
Variance = 0*0*0.797 + 1*1*0.179 + 2*2* 0.022 + 3* 3*0.002 - 0.229 * 0.229
= 0.233

6.79.
a. Possible values of *X*: -3, -1, 1, 3

b.

X	Elementary Outcomes
-3	TTT
-1	HTT, THT, TTH
1	HHT, HTH, THH
3	HHH

6.80. D = defective; G = good

a.

X = # defectives	Elementary outcomes
0	GG
1	GD, DG
2	DD

b. Probability distribution of *X*:

X = # defectives	f(x)
0	0.7*0.8333 = 0.583
1	0.7*0.1667 + 0.3*0.8333 = 0.367
2	0.3*0.1667 =0 .050

6.81.

W = # defectives	f(x)
0	0.78*0.776 = 0.605
1	0.22*0.806 + 0.78 * 0.224 = 0.350
2	0.22*0.204 = 0.045

6.82.

a.

X = # defectives	f(x)
1	(32/31)/2 = 0.516
2	(32/31)/4 = 0.258
3	(32/31)/8 = 0.129
4	(32/31)/16 = 0.065
5	(32/31)/32 = 0.032

b.

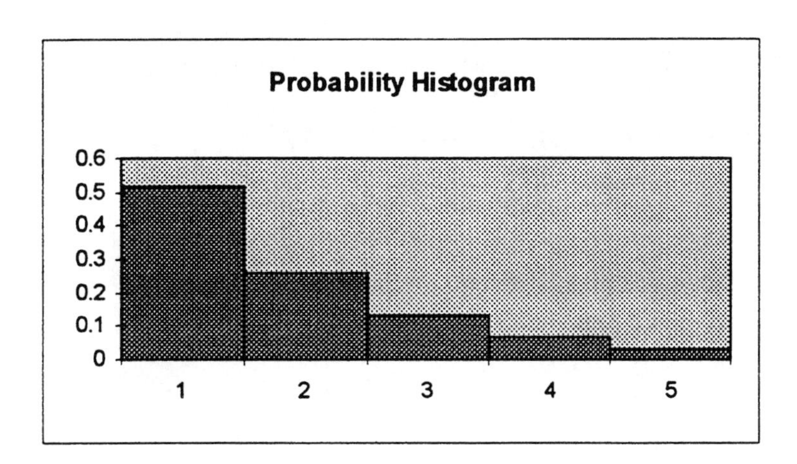

6.83.

a. Mean = 0*0.3 + 1*0.4 + 2*0.3 = 1

b. Variance = 0*0*0.3 + 1*1*0.4 + 2*2*0.3 - 1*1 = 0.6
Standard deviation = 0.775

c.

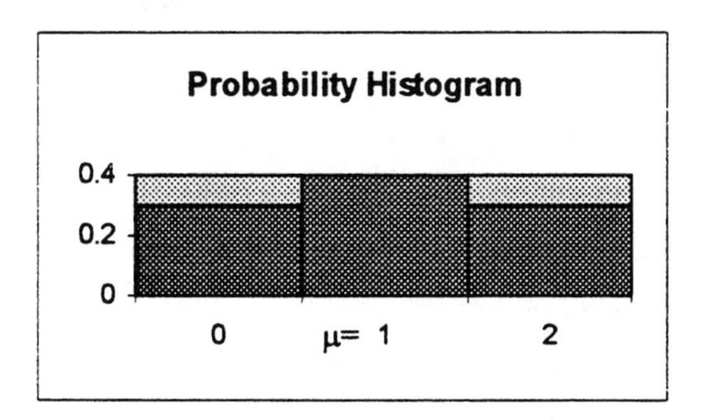

6.84.

a. Mean = 2*0.1 + 3*0.3 + 4*0.3 + 5*0.2 + 6*0.1 = 3.9

b. Variance = 2*2*0.1 + 3*3*0.3 + 4*4*0.3 + 5*5*0.2 + 6*6*0.1 - 3.9*3.9 = 1.29
Standard deviation = 1.136

c.

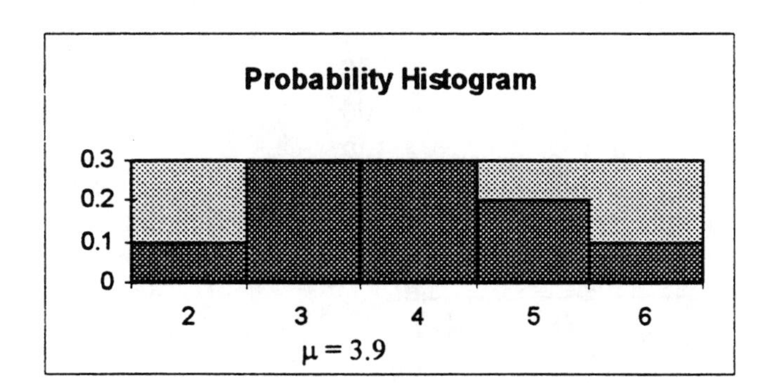

6.85.

a. $\mu + \sigma = 3.9 + 1.136 = 5.036$
$\mu - \sigma = 3.9 - 1.136 = 2.764$
Within this interval are: 3, 4, and 5
$P(\mu - \sigma \leq X \leq \mu + \sigma) = 0.3 + 0.3 + 0.2 = 0.8$

b. $\mu + 2\sigma = 3.9 + 2*1.136 = 6.172$
$\mu - 2\sigma = 3.9 - 2*1.136 = 1.628$
Within this interval are: 2, 3, 4, 5, 6
$P(\mu - 2\sigma \leq X \leq \mu + 2\sigma) = 1$

6.86.

a. P(win) = 2/1000 = 0.002
P(lose) = .998/1000 = 0.998

b. Expected gain = 0.002(260-1) + 0.998(-1) = $-0.48 (a loss)

6.87.

a. Expected gain = 0.3*15000 + 0.7*0 = $4500

b. Expected net gain = 0.3(15000-2500) + 0.7*(-2500) = $2000

6.88.

a. $P(X \leq 2) = 0.05 + 0.10 + 0.15 = 0.3$

b.

x	f(x)	x f(x)	$x^2 f(x)$
0	0.05	0.00	0.00
1	0.10	0.10	0.10
2	0.15	0.30	0.60
3	0.35	1.05	3.15
4	0.20	0.80	3.20
5	0.15	0.75	3.75
		3.00	10.80

Mean = 3.000
Variance = 1.800
Standard deviation = 1.342

6.89. a.

x	f(x)	x f(x)	$x^2 f(x)$
4000	0.0002	0.800	3200.000
1000	0.0006	0.600	600.000
100	0.0190	1.900	190.000
5	0.0850	0.425	2.125
0	0.8952	0	0
	1	3.725	3992.125

b. Mean = 3.725

c. P(lose money) = P(X = 0 or 5) = 0.085 + 0.8952 = 0.980

6.90. P(Win) = 18/38 = 0.474 ; P(Lose) = 20/38 = 0.526

a. and b.:

x	f(x)	f(x)	x*f(x)
-15	0.4737 **3	0.1064	-1.59543
-5	3*(0.4737^2) *0.5263	0.3543	-1.77146
5	3*(0.4737 *0.5263^2)	0.3936	1.968164
15	0.5263^3	0.1458	2.186711
		1.000067	0.78799

Mean = 0.788

c. The expected net gain alternating with black will be the same since the probability of black is the same as that of red.

6.91.

x	f(x)	f(x)	x f(x)	x^2f(x)
0	0.25*(.75)^0	0.2500	0	0
1	0.25*(.75)^1	0.1875	0.1875	0.1875
2	0.25*(.75)^2	0.1406	0.2813	0.5625
3	0.25*(.75)^3	0.1055	0.3164	0.9492
4	1 - 0.6836	0.3164	1.2656	5.0625
	Total:	1.0000	2.0508	6.7617

Mean = 2.051

Variance = 2.560

Standard deviation = 1.599

a. $f(4) = 1 - (0.2500 + 0.1875 + 0.1406 + 0.1055) = 0.316$

b. $P(X \geq 2) = 0.1406 + 0.1055 + 0.3164 = 0.5625$

c. $E(X) = 2.051$

d. $sd(X) = 1.599$

6.92.

a.

x	f(x)	F(x)
1	0.0700	0.07
2	0.1200	0.19
3	0.2500	0.44
4	0.2800	0.72
5	0.1800	0.90
6	0.1000	1
	1.0000	

b. $f(6) = F(6) - F(5) = 1\text{-}0.90 = 0.10$
$f(5) = F(5) - F(4) = 0.90 - 0.72 = 0.18$
$f(4) = F(4) - F(3) = 0.72 - 0.44 = 0.28$
$f(3) = F(3) - F(2) = 0.44 - 0.19 = 0.25$
$f(2) = F(2) - F(1) = 0.19 - 0.7 = 0.12$
$f(1) = F(1) = 0.07$

6.93.

a. $P(X = 0) = 0.9$
b. $P(X \geq 2) = 0.03 + 0.02 = 0.05$
c. $P(1 \leq X \leq 2) = 0.05 + 0.03 = 0.08$

6.94.

a. Expected profit = 1.9*300 + 0.5*800 = 970
b. Variance = 0.89 * 90,000 + 0.45 * 640,000 = 368,100
Standard deviation = 606.712

6.95.

This is not a Bernoulli trial since the trials are not independent, i.e., the probability of success changes since there is no replacement.

6.96.

a. This is not a Bernoulli trial since the presence of a cavity in one tooth means that the next tooth is more probable to have a cavity. Also, certain teeth are more likely to have cavities than others.

b. This is a set of Bernoulli trials.

c. This is a set of Bernoulli trials if the trials are independent, i.e., if it is as likely to be cloudy on one day of the first week of April as compared to another

d. This is not a set of Bernoulli trials; there is not a success vs. failure probability.

e. This is a set of Bernoulli trials assuming the trials are independent.

6.97. (Answers will vary.)

a. The students in a statistics class will each either pass or fail.

b. A door-to-door book salesman goes to each of 10 houses in a neighborhood where the residents know each other very well and call one another to warn of this salesman.

c. A marble is taken out of a bag of red and green marbles without replacement.

6.98.

a. This is binomial distribution if the student's chance of getting each question correct does not vary from question to question due to knowledge about the tested information.

b. This is not a binomial distribution since the probability of getting the questions correct varies from 0.2 for Part I to 0.25 for Part II.

c. This is not a binomial distribution since the probability varies as to whether or not the people jog.

6.99. P(first basket on third possession) = 0.6 * 0.6 * 0.4 = 0.144

6.100. $P(X \geq 1) = 1 - P(X = 0) = 1 - \binom{2}{0} .51^0 .49^2 = 1 - .2401 = 0.760$

6.101. $n = 20;\ p = 0.8$

P(12 or fewer agree) = 0.032

The claim is not very plausible since it is not very likely that 12 or fewer students agree.

6.102.

a. $P(4 \leq X \leq 9) = 0.858$

b. $P(4 < X \leq 9) = 0.703$

c. $P(4 < X < 9) = 0.663$

d. $E(X) = np = 14 * .4 = 5.600$

e. $sd(X) = (14 * .4 * .6)^{1/2} = 1.833$

6.103.

a.

x	f(x)	xf(x)	$x^2f(x)$
0	0.078	0.00	0.000
1	0.259	0.26	0.259
2	0.346	0.69	1.382
3	0.230	0.69	2.074
4	0.077	0.31	1.229
5	0.010	0.05	0.255
	1.0000	2.00	5.199

Mean =	2.000
Variation=	1.200
Standard deviation =	1.095

b.

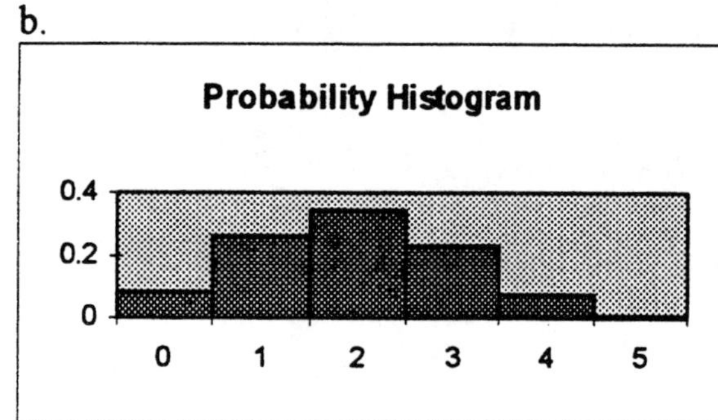

c. Expected value = 2
Variance = 1.200
Standard deviation = 1.095

d. E(X) = 2
var(X) = 1.2

6.104.

a. $P(X = 3) = 0.279$

b. $P(X = 7) = 0.189$

c. $P(X \leq 3) = 0.483$

d. $P(X > 12) = 0.450$

e. $P(8 \leq X \leq 13) = 0.840$

6.105.

a.

p	P(three or fewer successes):
0.1	0.974
0.2	0.795
0.3	0.493
0.4	0.225
0.5	0.073

b.

p	P(three or fewer successes):
0.1	0.902
0.2	0.501
0.3	0.165
0.4	0.033
0.5	0.004

6.106.

a. P($X \leq 3$) = 0.598

b. P(wrong at least 10 times) = P($X \leq 6$) = 0.973

6.107.

expected value = 16 * 0.2 = 3.2

standard deviation = (16 * 0.2 * 0.8)^0.5 = 1.6

7.1. a. This is a probability density function for a continuous random variable since it is continuous and the area under the curve is 1.

b. This is not a probability density function for a continuous random variable since the graph cannot go below the x axis. Negative probabilities are not allowed.

c. This is a probability density function for a continuous random variable since it is continuous and the area under the curve is 1.

d. This is not a probability density function for a continuous random variable since the area under the curve is not equal to 1.

7.2.

a. $P(0 < X < 0.5) = 0.5 * 0.5 = 0.25$

b. $P(.5 < X < 1) = 0.5 * 0.5 = 0.25$

c. $P(1.5 < X < 2) = 0.5 * 0.5 = 0.25$

d. $P(X = 1) = 0$

7.3. The interval $(1.5 < X < 2)$ is assigned a higher probability than $(0 < X < 0.5)$ since the area under the curve $(1.5 < X < 2)$ is larger.

7.4.

First quartile = 0.5

Median = Second quartile = 1

Third quartile = 0.75

7.5.

First quartile = 1

Median = Second quartile = 1.414

Third quartile = 1.732

7.6. Tenth percentile = 0.2

7.7.

a. The median is later than 1:20 pm.

b. The median cannot be determined. It can be less than, equal to, or greater than the mean. We are not given information about whether the graph is skewed or symmetric.

7.8.

a. Since the median is less than the mean, the third distribution of Fig. 7.3 is compatible with this statement.

b. Since the median is less than the mean, the third distribution of Fig. 7.3 is compatible with this statement also.

7.9.

a. $Z = (X - 15)/4$

b. $Z = (X + 9)/7$

c. $Z = (X - 151)/5$

7.10.

a. $Z = (X - 8)/2$

b. $Z = (X - 350)/25$

c. $Z = (X + 67)/10$

7.11.

a. $Z = (X - 70)/2.8$

b. $Z = (X - 65)/2.4$

c. $Z = (66 - 70)/2.8 = -1.429$

d. $Z = (66 - 65)/2.4 = 0.417$ This number is positive since the person is taller than the mean whereas the number in part c is negative since the person is shorter than the mean.

7.12.

a. $Z = (29 - 25)/2.2 = 1.82$

$P(X > 29) = P(Z > 1.82) = 0.0344$

b. $Z_1 = (24 - 25)/2.2 = -0.45$; $Z_2 = (27 - 25)/2.2 = 0.91$

$P(-0.45 < Z < 0.91) = 0.8186 - 0.3264 = 0.4922$

7.13.

$Z_1 = (62.5 - 64.3)/1.15 = -1.57$; $Z_2 = (67.7 - 64.3)/1.15 = 2.96$

$P(62.5 < Z < 67.7) = P(-1.57 < X < 2.96) = 0.9985 - 0.0582 = 0.940$

7.14. $Z = (4 - 5)/1.1 = -0.91$

$P(Z > 4) = P(X > -0.91) = 0.8186$

7.15.

a. $Z = (0.75 - 0.5)/0.2 = 1.25$

$P(X > 0.75) = P(Z > 1.25) = 0.1056$

b. $Z = (0.6 - 0.5) / 0.2 = 0.5$

$P(X > 0.6) = P(Z > 0.5) = 0.3085$

P(two rates of return are both larger than 0.6%) = 0.3085 ^ 2 = 0.095

c. 0.1056

7.16. $P(X_1 < X < X_2) = 0.8$

$P(0 < X < X_2) = 0.4$

$Z = \pm 1.28$

$X = Z\sigma + \mu = \pm 1.28 * 2.8 + 69 = 65.4$ to 72.6

7.17.

a. P(overstated weight) = 0.5

b. $Z = (2.8 - 0.05) / 1.5 = 1.83$

$P(|X| > 2.8) = P(|Z| > 1.83) = 0.0336 * 2 = 0.0672$

c. $P(Z < Z_0) = 0.80$
$Z_0 = 0.84$
$X = Z\sigma + \mu = 0.84 * 1.5 + 0.05 = 1.31$

7.18. a.
i. $Z = (22 - 17) / 3 = 1.67$
$P(X > 22) = P(Z > 1.67) = 0.048$

ii. $Z_1 = (13 - 17) / 3 = -1.33$; $Z_2 = (21 - 17) / 3 = 1.33$
$P(13 < X < 21) = P(-1.33 < Z < 1.33) = 0.9082 - 0.0918 = 0.816$

iii. $Z_1 = (15.5 - 17) / 3 = -0.5$; $Z_2 = (18.5 - 17) / 3 = 0.5$
$P(15.5 < X < 18.5) = P(-0.5 < Z < 0.5) = 0.6915 - 0.3085 = 0.383$

b. Highest probability time is centered around the mean:
$Z_1 = (16.5 - 17) / 3 = -0.17$
$P(16.5 < X < 17.5) = P(-0.17 < Z < 0.17) = 0.5675 - 0.4235 = 0.144$

7.19.
a. $Z = (25 - 32) / 4 = -1.75$
$P(X < 25) = P(Z < -1.75) = 0.040$

b. $Z = (35 - 32) / 4 = 0.75$
$P(X < 35) = P(Z < .75) = 0.773$

7.20.
a. $P(X = 17) = 0.120$
$P(11 \leq X \leq 18) = 0.926 - 0.034 = 0.892$
$P(11 < X < 18) = 0.846 - 0.078 = 0.768$

b. Mean = np = 25 * 0.6 = 15
Standard deviation = (np(1-p))^0.5 = (25 * 0.6 * 0.4)^0.5 = 2.449
$P(X = 17) = P(16.5 < X < 17.5) = P((16.5 - 15)/2.449) < Z < (17.5 - 15) / 2.449)$
$= P(0.61 \leq Z \leq 1.02) = 0.8461 - 0.7291 = 0.117$

$P(11 \leq X \leq 18) = P((11 - 15)/2.449 < Z < (18 - 15) / 2.449)$
$= P(-1.63 \leq Z \leq 1.22) = 0.8888 + 0.0516 = 0.837$

$P(12 \leq X \leq 17) = P((12 - 15) / 2.449) < Z < (17 - 15) / 2.449)$
$= P(-1.22 \leq Z \leq 0.82) = 0.7939 - 0.1112 = 0.6827$

7.21.
a. $P(X = 15) = 0.091$
$P(13 \leq X \leq 19) = 0.807 - 0.017 = 0.790$
$P(13 < X < 19) = 0.659 - 0.044 = 0.615$

b. Mean = np = 25 * 0.7 = 17.5
Standard deviation = (np(1-p))^.5 = (25 * 0.7 * 0.3)^0.5 = 2.2913
$P(X = 15) = P(14.5 < X < 15.5) = P((14.5 - 17.5)/2.2913) < Z < (15.5 - 17.5)/2.2913)$
$= P(-1.31 < Z < -0.87) = 0.1922 - 0.0951 = 0.097$

$P(13 \leq X \leq 19) = P((13 - 17.5)/2.2913 < Z < (19 - 17.5)/2.2913)$
$= P(-1.96 \leq Z \leq 0.65) = 0.7422 - 0.0250 = 0 .717$

$P(13 < X < 19) = P(14 \leq X \leq 18)$
$= P((14 - 17.5) / 2.2913 \leq Z \leq (18 - 17.5)/2.2913) = P(-1.53 \leq Z \leq 0.22)$
$= 0.5871 - 0.0630 = 0.524$

7.22. Mean = np = 300 * 0.25 = 75
Standard deviation = (np(1-p))^0.5 = (300 * 0.25 * 0.75)^0.5 = 7.5

a. $P(X = 80) = P(79.5 < X < 80.5) = P((79.5 - 75)/7.5 < Z < (80.5 - 75)/7.5)$
$= P(0.6 < Z < 0.73) = 0.7673 - 0.7257 = 0.042$

b. $P(X \leq 65) = P(Z < (65 - 75)/7.5) = P(Z < -1.33) = 0.092$

c. $P(68 \leq X \leq 89) = P((68 - 75) / 7.5 < Z < (89 - 75)/7.5)$
$= P(-0.93 < Z < 1.87) = 0.9693 - 0.1762 = 0.793$

7.23. Mean = np = 200 * 0.75 = 150
Standard deviation = (np(1-p))^0.5 = (200 * 0.75 * 0.25)^0.5 = 6.124

a. $P(X = 140) = P(139.5 < X < 140.5)$
$= P((139.5 - 150)/6.124 < Z < (140.5 - 150)/6.124)$
$= P(-1.71 < Z < -1.55) = 0.0606 - 0.0436 = 0.017$

b. $P(X \leq 160) = P(Z \leq (160 - 150)/6.124) = P(Z \leq 1.63) = 0.948$

c. $P(137 \leq X \leq 162) = P(137 < X < 162)$
$= P((137 - 150)/6.124 < Z < (162 - 150)/6.124)$
$= P(-2.12 < Z < 1.96) = 0.9750 - 0.0170 = 0.958$

7.24. The normal approximation is appropriate if np > 15 and n(1-p) > 15.

a. np = 90 * 0.23 = 20.7; n(1-p) = 90 * 0.77 = 69.3; normal approximation is appropriate.

b. np = 100 * 0.02 = 2; n(1-p) = 100 * 0.98 = 98; normal approximation is not appropriate.

c. $np = 71 * 0.4 = 28.4$; $n(1\text{-}p) = 71 * 0.6 = 42.6$; normal approximation is appropriate.

d. $np = 120 * 0.97 = 116.4$; $n(1\text{-}p) = 120 * 0.03 = 3.6$; normal approximation is not appropriate.

7.25.

	n	p	np	n(1-p)	Normal approximation appropriate?
a.	500	0.33	165	335	Yes
b.	10	0.50	5	5	No
c.	400	0.01	4	396	No
d.	200	0.98	196	4	No
e.	100	0.61	61	39	Yes

7.26.

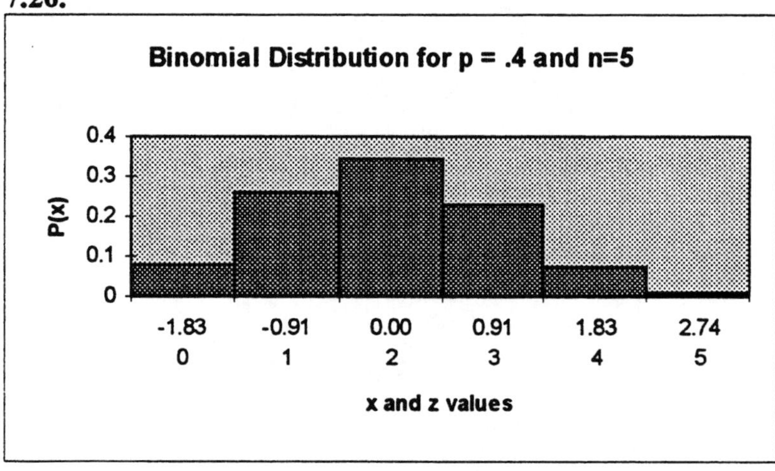

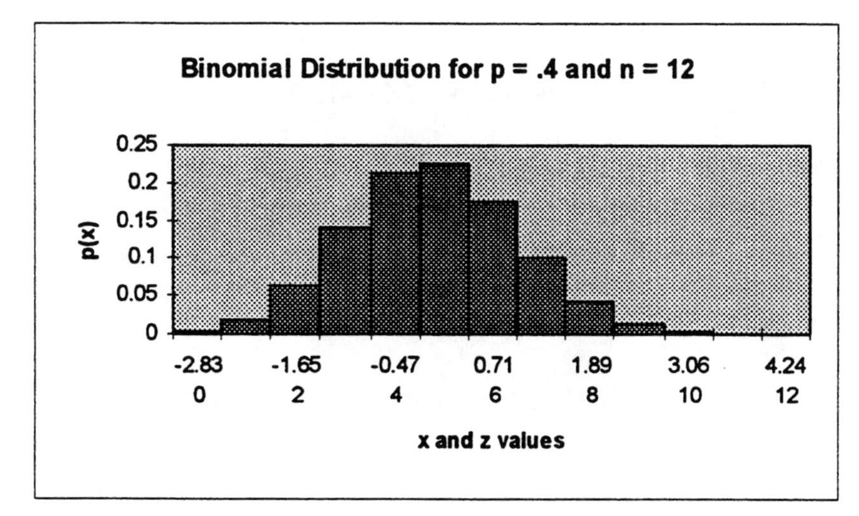

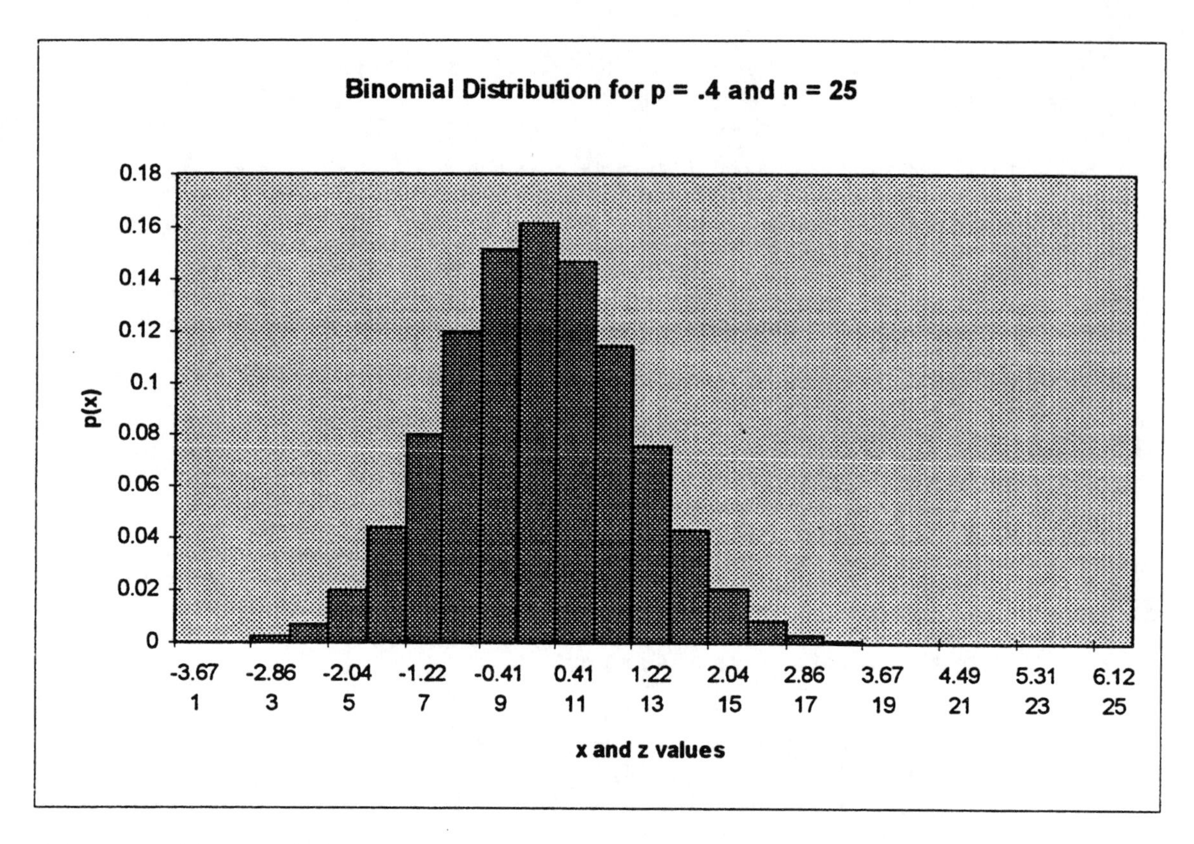

7.27. $n = 200$; $p = 0.5$; mean = np = 0.5 * 200 = 100
standard deviation = (0.5 * 0.5 * 200) ^ 0.5 = 7.071

$P(X \geq 110) = P(Z \geq (110-100)/7.071) = P(Z \geq 1.41) = 0.079$

7.28. $n = 300$; $p = 0.079$; mean = np = 300 * 0.079 = 23.7;
standard deviation = (300 * 0.079 * (1 - 0.079))^0.5 = 4.672

a. $P(X < 18) = P(X \leq 17) = P(Z < (17 - 23.7)/4.672) = P(Z < -1.43) = 0.076$

b. $P(X > 30) = P(X \geq 31) = P(Z \geq (31 - 23.7)/4.672) = P(Z \geq 1.56) = 0.059$

7.29. mean = 70 * 0.25 = 17.5; standard deviation = (70 * 0.25 * 0.75)^0.5 = 3.623

$P(X > 20) = P(X \geq 21) = P(Z \geq (21-17.5)/3.623) = P(Z \geq 0.97) = 0.166$

7.30. $P(X > 3000) = P(X \geq 3001) = P(Z > (3001 - 2500)/400) = P(Z > 1.25) = 0.106$

7.31. $P(X \geq X_0) = 0.05$
$Z = 1.645$
$X_0 = 1.645 * 400 + 2500 = 3158$

7.32. mean = 200 * 0.3 = 60; standard deviation = (200 * 0.3 * 0.7)^0.5 = 6.481

$P(50 \leq X \leq 75) = P((50 - 60)/6.481 \leq Z \leq (75 - 60)/6.481)$
$= P(-1.54 \leq Z \leq 2.31) = 0.9896 - 0.0618 = 0.9278$

7.33. mean = 100 * 0.352 = 35.2; standard deviation = (100 * 0.352 * 0.648)^0.5 = 4.776

$P(X \geq 41) = P(Z \geq (41 - 35.2)/4.776) = P(Z \geq 1.21) = 0.113$

7.34. mean = 300 * 0.2 = 60; standard deviation = (300 *0 .2 * 0.8)^.5 = 6.928

a. $P(49 \leq X \leq 71) = P((49 - 60)/6.928 \leq Z \leq (71 - 60)/6.928)$
$= P(-1.59 \leq Z \leq 1.59) = 0.9441 - 0.0559 = 0.888$

b. $P(X = 72) = P(71.5 < X < 72.5) = P((71.5 - 60)/6.928 < Z < (72.5 - 60)/6.928)$
$= P(1.66 < Z < 1.80) = 0.9641 - 0.9515 = 0.013$

Although it is not likely that 72 claims were incorrectly documented, it is slightly possible that this event occurs.

7.35.

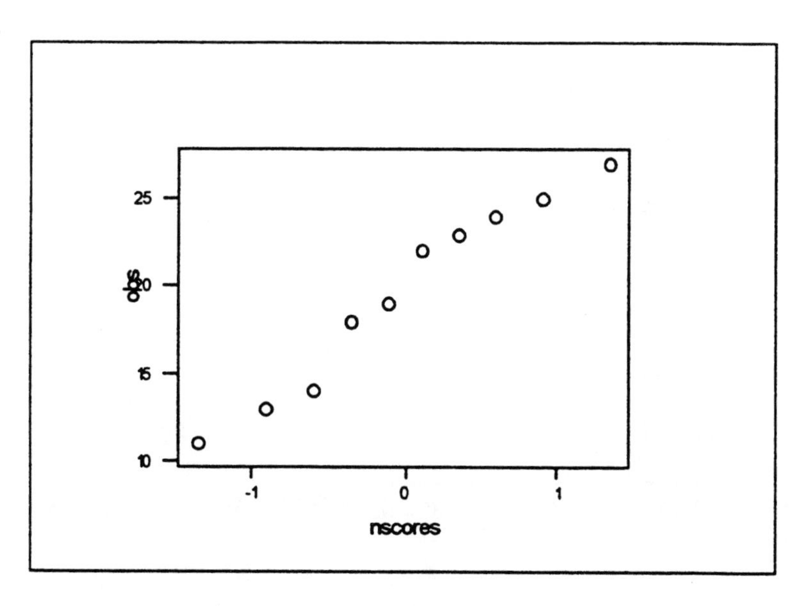

7.36.

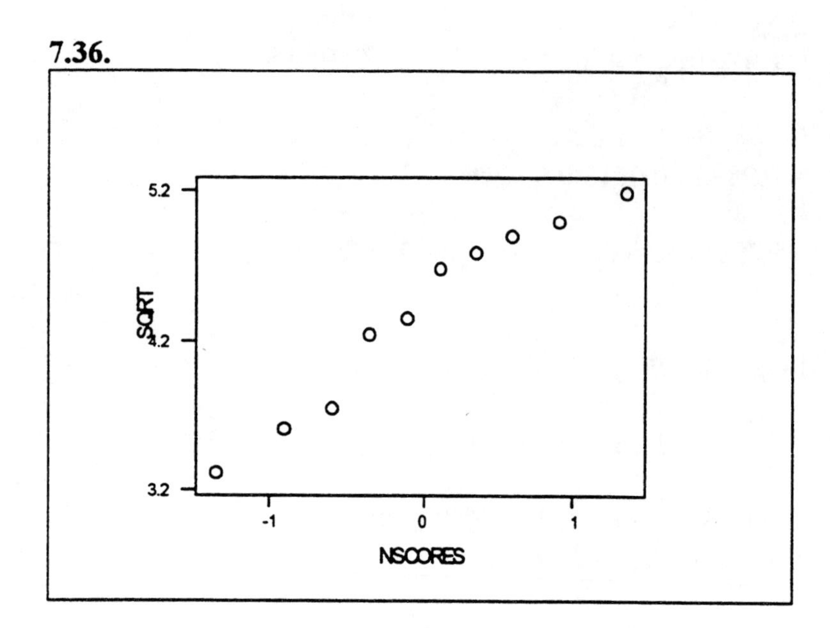

This data are near normal since the plot is nearly a straight line.

7.37.

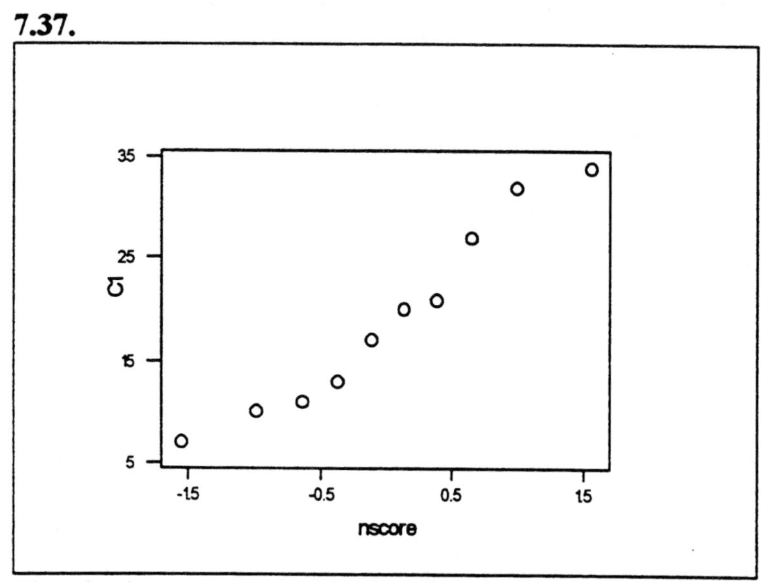

The plot is not very straight, but the sample size is small, so it is difficult to argue against normality.

7.38.

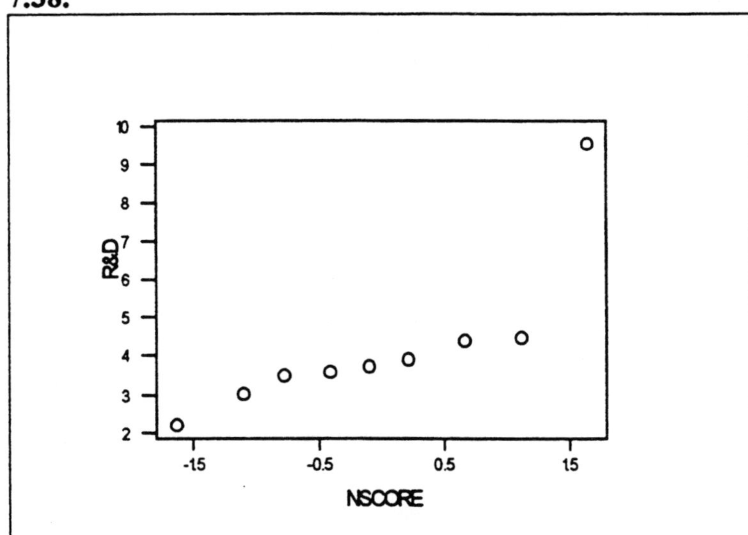

The data are non-normal since there is one clearly outlying observation.

7.39.

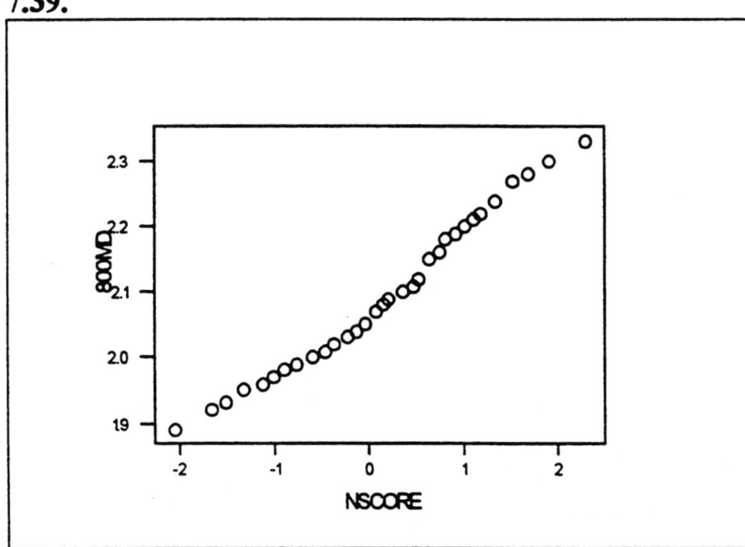

This graph is nearly straight, therefore the data are nearly normal .

7.40.

a.

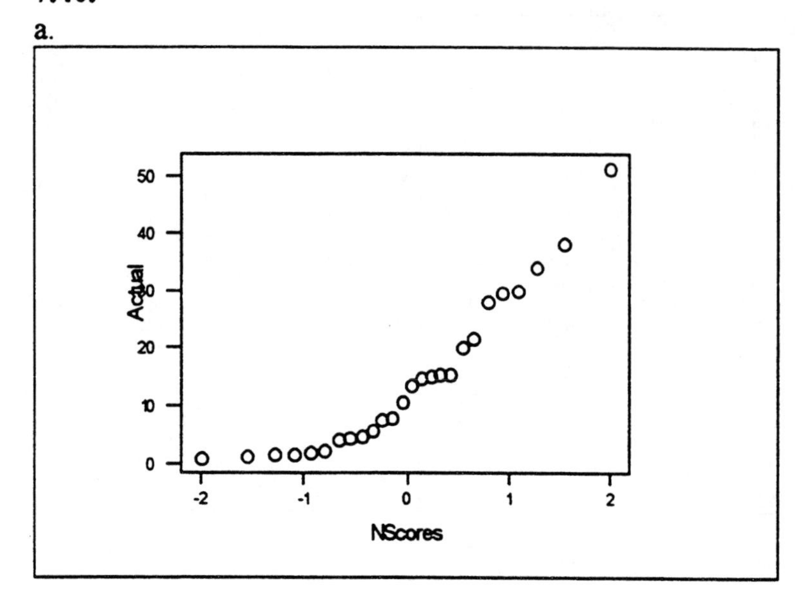

The normal scores plot curves upward, therefore the data are not normal.

b.

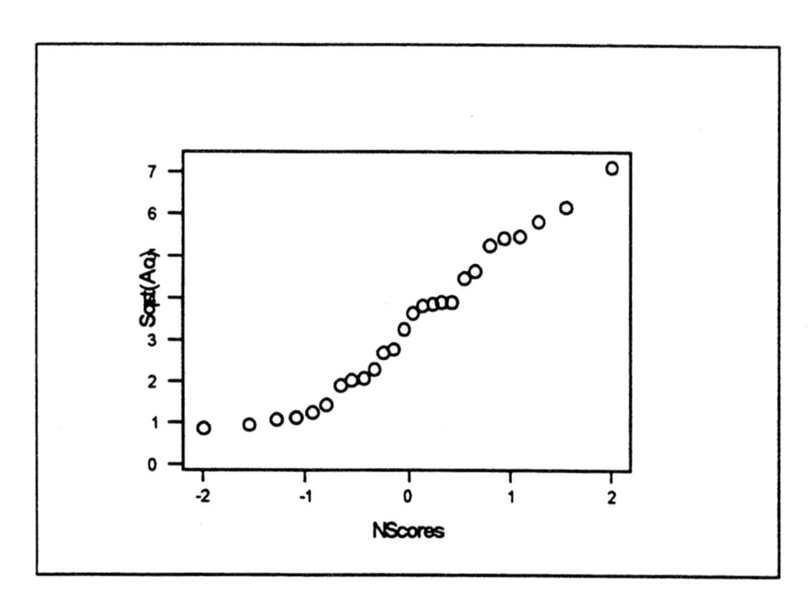

The graph is nearly straight, therefore the data are near normal.

c.

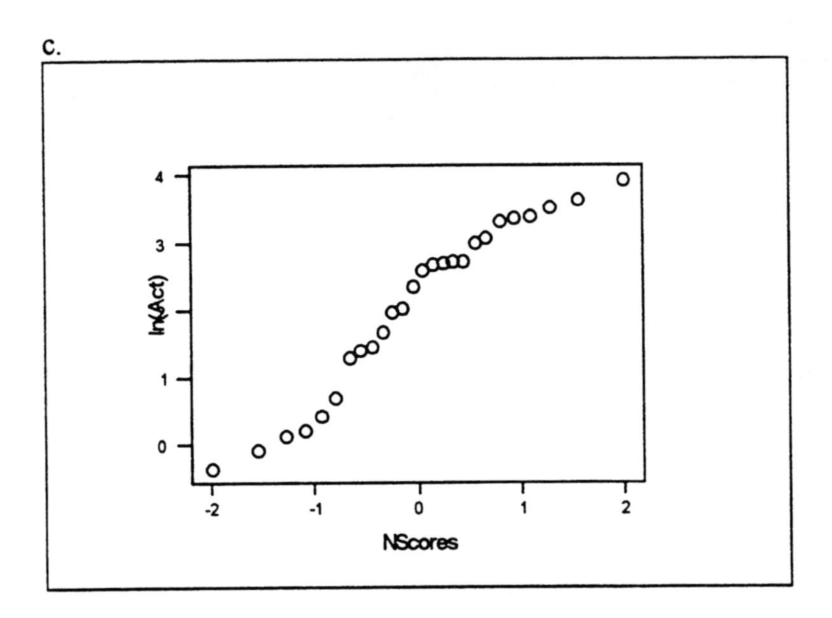

The graph curves more than the graph in part b. The data are not as close to being normal.

d. The graph in part 'b' is the straightest and therefore the closest to normal.

7.41.

a. Square-root transformation:

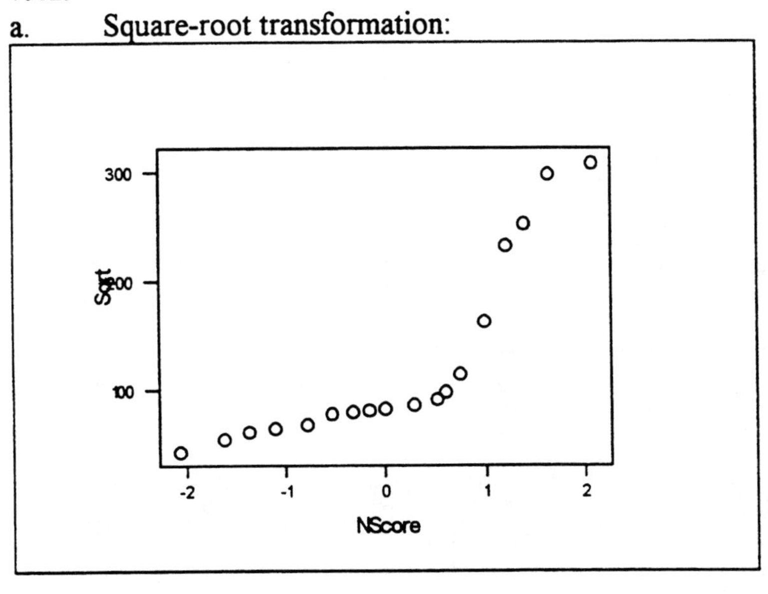

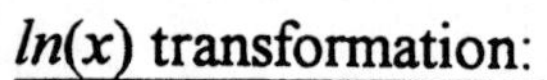

ln(*x*) transformation:

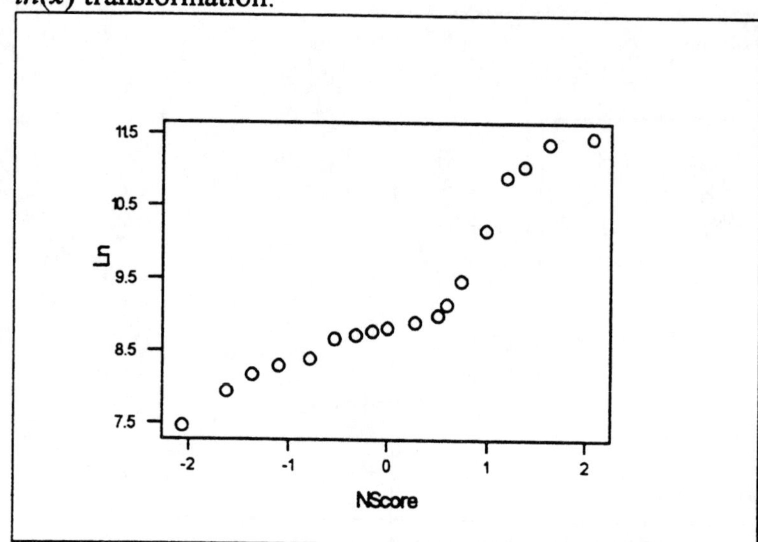

One divided by the square-root of (x) transformation:

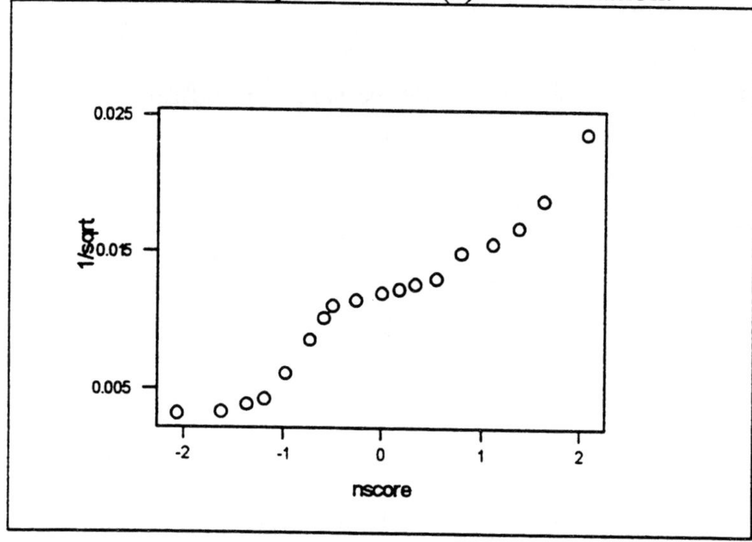

Fourth root of x transformation:

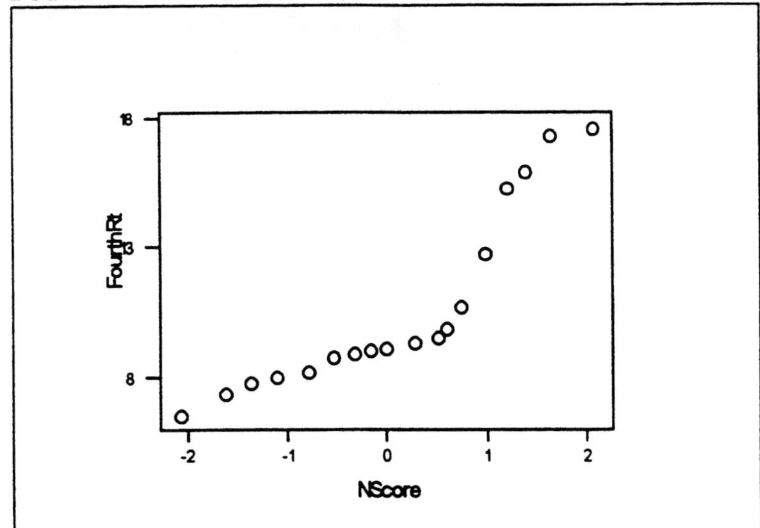

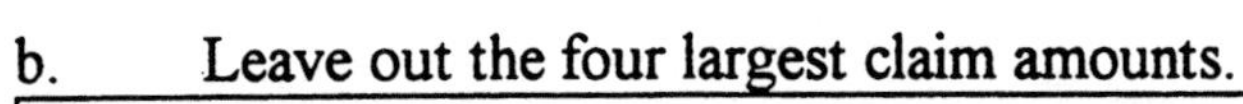
b. Leave out the four largest claim amounts.

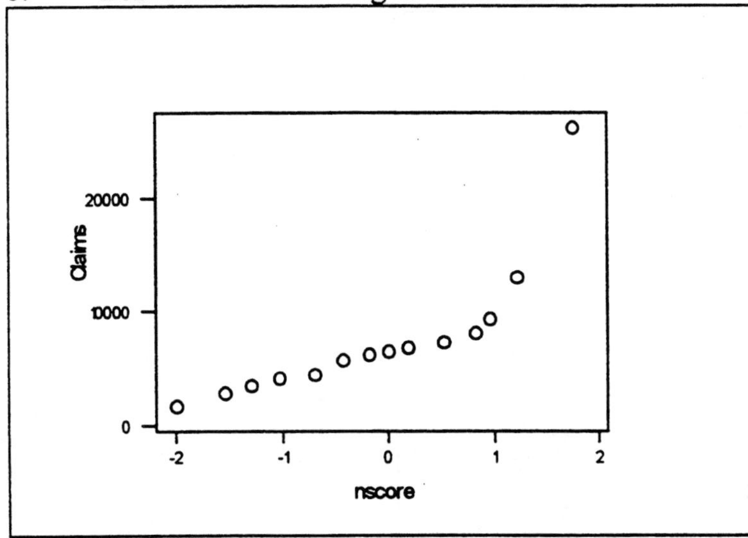

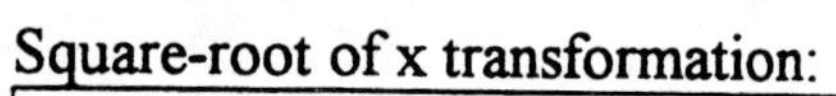

Square-root of x transformation:

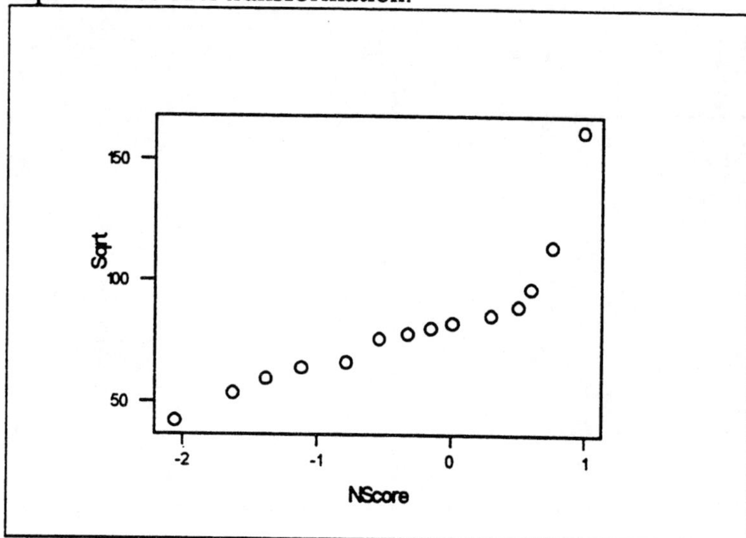

Ln(x) transformation:

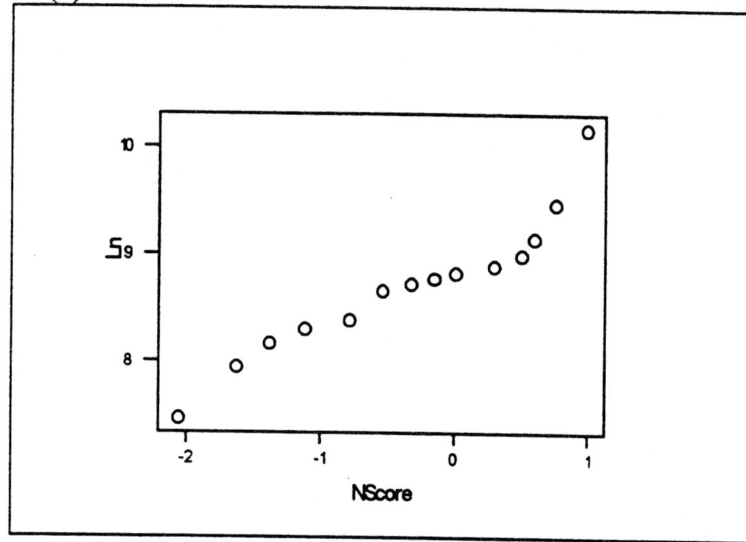

One divided by the square-root of x transformation:

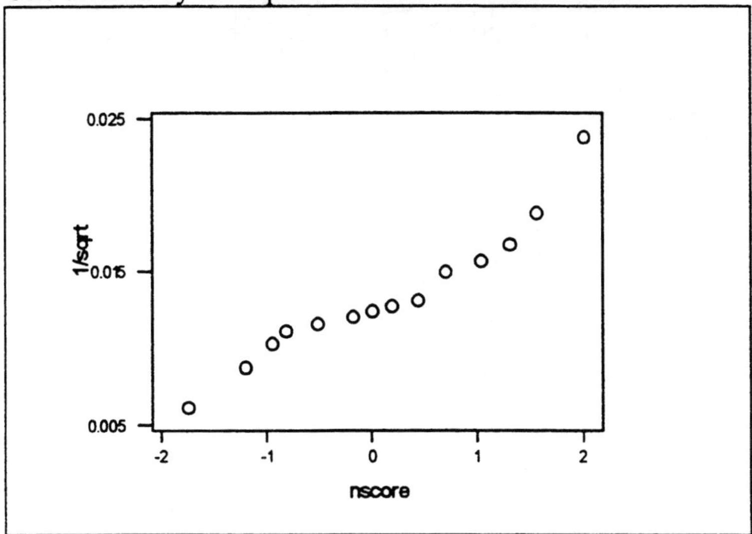

7.42.

No transformation:

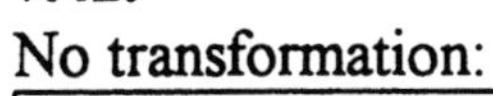

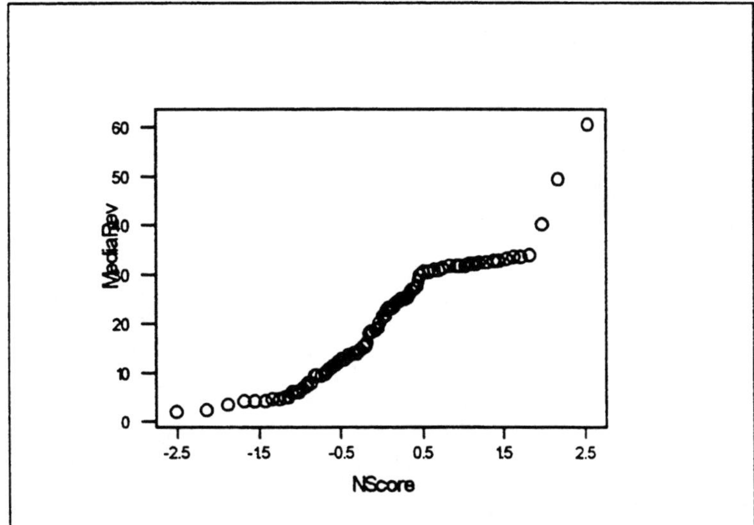

These data are non-normal since the normal scores plot curves upward.

Square-root of x transformation:

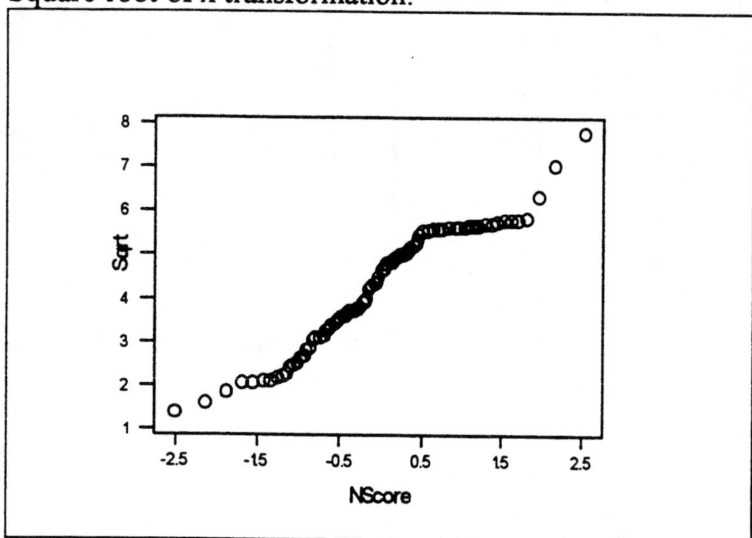

These data are closer to normal, but still the normal scores plot deviates from a straight line.

Ln(x) transformation:

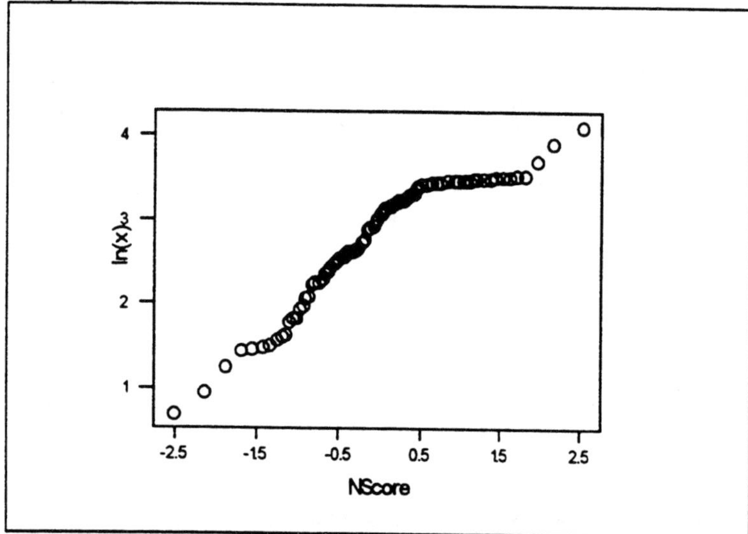

The normal scores plot is not a very straight line; these data are non-normal.

One divided by square-root of x transformation:

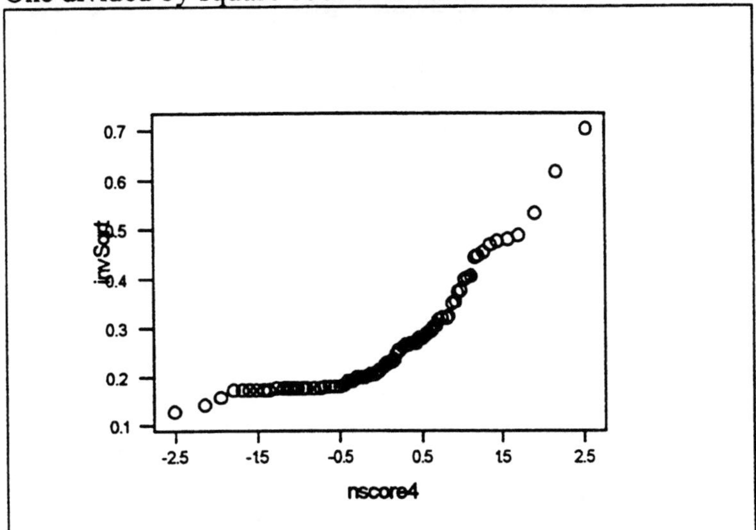

These data are non-normal since the normal scoress plot is not a straight line.

7.43.

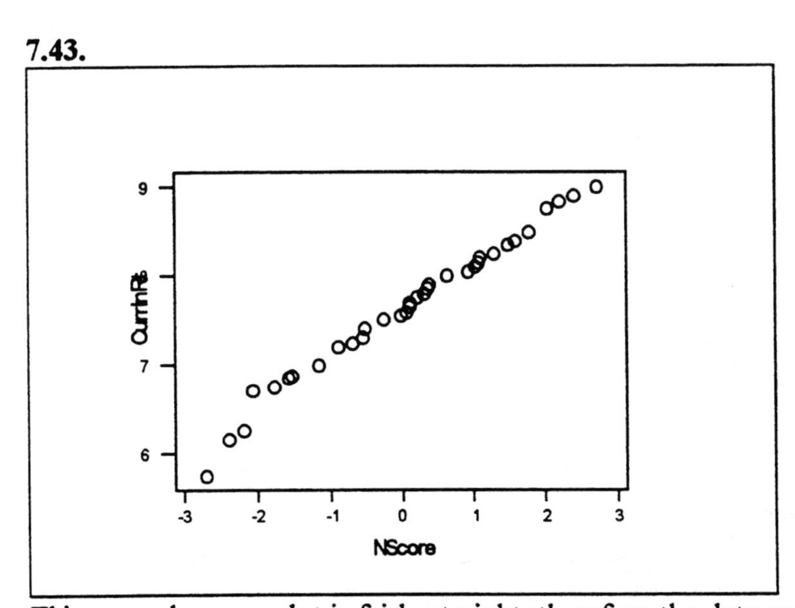

This normal scores plot is fairly straight; therefore the data are normal.

7.44.

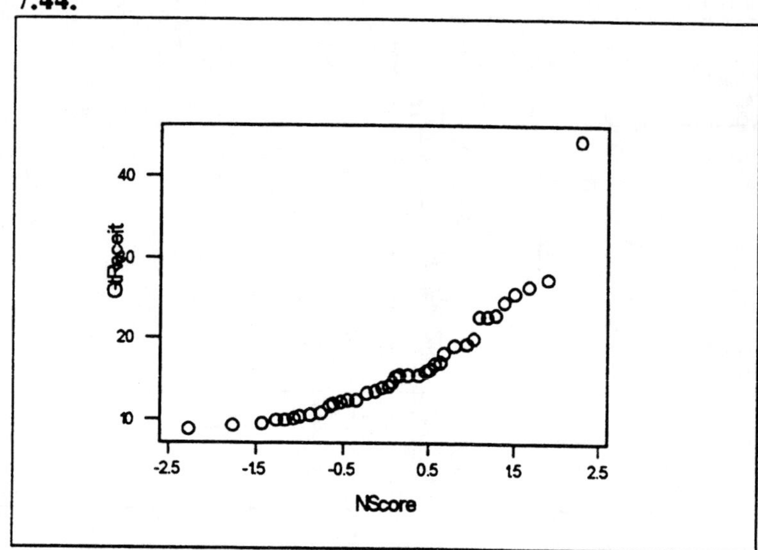

Square-root transformation:

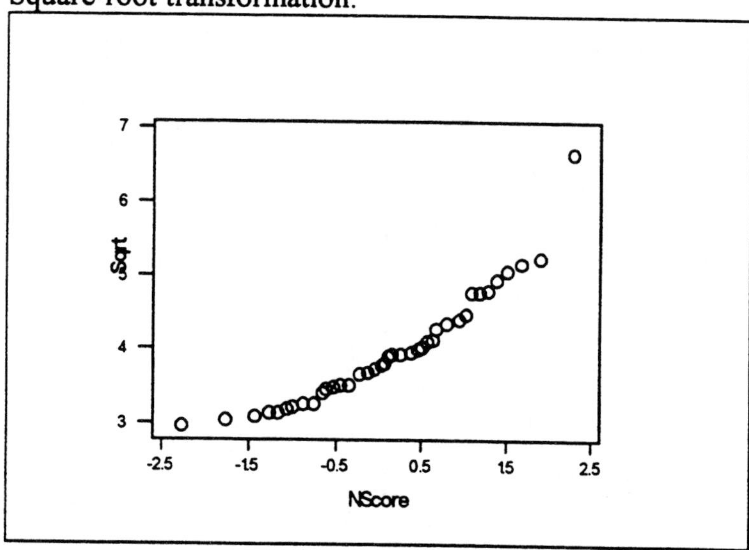

Natural log transformation:

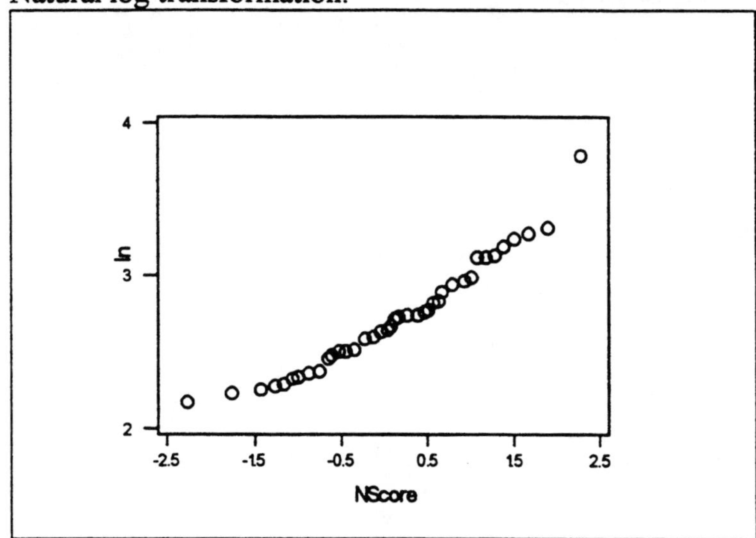

One divided by the square-root of x transformation:

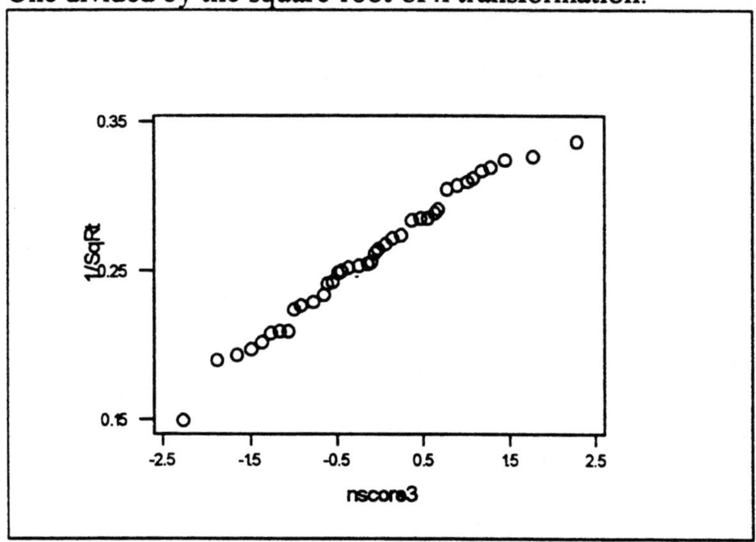

One divided by the fourth root of x transformation:

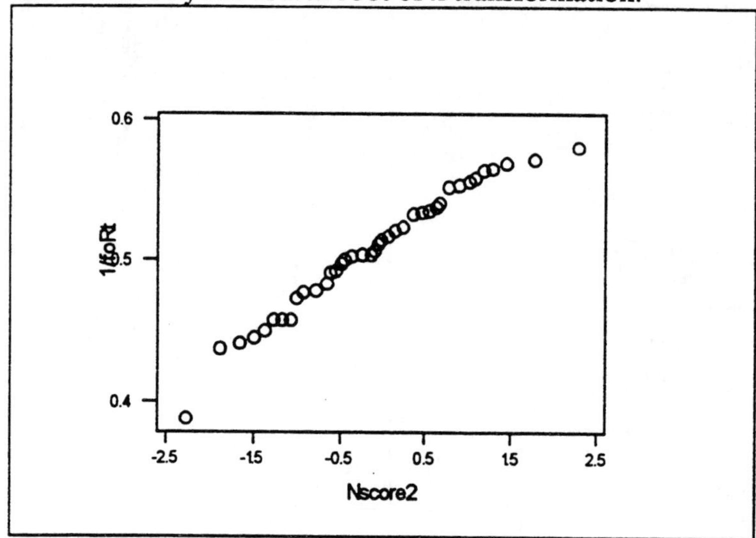

7.45. A normal-scores plot:

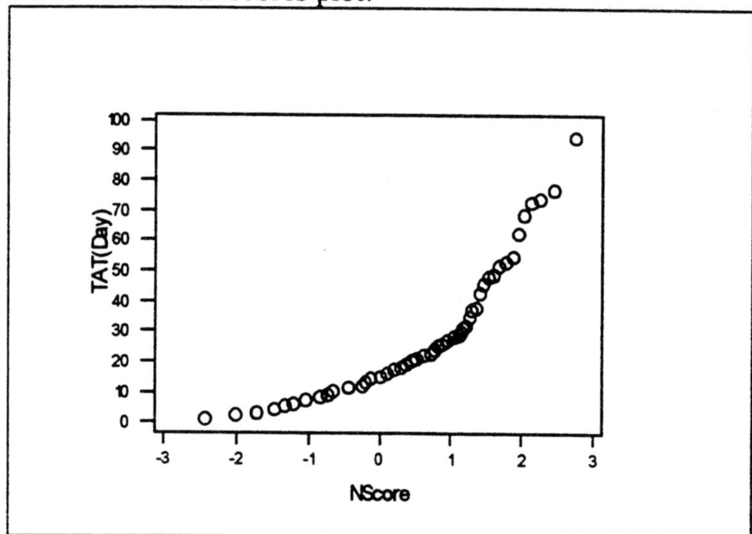

The non-normality of the full data set is related to the four largest numbers. Without these numbers the normal scores plot as shown above is closer to a straight line indicating that these data are closer to normal.

7.46.

a. Graph of ln(turnaround time):

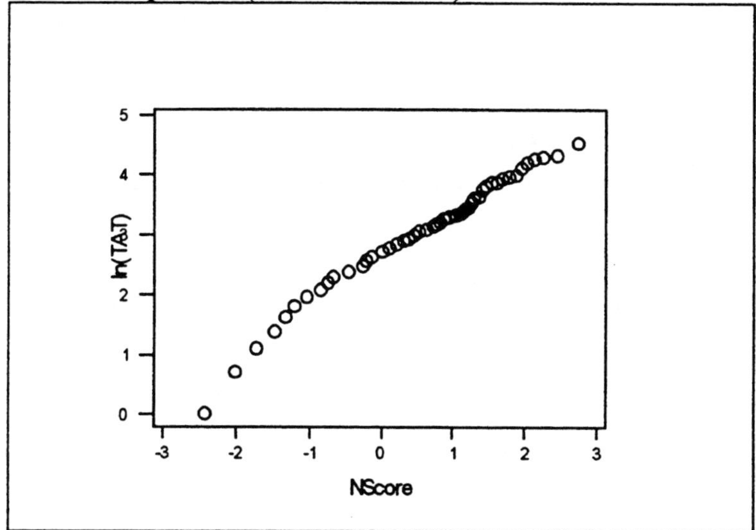

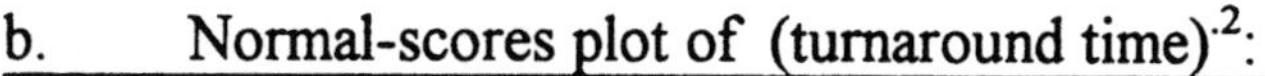

b. Normal-scores plot of (turnaround time)$^{.2}$:

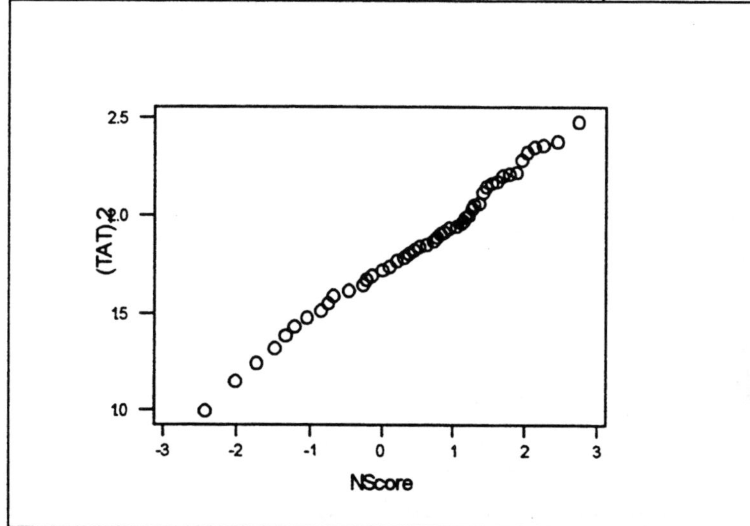

The transformation in part (b) produces nearly normal data since the normal scores plot is close to a straight line.

7.47.

$P(X > (14)^{0.2}) = P(Z > (1.695 - 1.719)/0.263) = P(Z > -0.09) = 0.536$

This probability is closer to the correct probability since the four largest numbers have been removed.

7.48.

a. The sample standard deviation is a statistic.

b. The sample range is a statistic.

c. The population 10th percentile is a parameter.

d. The sample 1st quartile is a statistic.
e. The population median is a parameter.

7.49.

a. The 11 persons who served on the Supreme Court are the population.
b. The fact that 185 out of 1000 were out of work is a statistic.
c. The 46 out of 100 who read mail advertisements is a statistic.

7.50.

a.

Sample	Sample mean
1,1	1
1,3	2
1,5	3
3,1	2
3,3	3
3,5	4
5,1	3
5,3	4
5,5	5

b.

Value of $\overline{X}$	Probability
1	1/9
2	2/9
3	3/9 = 1/3
4	2/9
5	1/9

7.51.

a.

Sample	Sample mean	Sample variation
2,2	2	0
2,4	3	2
2,6	4	8
4,2	3	2
4,4	4	0
4,6	5	2
6,2	4	8
6,4	5	2
6,6	6	0

b.

Value of $\overline{X}$	Probability
2	1/9
3	2/9
4	3/9 = 1/3
5	2/9
6	1/9

c.

Value of s^2	Probability
0	3/9 = 1/3
2	4/9
8	2/9

7.52. The berries taken from the top are not a random sample since there may be variation in berries closer to the bottom of the basket. Workers may have placed the nicer looking berries at the top of the basket.

7.53. To get a random sample, the times should be chosen randomly instead of at regular time periods.

7.54.

a.

Mean of population: 37.092

Standard deviation of the population: 18.892

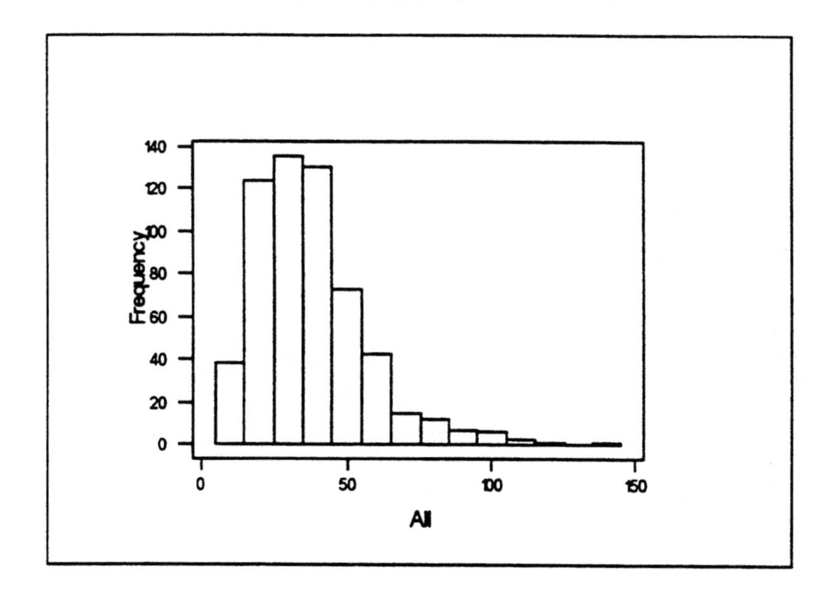

b.

Mean of sample means: 36.86
Standard deviation of the sample means: 5.54

The mean of the sample means is very close to the mean of the population.
The standard deviation of the sample means is about a third of the population standard deviation.

c.

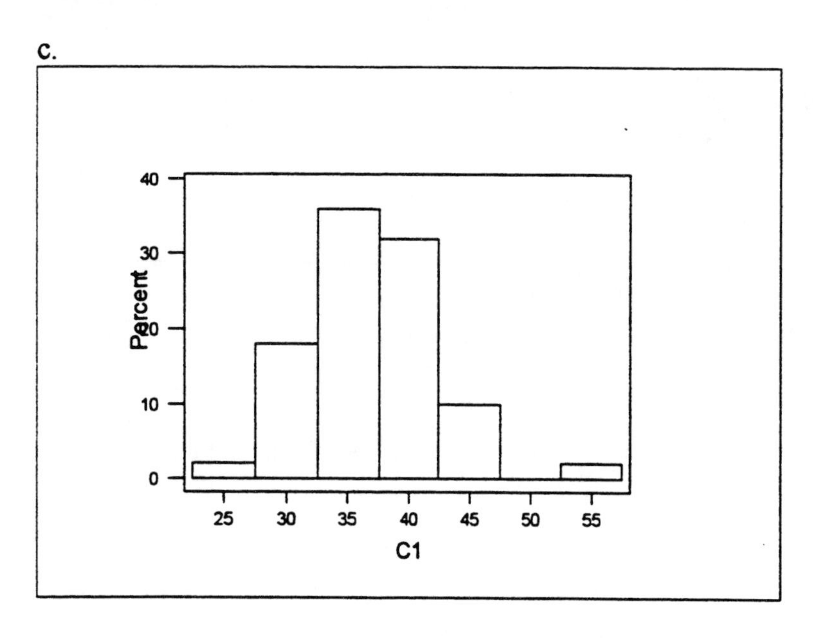

d. The general shapes of these two histograms are similar, however, the histogram of sample means is more symmetric. Both are centered on the upper 30's values. The spread of the first histogram is much larger than the spread of the second histogram.

7.55.

a. $E(\overline{X}) = \mu = 99$
$\text{sd}(\overline{X}) = \sigma/\sqrt{n} = 7/2 = 3.5$

b. $E(\overline{X}) = \mu = 99$
$\text{sd}(\overline{X}) = \sigma/\sqrt{n} = 7/5 = 1.4$

7.56.

a. $E(\overline{X}) = \mu = 250$
$\text{sd}(\overline{X}) = \sigma/\sqrt{n} = 12/1.732 = 6.928$

b. $E(\overline{X}) = \mu = 250$
$\text{sd}(\overline{X}) = \sigma/\sqrt{n} = 12/4 = 3$

7.57.

a. $sd(\overline{X}) = \sigma/\sqrt{n}$ = 10/5 = 2

b. $sd(\overline{X}) = \sigma/\sqrt{n}$ = 10/10 = 1

c. $sd(\overline{X}) = \sigma/\sqrt{n}$ = 10/20 = 0.5

7.58.

a. $sd(\overline{X}) = \sigma/\sqrt{n}$ = 84/6 = 14

b. $sd(\overline{X}) = \sigma/\sqrt{n}$ = 84/12 = 7

7.59.

$\mu = 3$

$E(\overline{X}) = 27/9 = 3 = \mu$

$sd(\overline{X}) = 1.633/1.414 = 1.155 = \sigma/\sqrt{n}$

7.60.

$\mu = 4$

$E(\overline{X}) = 36/9 = 4$

$sd(\overline{X}) = 1.633/1.414 = 1.155$

7.61.

a. $E(\overline{X}) = 27$

b. $sd(\overline{X}) = \sigma/\sqrt{n}$ = 3/2 = 1.5

c. The distribution of $\overline{X}$ is approximately normal.

7.62.

a. $E(\overline{X}) = 20$

b. $sd(\overline{X}) = \sigma/\sqrt{n}$ = 5 / (6^0.5) = 2.041

c. The distribution of $\overline{X}$ is approximately normal.

7.63.

$sd(\overline{X}) = \sigma/\sqrt{n}$ = 22/(5^0.5) = 9.839

$P(\overline{X} > 140) = P(Z > (140 - 134)/9.839) = P(Z > 0.61) = 0.271$

7.64.

a. $P(\overline{X} < 16.0) = P(Z < (16\text{-}16.08)/(0.122/3)) = P(Z < -1.97) = 0.024$

b. In the long run, 2.4% of the bags will have less than 16 ounces of chips in them.

7.65.

a. $P(\overline{X} < 16) = 0.02$

$Z = -2.055$

$\mu = -Z * \sigma/\sqrt{n} + 16 = 2.055 * (0.122/3) + 16 = 16.084$

b. In the long run, 2% of the bags will have less than 16 ounces of chips in them.

7.66.

a. The distribution of $\overline{X}$ is normal. It is an approximate distribution.

b.

i. $P(\overline{X} > 3.7) = P(Z > (3.7 - 3.0)/(0.5/4)) = P(Z > 5.6) = 0$

ii. $P(2.84 < \overline{X} < 3.16) = P((2.84 - 3.0)/(0.5/4) < Z < (3.16 - 3.0)/(0.5/4))$
$= P(-1.28 < Z < 1.28) = 0.8997 - 0.1003 = 0.799$

7.67.

a. Since the sample size is large, the Central Limit Theorem states that the population distribution of the sample mean is approximately normal.

b. $P(\overline{X} > 31{,}500) = P(Z > (31{,}500\text{-}31{,}000)/(5000/10)) = P(Z > 1) = 0.159$

7.68.

a. Since the sample size is large, the Central Limit Theorem states that the population distribution of the sample mean is approximately normally distributed.

b. $P(\overline{X} > 20.75) = P(Z > (20.75 - 20) / (5/10)) = P(Z > 1.5) = 0.067$

7.69. $P(\overline{X} < 5) = P(Z < (5 - 6) / (2 / 3.464)) = P(Z < -1.73) = 0.0418$

7.70. $P(\overline{X} > 72) = P(Z \geq (72 - 70)/(2.8/2.236)) = P(Z > 1.60) = 0.0548$

7.71.

$P(34.1 < \overline{X} < 35.2) = P((34.1 - 34.5)/(1.3/2.449) < Z < (34.2 - 34.5)/(1.3/2.449))$
$= P(-0.75 < Z < -0.57) = 0.2843 - 0.2266 = 0.058$

7.72.

$P(0.048 < \overline{X} < 0.053) = P((0.048 - 0.05)/(0.015/10) < Z < (0.053 - 0.05)/(0.015/10))$
$= P(-1.33 < Z < 2) = 0.9772 + 0.0918 = 0.885$

7.73.

a. Medians: 6 7 4 6 4 4 4 7 6 6 5 3 5 6 5 4 4 4 4 6 2 2 5 6 2 3 5 8 4 5 4 6 4 7 4
4 2 5 5 4 6 4 4 4 2 9 4 5 4 2 7 2 4 4 5 4 4 6 8 7 6 4 7 6 5 5 4 5 4 4 1 5 6 3 5 6 1 6 7 9 3 7
2 4 4 7 3 4 3 6 6 6 5 1 5 8 2 3 2

b.

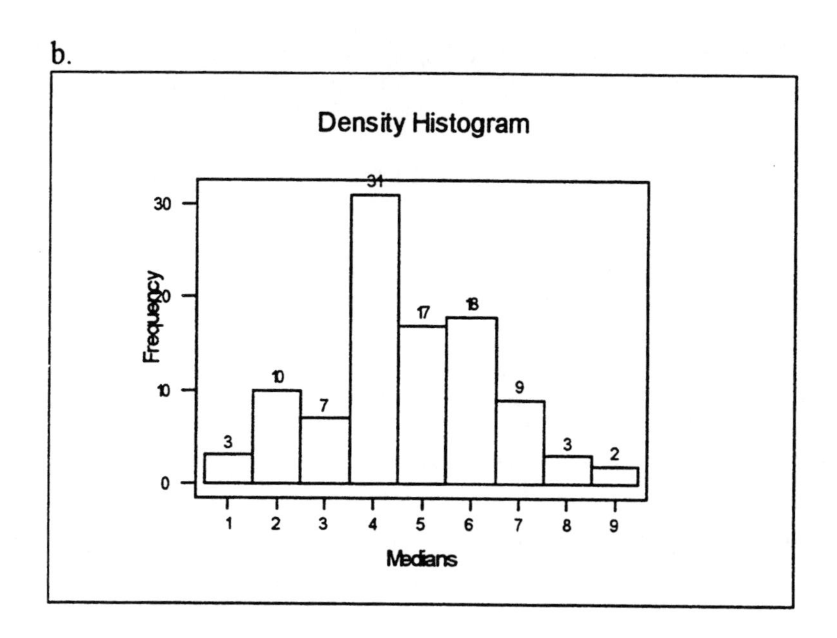

c. The sampling distribution of the sample mean has a variance smaller than that of the sample median.

7.74. The Central Limit Theorem expresses the fact that the probability distribution of $\overline{X}$ is approximately normal and concentrates more and more on the population mean as the sample size increases since the standard deviation grows smaller as the sample size gets larger.

7.75.

a. Median = 0.5

b. First quartile = 0.25
Third quartile = 0.75

7.76.

a. $P(X > 0.8) = 1 - 0.8 = 0.2$

b. $P(0.5 \leq X \leq 0.8) = 0.8 - 0.5 = 0.3$

c. $P(0.5 < X < 0.8) = 0.8 - 0.5 = 0.3$

7.77.

a. $P(Z < 1.31) = 0.905$

b. $P(Z > 1.205) = 0.114$

c. $P(0.67 < Z < 1.98) = 0.9761 - 0.7486 = 0.228$

d. $P(-1.32 < Z < 1.055) = (0.8531 + 0.8554)/2 - 0.0934 = 0.761$

7.78.

a. $Z = (696 - 582)/75 = 1.52$

b. $X = Z\sigma + \mu = -0.8 * 75 + 582 = 522$

c. $380 < X < 560 = (380-582) / 75 < Z < (560 - 582) / 75 = -2.69 < Z < -0.29$

d. $-1.2 < Z < 1.2 = -1.2 * 75 + 582 < X < 1.2 * 75 + 582 = 492 < X < 672$

7.79.

a. $P(X < 107) = P(Z < (107 - 100)/8) = P(Z < 0.875) = (0.8106 + 0.8078)/2$
$= 0.809$

b. $P(X < 97) = P(Z < (97 - 100)/8) = P(Z < -0.375) = (0.3557 + 0.3520)/2$
$= 0.354$

c. $P(X > 110) = P(Z > (110 - 100)/8) = P(Z > 1.25) = 0.106$

d. $P(X > 90) = P(Z > (90 - 100)/8) = P(Z > -1.25) = 0.894$

e. $P(95 < X < 106) = P((95 - 100)/8 < Z < (106 - 100)/8) = P(-0.625 < Z < 0.75)$
$= 0.7734 - 0.2660 = 0.507$

f. $P(103 < X < 114) = P((103 - 100)/8 < Z < (114 - 100)/8) = P(0.38 < Z < 1.75)$
$= 0.9599 - 0.6480 = 0.312$

g. $P(88 < X < 100) = P((88 - 100)/8 < Z < (100 - 100)/8) = P(-1.5 < Z < 0)$
$= 0.5 - 0.0668 = 0.433$

h. $P(60 < X < 108) = P((60 - 100)/8 < Z < (108 - 100)/8) = P(-5 < Z < 1) = 0.841$

7.80.

a. $P(X < b) = 0.670; \quad b = 0.44 * \sigma + \mu = 0.44 * 5 + 200 = 202.2$

b. $P(X > b) = 0.011; \quad b = 2.29 * 5 + 200 = 211.45$

c. $P(|X - 200| < b) = 0.966; \; P(|Z| < b/5) = 0.966; \; P() < Z < b/5) = 0.483$;
$b/5 = 2.12; \quad b = 10.6$

7.81. $P(X \le 98) = P(Z \le (98 - 100)/8) = P(Z \le -0.25) = 0.4013$

7.82. $P(X < 58) = P(Z < (58 - 70)/8) = P(Z < -1.5) = 0.067$

$P(58 < X < 66) = P((58 - 70)/8 < Z < (66 - 70)/8) = P(-1.5 < Z < -0.5) = 0.3085 - 0.0668$
$= 0.242$

$P(66 < X < 74) = P((66-70)/8 < Z < (74 - 70)/8) = P(-0.5 < Z < 0.5) = 0.6915 - 0.3085$
$= 0.383$

$P(74 < X < 82) = P((74 - 70)/8 < Z < (82 - 70)/8) = P(0.5 < Z < 1.5) = 0.9332 - 0.6915$
$= 0.242$

$P(X > 82) = P((82 - 70)/8) = P(Z > 1.5) = 0.067$

7.83. $P(X \geq 730) = P(Z \geq (730 - 530)/100) = P(Z > 2) = 0.023$

7.84.

a. $P(X < 80) = P(Z < (80 - 90)/20) = P(Z < -0.5) = 0.309$

30.9 % of the candidates will pass the test.

b. $P(X > b) = 0.05;$ $b = 1.645 * 20 + 90 = 122.9$ minutes

7.85.

a.

i. $P(X \leq 5) = 0.029$

ii. $P(11 \leq X \leq 17) = 0.965 - 0.048 = 0.917$

iii. $(X \geq 11) = 1 - 0.895 = 0.105$

b.

i. $\mu = 25 * 0.4 = 10;$ $\sigma = (25 * 0.4 * 0.6)^\wedge 0.5 = 2.449$

$P(X \leq 5) = P(Z \leq (5 - 10)/2.449) = P(Z \leq -2.04) = 0.021$

ii. $\mu = 20 * 0.7 = 14;$ $\sigma = (20 * 0.7 * 0.3)^\wedge 0.5 = 2.049$

$P(11 \leq X \leq 17) = P((11 - 14)/(2.049) \leq Z \leq (17 - 14)/(2.049))$

$= P(-1.46 \leq Z \leq 1.46) = 0.9279 - 0.0721 = 0.856$

iii. $\mu = 16 * 0.5 = 8;$ $\sigma = (16 * 0.5 * 0.5)^\wedge 0.5 = 2$

$P(X \geq 11) = P(Z \geq (11 - 8)/(2)) = P(Z \geq 1.5) = 0.067$

7.86. $\mu = 20{,}000 * 0.09 = 1800;$ $\sigma = (20{,}000 * 0.09 * 0.91)^\wedge 0.5 = 40.472$

a. $P(X < 1750) = P(X \leq 1749) = P(Z \leq (1749 - 1800)/40.472) = P(Z < -1.26)$
$= 0.104$

b. $P(X \geq 2000) = P(Z \geq (2000 - 1800)/40.472) = P(Z \geq 4.94) = 0$

7.87. $\mu = 400 * 0.3 = 120;$ $\sigma = (400 * 0.3 * 0.7)^\wedge 0.5 = 9.165$

a. $P(X \leq 104) = P(Z \leq (104 - 120)/9.165) = P(Z < -1.75) = 0.040$

b. If the number of viewers is actually less than 105, the suspicion that the population percentage has decreased will be strongly supported.

7.88.

a. $np = 400 * 0.28 = 112;$ $n(1-p) = 400 * 0.72 = 288$

The normal approximation is appropriate for this situation.

b. $np = 20 * 0.04 = 0.8$; $n(1\text{-}p) = 20 * 0.96 = 19.2$
The normal approximation is not appropriate for this situation.

c. $np = 90 * 0.99 = 89.1$; $n(1\text{-}p) = 90 * 0.01 = 0.9$
The normal approximation is not appropriate for this situation.

7.89. mean = 400 * 0.1 = 40; standard deviation = (400 * 0.1 * 0.9) ^0.5 = 6

a. $P(X > 370) = P(X \geq 371) = P(Z \geq (371 - 400)/6) = P(Z \geq -4.83) = 1.00$

b. $P(X < 350) = P(X \leq 349) = P(Z \leq (349 - 400)/6) = P(Z \leq -8.5) = 0$

7.90.

a.

Samples	Sample Mean
2, 2	2
2, 4	3
2, 6	4
2, 8	5
4, 2	3
4, 4	4
4, 6	5
4, 8	6
6, 2	4
6, 4	5
6, 6	6
6, 8	7
8, 2	5
8, 4	6
8, 6	7
8, 8	8

b.

Value of sample mean	Probability
2	1/16
3	2/16 = 1/8
4	3/16
5	4/16 = 1/4
6	3/16
7	2/16 = 1/8
8	1/16

c. Population distribution:

X	Probability of X
2	0.25
4	0.25
6	0.25
8	0.25

Population: $\mu = 5$; $\sigma = 2.236$

d.
Sampling distribution:
sample mean = 5
sample standard deviation = 1.581 = σ/1.414

7.91.

a.

Samples	Sample Range
2, 2	0
2, 4	2
2, 6	4
2, 8	6
4, 2	2
4, 4	0
4, 6	2
4, 8	4
6, 2	4
6, 4	2
6, 6	0
6, 8	2
8, 2	6
8, 4	4
8, 6	2
8, 8	0

b.

Possible range	Probability of sampling distribution of R
0	4/16 = 1/4
2	6/16 = 3/8
4	4/16 = 1/4
6	2/16 = 1/8

7.92.

a. $E(\overline{X}) = 550$

$\text{Sd}(\overline{X}) = \sigma/\sqrt{n} = 70/4 = 17.5$

b. $E(\overline{X}) = 550$
$Sd(\overline{X}) = \sigma/\sqrt{n} = 70 / 12.649 = 5.534$

7.93.

a. 16
b. 64
c. 44.444

7.94.

a. $E(\overline{X}) = 80$
$Var(\overline{X}) = \sigma/\sqrt{n} = 10/3 = 3.333$

b. The distribution of $\overline{X}$ is approximately normal.

c. $P(76 < \overline{X} < 84) = P((76\text{-}80)/3.333 < Z < (84 - 80)/3.333) = P(-1.2 < Z < 1.2)$
$0.8849 - 0.1151 = 0.770$

7.95.

a. $P(0.28 < X < 0.34) = P((0.28 - 0.32)/ 0.08 < Z < (0.34 - 0.32)/ 0.08)$
$= P(-0.5 < Z < 0.25) = 0.5987 - 0.3085 = 0.290$

b. $P(0.28 < \overline{X} < 0.34) = P((0.28 - 0.32)/(0.08 / 2) < Z < (0.34 - 0.32)/(0.08 / 2))$
$= P(-1 < Z < 0.5) = 0.6915 - 0.1587 = 0.533$

7.96.

a. The sample is large enough to assume that it is approximately normal.

b. $P(59 < \overline{X} < 61) = P((59 - 60)/(8 / 12.247) < Z < (61 - 60)/(8 / 12.247))$
$= P(-1.53 < Z < 1.53) = 0.9370 - 0.0630 = 0.874$

c. $P(\overline{X} > 62) = P(Z > (62 - 60) / (8 / 12.247)) = P(Z > 3.06) = 0.001$

7.97.

a. $E(\overline{X}) = 94$
$Var(\overline{X}) = \sigma/\sqrt{n} = 10 / 9 = 1.111$

b. The distribution of $\overline{X}$ is approximately normal by the Central Limit Theorem.

7.98.

a. $P(\overline{X} > 96) = P(Z > (96 - 94)/1.111) = P(Z > 1.80) = 0.036$

b. $P(92.3 < \overline{X} < 96.0) = P((92.3 - 94) / 1.111 < Z < (96.0 - 94) / 1.111)$
$= P(-1.53 < Z < 1.8) = 0.9641 - 0.0630 = 0.901$

c. $P(\overline{X} < 95) = P(Z < (95 - 94)/1.111) = P(Z < 0.90) = 0.816$

7.99. $sd(\overline{X}) = 7/(40)$^0.5 $= 1.107$

a. $P(54 < \overline{X} < 56) = P((54 - 55)/1.107 < Z < (56 - 55)/1.107)$
$= P(-0.90 < Z < 0.90) = 0.8159 - 0.1841 = 0.632$

b. $0.95/2 = 0.475; 0.5 + 0.475 = 0.975; Z = 1.96; X = 1.96(1.107) + 55 = 57.170$
$P(-57.170 < \overline{X} < 57.170) = 0.95$

7.100. $sd(\overline{X}) = 20 / 10 = 2$

a. $P(-2 \le \overline{X} - \mu \le 2) = P(-2/2 \le Z \le 2/2) = P(-1 \le Z \le 1) = 0.8413 - 0.1587 = 0.683$

b. $0.90 / 2 = 0.45; 0.50 + 0.45 = 0.95; P(-1.645 \le Z \le 1.645) = 0.90$
$k/2 = 1.645; k = 3.290$

c. $P(|\overline{X} - \mu| > 4) = P(|Z| > 4/2) = 2 * P(Z > 2) = 2 * 0.0228 = 0.046$

7.101. $sd(\overline{X}) = 0.6/(50)$^0.5 $= 0.085$
$P(1.8 < \overline{X} < 2.25) = P((1.8 - 2)/ 0.085 < Z < (2.25 - 2)/ 0.085)$
$= P(-2.35 < Z < 2.94) = 0.9984 - 0.0094 = 0.989$

7.102. Answers will vary.

7.103. Answers will vary.

7.104. Answers may vary.

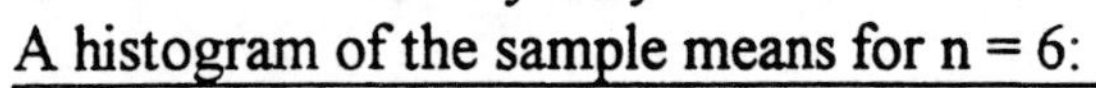

A histogram of the sample means for n = 6:

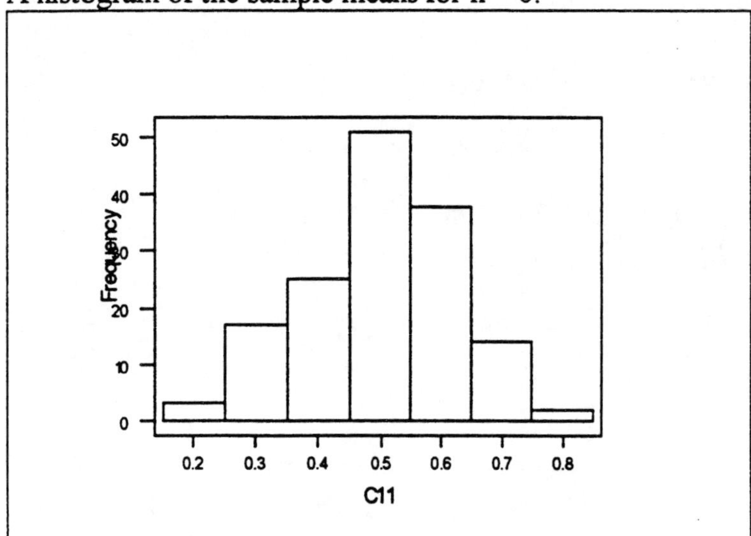

A normal scores plot for n = 6:

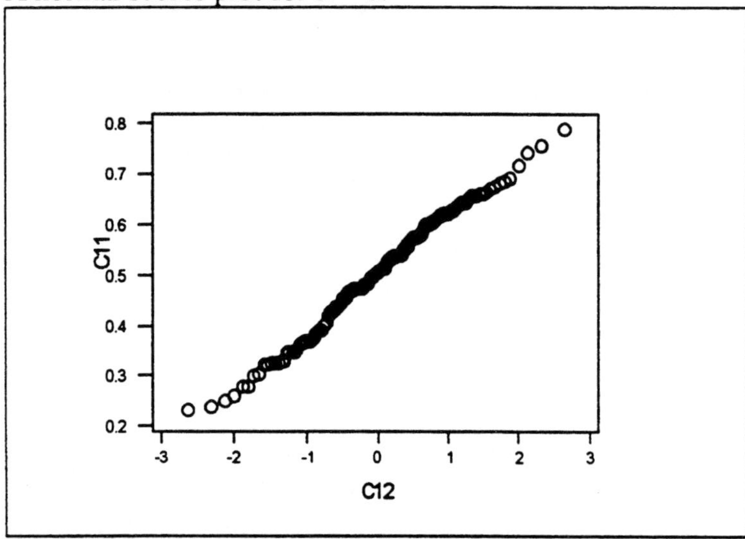

The distribution of the sample means appears to be nearly normal.

A histogram of sample means for n = 20:

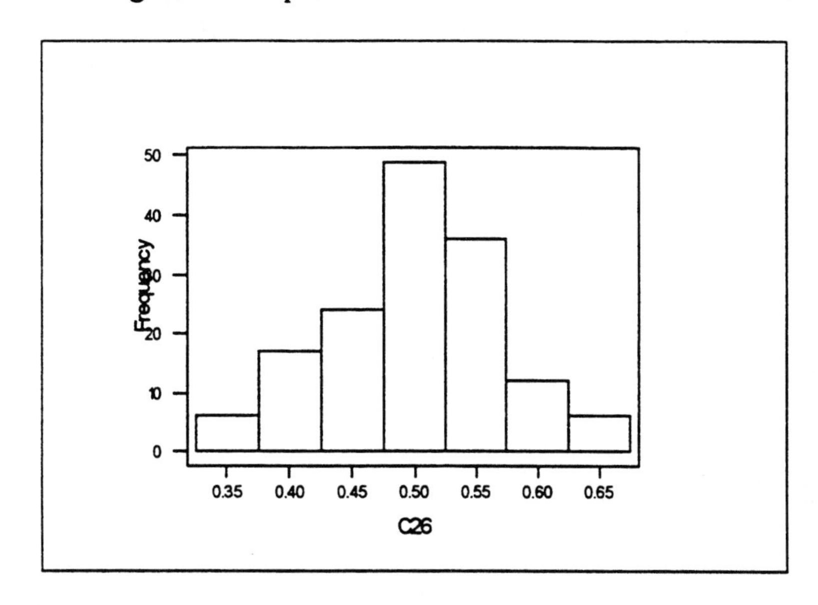

A normal-scores plot for n = 20:

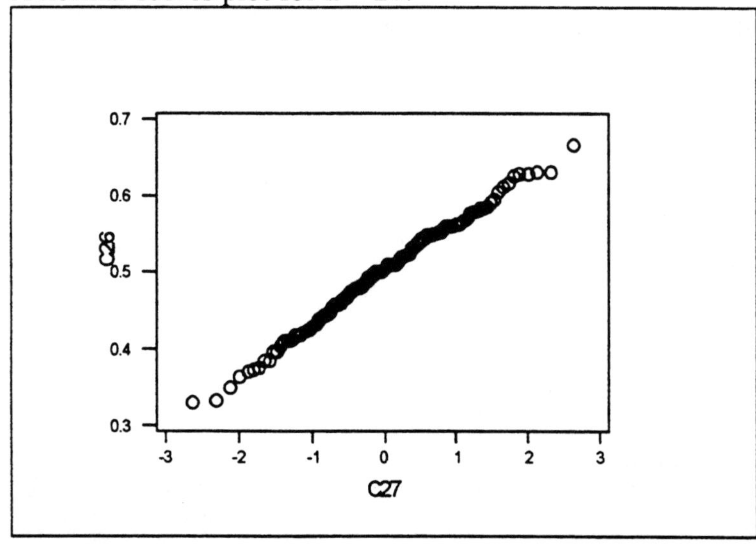

The larger sample size makes the shape of the distribution of the sample means closer to a normal distribution.

A histogram of sample means for n = 6 using the discrete distribution on the integers 0, 1, ...,9.

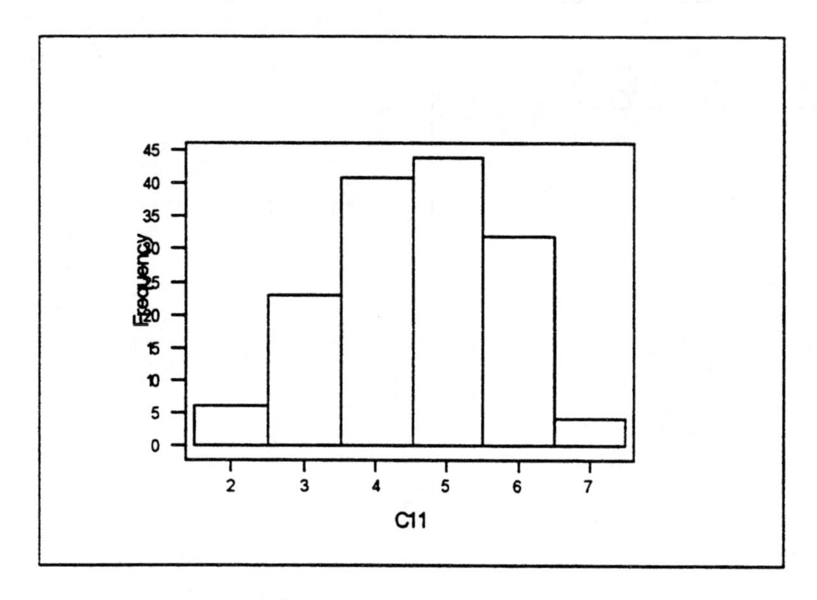

A normal-scores plot of the sample means from a discrete distribution:

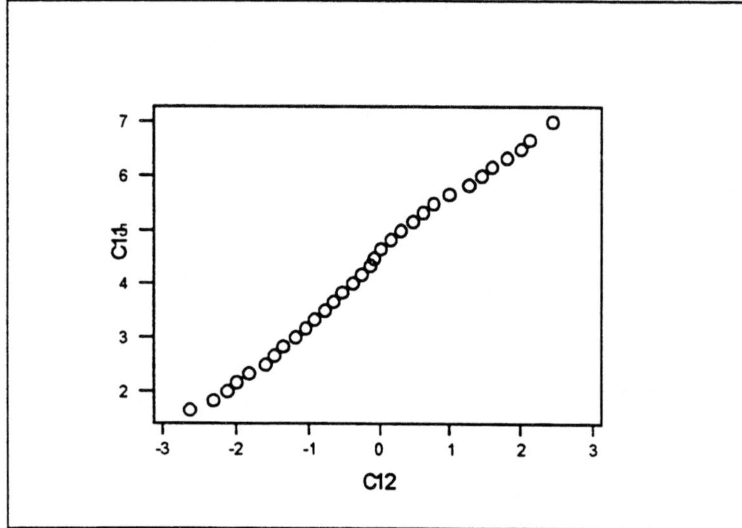

7.105.

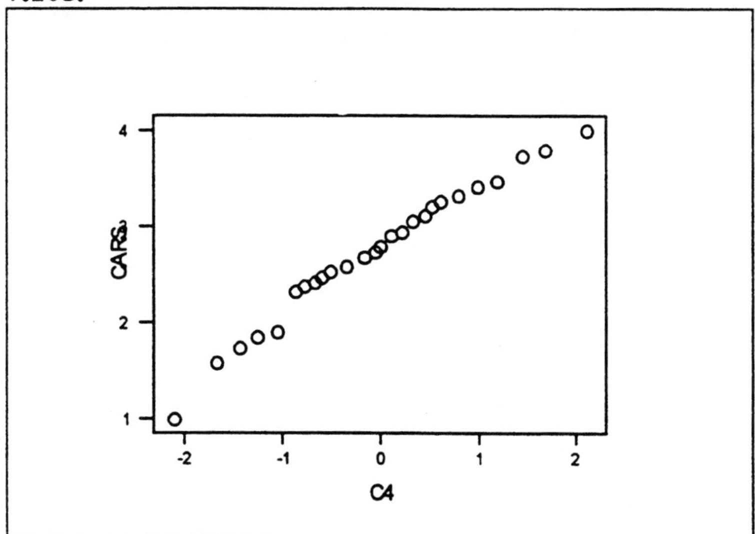

This normal scores plot is nearly a straight line, so the data are nearly normal.

7.106.

a. $P(X > 14) = 1 - P(X \leq 13) = 1 - (1 - e^{-13/18.65}) = e^{-0.697} = 0.498$

b. $P(X > b) = 0.5$

$e^{-b/18.65} = 0.5$

$b = 12.927$

7.107. Dropping the four largest observations changes the distribution of turnaround times a little. In problem 7.106. part a, the answer changed from 0.541 to 0.498, a difference of 0.043.

8.1.

	n	σ	Standard error of $\overline{X}$	100(1-α)% error margin
a.	175	20	1.512	2.963
b.	103	5.4	0.532	0.931
c.	200	63	4.455	11.471

8.2.

	n	$\bar{x}$	s	1-α	Point estimate of μ	Estimated standard error	100(1-α)% error margin
a.	75	83.6	31	0.98	83.6	3.5796	8.340
b.	64	427	3.9	0.975	427	0.4875	1.092
c.	82	0.724	7.7	0.90	0.724	0.8503	1.399

8.3.

	n	$\sum x_i$	$\sum (x_i - \bar{x})^2$	Point estimate of μ	Estimated standard error
a.	68	1506	327	22.147	2.209
b.	52	792	102	15.231	1.414
c.	73	1124	209	15.397	1.704

8.4.

a. Error margin = 4.418
b. Error margin = 2.828
c. Error margin = 3.408

8.5.

Population mean weekly earnings = 624
Error margin = 12.196

8.6.

a. s.e. = 4.83 / 1.96 = 2.464
b. 90% error margin = 1.645 * 2.464 = 4.053

8.7.

a. $n = (1.96 * 4.7 / 0.85)^2 = 117.45 \approx 118$
b. $n = (2.17 * 315 / 10)^2 = 4672.4 \approx 4673$
c. $n = (1.28 * 0.44 / 0.02)^2 = 792.99 \approx 793$

8.8.

a. Error margin = $2 * (\sigma/\sqrt{n}) = \sigma / 6$
$n = 12^2 = 144$

b. Error margin = $2 * (\sigma/\sqrt{n}) = 0.12 * \sigma$
$n = (2 / 0.12)^2 = 278$

8.9. $n = (2.33 * 44/ 4.5)^2 = 519.03 \approx 520$

8.10.

a. Estimate of the mean number of requests per hour = 12.7
Error margin = $1.645 * 5.2/ (48)^{.5} = 1.235$

b. Estimate of the mean number of requests per hour = 25.4
Error margin = $1.645 * 5.2/ (96)^{.5} = 0.873$

c. The error margin is smaller in part b. This result is due to that fact that the larger sample size in part b creates a smaller variance.

8.11.

a. Estimate of the mean number of copies per day = 382.7
Error margin = $1.645 * 61.28/ (56)^{.5} = 13.471$

b. Estimate of the mean number of copies per 5 days = 5 * 382.7 = 1913.5
Error margin = $1.645 * 61.28 / (56*5)^{.5} = 6.024$

c. The error margin is smaller in part b. This result is due to that fact that the larger sample size in part b creates smaller variance.

8.12. Estimate of the mean value = \$1790
Error margin = $1.96 * 635 / (340)^{.5} = 67.498$

8.13. $P(|(\overline{X} - \mu)| < \sigma/\sqrt{n}) = P(|(\overline{X} - \mu) /\sigma| < 1/\sqrt{n}) = P(-1/\sqrt{n} < (\overline{X} - \mu) /\sigma < 1/\sqrt{n})$
Since $n \geq 1$, this interval is no larger than $P(-1 < Z < 1) = 0.6826$.

8.14.

a. Point estimate of the population mean = 1.73
(40)^0.5 = 6.325
Standard error = 0.62/ 6.325 * [(350 - 40)/(350-1)]^0.5 = 0.092

b. Point estimate of the population mean = 0.19
(20)^0.5 = 4.472
Standard error = 0.06 / 4.472 * [(82 - 20)/(82-1)]^0.5 = 0.012

8.15. $38.4 \pm 1.645 * 6.1 / 7 = 38.4 \pm 1.4335 = (36.97 , 39.83)$

8.16. $7.3 \pm 1.96 * 2.8 / 8 = 7.3 \pm 0.686 = (6.61 , 7.99)$

8.17. 90% * 400 = 360 of the intervals will cover the true mean.

8.18. 95% * 300 = 285 of the intervals will cover the true mean.

8.19. 0.95 * 365 = 347 of the intervals will cover the true mean.

8.20.

a. Yes, the statement is correct.

b. We do not know whether this interval covers the true mean. 95% of the intervals constructed in this manner based on the means of random samples will cover the true mean.

8.21.

a. Yes, the statement is correct.

b. We do not know whether this interval covers the true mean. 90% of the intervals constructed in this manner based on the means of random samples will cover the true mean.

8.22. $2.83 \pm 1.96 * 1.2 / 6.9282 = 2.83 \pm 0.33948 = (2.49 , 3.17)$

8.23. $2.7 \pm 1.96 * 0.92 / 7.61577 = 2.7 \pm 0.236777 = (2.46 , 2.94)$

8.24. $624 \pm 2.33 * 44 / 7.07 = 624 \pm 14.5 = (609.5 , 638.5)$

8.25. Error margin = $1.96 * \sigma/\sqrt{n} = 75.2$
$\sigma/\sqrt{n} = 75.2 / 1.96 = 38.367$

a. $1234 \pm 1.96 * 38.367 = 1234 \pm 75.2 = (1158.80 , 1309.20)$
b. $1234 \pm 1.645 * 38.367 = 1234 \pm 63.1137 = (1170.89 , 1297.11)$

8.26. Error margin = $1.96 * \sigma/\sqrt{n} = 1.2$
$\sigma/\sqrt{n} = 1.2 / 1.96 = 0.61224$

a. $3.4 \pm 1.96 * 0.61224 = 3.4 \pm 1.19999 = (2.20 , 4.60)$

b. $3.4 \pm 1.645 * 0.61224 = 3.4 \pm 1.007 = (2.39 , 4.41)$

8.27.

a. Point estimate for the population mean = (102.59 + 23.41) / 2 = 63
Error margin = (102.59 - 23.41) / 2 = 39.59

b. $\sigma/\sqrt{n} = 39.59 / 1.96 = 20.199$
$63 \pm 1.645 * 20.199 = 63 \pm 33.227 = (29.773 , 96.227)$

8.28.

a. Point estimate for the population mean = (43.7 + 162.3) / 2 = 103
Error margin = (162.3 - 43.7) / 2 = 59.3

b. $\sigma/\sqrt{n}$ = 59.3 / 1.96 = 30.255
103 ± 1.645 * 30.255 = 103 ± 49.77 = (53.23 , 152.77)

8.29. $0.13s = z * s/\sqrt{n}$; $z = 0.13 * 12 = 1.56$
Confidence level = 0.881

8.30.

a. Cannot tell since only 95% of confidence intervals constructed about the means of random samples contain the population mean.
b. Yes, the sample mean is contained within the interval since the interval was built around the sample mean.
c. Cannot tell since the means will differ.
d. No, the 95% applies to the number of confidence intervals which contain the population mean.
e. No, higher confidence levels produce larger confidence intervals.

8.31.

a. Cannot tell since only 90% of confidence intervals constructed about the means of random samples contain the population mean.
b. Yes, the sample mean is contained within the interval since the interval was built around the sample mean.
c. Cannot tell since the means will differ.
d. No, the 90% applies to the number of confidence intervals which contain the population mean.
e. Yes, lower confidence levels produce narrower confidence intervals.

8.32.

a. $H_0 : \mu = 24$ versus $H_1 : \mu > 24$
b. $H_0 : \mu = 2000$ versus $H_1 : \mu < 2000$
c. $H_0 : \mu = 31.80$ versus $H_1 : \mu \neq 31.80$
d. $H_0 : \mu = 4.7$ versus $H_1 : \mu < 4.7$

8.33.

a. $H_0 : \mu = 9.5$ versus $H_1 : \mu < 9.5$
b. $H_0 : \mu = 7.40$ versus $H_1 : \mu > 7.40$
c. $H_0 : \mu = 3.4$ versus $H_1 : \mu \neq 3.4$
d. $H_0 : \mu = 35$ versus $H_1 : \mu > 35$

8.34.

a. The decision to reject the null hypothesis is correct when the average score is greater than 24.
A Type I error occurs when the mean score is 24 and we, incorrectly, reject the null hypothesis.

b. The decision to reject the null hypothesis is correct when the mean bill is less than $2000.
A Type I error occurs when the mean bill is $2000 and we, incorrectly, reject the null hypothesis.

c. The decision to reject the null hypothesis is correct when the mean expense is different from $31.80.
A Type I error occurs when the mean expense is $31.80 and we, incorrectly, reject the null hypothesis.

d. The decision to reject the null hypothesis is correct when the mean is less than 4.7.
A Type I error occurs when the mean is 4.7 and we, incorrectly, reject the null hypothesis.

8.35.

a. The decision to reject the null hypothesis is correct when the mean is less than 9.5 days.
A Type I error occurs when the mean is 9.5 and we, incorrectly, reject the null hypothesis.

b. The decision to reject the null hypothesis is correct when the mean is more than $7.40.
A Type I error occurs when the mean is $7.40 and we, incorrectly, reject the null hypothesis.

c. The decision to reject the null hypothesis is correct when the mean is different from 3.4.
A Type I error occurs when the mean is 3.4 and we, incorrectly, reject the null hypothesis.

d. The decision to reject the null hypothesis is correct when the mean is more than 35 days.
A Type I error occurs when the mean is 35 and we, incorrectly, reject the null hypothesis.

8.36.

a. The rejection region is one-sided to the right.
b. The rejection region is one-sided to the left.
c. The rejection region is two-sided.

d. The rejection region is one-sided to the left.

8.37.

a. The rejection region is one-sided to the left.
b. The rejection region is one-sided to the right.
c. The rejection region is two-sided.
d. The rejection region is one-sided to the right.

8.38.

1. Hypotheses: $H_0 : \mu = 1.51$ versus $H_1 : \mu < 1.51$
2. Test statistic: $z = (\bar{x} - 151) / (s / (45)^{\wedge} 0.5)$
3. Rejection region: $z < -1.96$
4. Calculation: $z = (147 - 151) / (12.4 / (45)^{\wedge}.5) = -2.16$
5. Conclusion: Reject the null hypothesis since the test statistic is within the rejection region.
6. P-value = P(Z < -2.16) = 0.015

8.39.

Hypotheses: $H_0 : \mu = 1.51$ versus $H_1 : \mu \neq 1.51$

P-value = 2 * $P(Z < -2.16)$ = 2 * .0154 = 0.031
Since it is a two-tailed test, this value is two times the P-value for a one-tailed test.

8.40.

a. Select the test statistic: $z = (\bar{x} - 20) / (s / 8.0623)$
Rejection regions: $|z| > 1.96$

b. Calculate test statistic: $z = (17.14 - 20) / (15.6 / 8.0623) = -1.48$
Do not reject the null hypothesis since the test statistic is not in the rejection region.

c. P-value = 2 * P(Z < -1.48) = 2 * 0.0694 = 0.139
Do not reject the null hypothesis since the P-value is not less than α.

8.41.

1. Hypotheses: $H_0 : \mu = 15$ versus $H_1 : \mu \neq 15$
2. Select the test statistic: $z = (\bar{x} - 15) / (s / (30)^{\wedge}0.5)$
3. Rejection regions: $|z| > 1.645$
4. Calculate test statistic: $z = (10.76 - 15) / (13.71 / (30)^{\wedge}0.5) = -1.69$
5. Conclusion: Reject the null hypothesis since the test statistic is within the rejection region.

8.42.

1. Hypotheses: $H_0 : \mu = 2.0$ versus $H_1 : \mu \neq 2.0$
2. Select the test statistic: $z = (\bar{x} - 2.0) / (s / (3.2)^{\wedge}0.5)$
3. Rejection regions: $|z| > 1.96$
4. Calculate test statistic: $z = (2.8 - 2.0) / (3.2 / (50)^{\wedge}0.5) = 1.77$
5. Conclusion: Do not reject the null hypothesis since the test statistic is not within the rejection region.

8.43.

1. Hypotheses: $H_0 : \mu = 7.5$ versus $H_1 : \mu > 7.5$
2. Select the test statistic: $z = (\bar{x} - 7.5) / (s / (80)^{\wedge}0.5)$
3. Calculate test statistic: $z = (8.1 - 7.5) / (1.3 / (80)^{\wedge}0.5) = 4.13$
4. P-value = P($Z > 4.13$) = 0

Since the P-value is very small, the evidence against the null hypothesis is strong.

8.44.

a. Test statistic: $z = (70.3 - 64) / (15.9 / 6) = 2.377$
P-value = P($Z > 2.377$) = 0.009
The evidence against the null hypothesis is strong since the P-value is very small.

b. Test statistic: $z = (70.3 - 75) / (15.9 / 6) = -1.77$
P-value = P($Z < -1.77$) = 0.0384
Reject the null hypothesis for $\alpha > 0.038$; do not reject the null hypothesis for $\alpha < 0.038$.

c. Test statistic: $z = (70.3 - 66) / (15.9 / 6) = 1.62$
P-value = 2 * P($Z > 1.62$) = 2 * 0.0526 = 0.105
The evidence against the null hypothesis is weak since the P-value is larger than 0.10.

8.45.

a. Yes
b. Can't tell
c. No

8.46.

$\beta = P(Z \geq -1.645) = P((\bar{X} - 8.3)/ (3.1*(60)^{\wedge}0.5) \geq -1.645)$
$= P(\bar{X} \geq 8.3 - 3.1 / (60)^{\wedge}0.5 * 1.645)$
$= P(\bar{X} - 7.1 \geq 8.3 - 7.1 - 3.1 / (60)^{\wedge}0.5 * 1.645)$
$= P((\bar{X} - 7.1) / (3.1/(60)^{\wedge}0.5) \geq 3.00 - 1.645)$
$= 1 - P((\bar{X} - 7.1) / (3.1 / (60)^{\wedge}0.5) < -1.345) = 1 - 0.0885 = 0.9115$

8.47.

a. Reject the null hypothesis since -4 is not in the confidence interval.

b. Do not reject the null hypothesis since 1 is in the confidence interval.

8.48.

a. Do not reject the null hypothesis since 7.5 is in the confidence interval.
b. Reject the null hypothesis since 7 is not in the confidence interval.

8.49. Yes, the population mean, 16, would fall in the confidence interval since the null hypothesis was not rejected.

8.50.

a. 2.015
b. -2.101
c. -2.718
d. 1.337

8.51.

a. 1.740
b. 1.325
c. ±2.052
d. -1.729

8.52.

a. 1.363
b. 4.541
c. -1.706
d. ±0.688

8.53.

a. $P(t < -1.740) = 0.05$
b. $P(|t| > 3.143) = 0.01 * 2 = 0.02$
c. $P(-1.330 < t < 1.330) = 1 - 2 * 0.10 = 0.8$
d. $P(-1.372 < t < 2.764) = 10 - 0.10 - 0.01 = 0.89$

8.54.

a. $P(t < b) = 0.95$; $P(t > b) = 0.05$; $b = 1.943$
b. $P(-b < t < b) = 0.95$; $P(t > b) = 0.025$; $b = 2.131$
c. $P(t > b) = 0.01$; $b = 2.896$
d. $P(t > b) = 0.99$; $P(t < -b) = 0.01$; $b = -2.718$

8.55.

d.f.	5	10	15	20	29
$t_{.05}$ values	2.015	1.812	1.753	1.725	1.699

The percentile decreases as the degrees of freedom increase (for the same table value).

8.56.

a. P($t > 2.6$) is between 0.01 and 0.025 because 2.6 lies between $t_{.025} = 2.365$ and $t_{.01} = 2.998$.

b. P($t > 1.9$) is between 0.025 and 0.05 because 1.9 lies between $t_{.025} = 2.131$ and $t_{.05} = 1.753$.

c. P($t < -1.5$) is between 0.05 and 0.1 because 1.5 lies between $t_{.05} = 1.740$ and $t_{.1} = 1.333$.

d. P($|t| > 1.9$) is between 0.05 and 0.1 (i.e., 2 * 0.025 and 2 * 0.05) because 1.9 lies between $t_{.025} = 2.131$ and $t_{.05} = 1.753$.

e. P($|t| < 2.8$) is between 0.01 and 0.01250 (i.e., 2 * 0.005 and 2 * 0.00625) because 2.8 lies between $t_{.005} = 2.845$ and $t_{.00625} = 2.744$.

8.57.

a. c is between 2.015 and 2.571 because $t_{.05} = 2.015$ and $t_{.025} = 2.571$.

b. c is between 2.262 and 2.821 because $t_{.025} = 2.262$ and $t_{.01} = 2.821$.

c. c is greater than 3.055 because $t_{.005} = 3.055$.

d. $P(t > c) = 0.015$; c is between 2.571 and 3.365 because $t_{.025} = 2.571$ and $t_{.01} = 3.365$.

e. $P(t > c) = 0.02$; c is between 2.080 and 2.518 because $t_{.025} = 2.080$ and $t_{.01} = 2.518$.

8.58.

Sample mean = 25
Standard deviation = 3.937
95% C.I.: $25 \pm 2.776 * 3.937 / 2.236 = 25 \pm 4.888 = (20.11 , 29.89)$

8.59.

Sample mean = 11
Standard deviation = 3.916
90% C.I.: $11 \pm 2.353 * 3.916 / 2 = 11 \pm 4.607 = (6.39 , 15.61)$

8.60.

95% C.I.: $35.7 \pm 2.201 * 4.2 / 3.464 = 35.7 \pm 2.6686 = (33.03 , 38.37)$

8.61.

90% C.I.: $35.7 \pm 1.796 * 4.2 / 3.464 = 35.7 \pm 2.1776 = (33.52 , 37.88)$

8.62.
95% C.I.: $21.4 \pm 2.365 * 7.3 / 2.828 = 21.4 \pm 6.1048 = (15.30 , 27.50)$

8.63.
90% C.I.: $21.4 \pm 1.895 * 7.3 / 2.828 = 21.4 \pm 4.892 = (16.51 , 26.29)$

8.64.
Mean = 151.10
Standard deviation = 76.557
95% C.I.: $151.10 \pm 2.262 * 76.557 / 3.1623 = 151.10 \pm 54.76 = (96.34 , 205.86)$

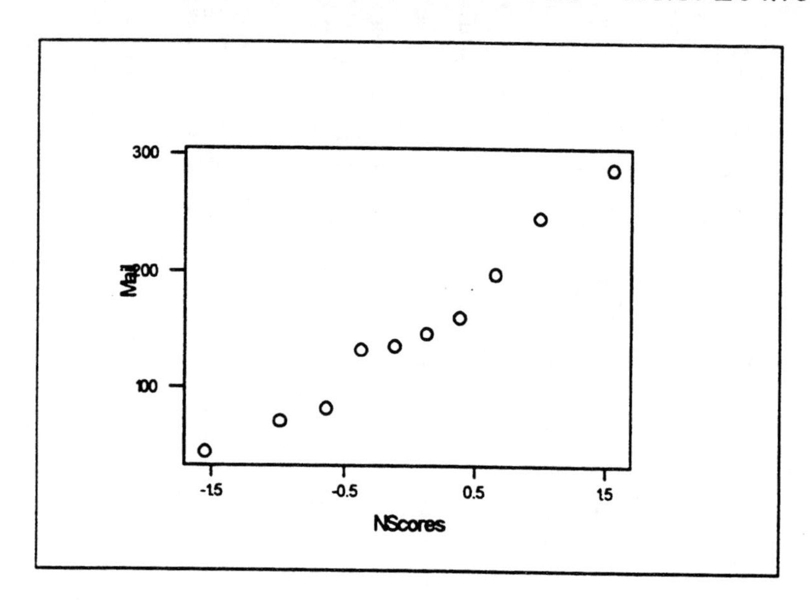

Normality is a reasonable assumption, particularly for a sample of this size. The normal scores plot is relatively straight.

8.65.
Confidence Intervals

Variable	N	Mean	StDev	SE Mean	95.0 % C.I.
C1	12	77.83	33.51	9.67	(56.54, 99.13)

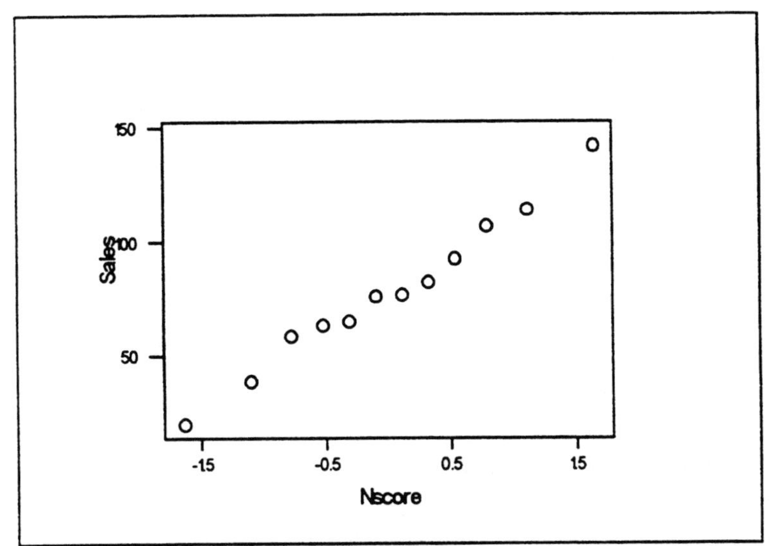

Normality is a reasonable assumption. The normal scores plot is relatively straight.

8.66.
Ratio of lengths of intervals: 1.833 / 2.262 = 0.81

8.67. 95% C.I.: $8.2 \pm 2.262 * 4.1 / 3.1623 = 8.2 \pm 2.9327 = (5.27, 11.13)$
We assumed that the original population was normal.

8.68.
H_o: $\mu = 6.3$ versus H_1: $\mu \neq 6.3$

8.69.
1. Hypotheses: H_o: $\mu = 335$; H_1: $\mu < 335$
2. Select test statistic: $t = (\bar{x} - 335)/(s / 3)$
3. Determine the rejection region: $t < -1.860$
4. Calculate the test statistic: $t = (324.5 - 335)/(15.2 / 3) = -2.07$
5. Draw a conclusion: Reject the null hypothesis.
6. *P*-value = $P(t < -2.07)$ is between 0.025 and 0.05.

The evidence against the null hypothesis is strong.

8.70.

1. Hypotheses: H_o: $\mu = 3.5$; H_1: $\mu > 3.5$
2. Select test statistic: t = $(\bar{x} - 3.5)/(s / 3.873)$
3. Determine the rejection region: t > 2.145
4. Calculate the test statistic: t = $(4.32 - 3.5)/(1.63 / 3.873) = 1.948$
5. Draw a conclusion: Do not reject the null hypothesis.
6. P-value = P(t > 1.948) is between 0.025 and 0.05.

The evidence against the null hypothesis is weak at $\alpha = 0.025$, but stronger at $\alpha = 0.050$.

8.71.

1. Hypotheses: H_o: $\mu = 3.5$; H_1: $\mu \neq 3.5$
2. Select test statistic: t = $(\bar{x} - 3.5)/(s / 3.873)$
3. Determine the rejection region: $|t| > 2.145$
4. Calculate the test statistic: t = $(4.32 - 3.5)/(1.63 / 3.873) = 1.948$
5. Draw a conclusion: Do not reject the null hypothesis.
6. *P*-value = 2 * P(t > 1.948) is between 0.05 and 0.1.

The evidence against the null hypothesis is moderately strong.

8.72.

1. Hypotheses: H_o: $\mu = 8\%$; H_1: $\mu < 8\%$
2. Select test statistic: t = $(\bar{x} - 8)/(s / 2.6458)$
3. Calculate the test statistic: t = $(7.1- 8)/(1.3 / 2.6458) = -1.83$
4. *P*-value = P(t < -1.83) is between 0.05 and 0.1.
5. $\alpha = .05$ level test result: Do not reject the null hypothesis.

8.73.

1. Hypotheses: H_o: $\mu = 2.6$; H_1: $\mu > 2.6$
2. Select test statistic: t = $(\bar{x} - 2.6)/(s / 4.2426)$
3. Calculate the test statistic: t = $(3.2 - 2.6)/(1.6 / 4.2426) = 1.59$
4. *P*-value = P(t > 1.59) is between 0.05 and 0.1.
5. $\alpha = .025$ level test result: Do not reject the null hypothesis.

8.74.

1. Hypotheses: H_o: $\mu = 2.6$; H_1: $\mu \neq 2.6$
2. Select test statistic: t = $(\bar{x} - 2.6)/(s / 4.2426)$
3. Calculate the test statistic: t = $(3.2 - 2.6)/(1.6 / 4.2426) = 1.59$
4. *P*-value = 2 * P(t > 1.59) is between 0.1 and 0.2.
5. $\alpha = .05$ level test result: Do not reject the null hypothesis.

8.75.

Confidence interval: $\left(\bar{x} - t_{\alpha/2}\frac{s}{\sqrt{n}}, \bar{x} + t_{\alpha/2}\frac{s}{\sqrt{n}}\right)$

Two-sided rejection region: $\frac{\bar{x} - \mu_0}{s/\sqrt{n}} > t_{\alpha/2}$ or $\frac{\bar{x} - \mu_0}{s/\sqrt{n}} < -t_{\alpha/2}$

Acceptance region: $-t_{\alpha/2} < \frac{\bar{x} - \mu_0}{s/\sqrt{n}} < t_{\alpha/2}$

or: $\bar{x} - t_{\alpha/2}\frac{s}{\sqrt{n}} < \mu_0 < \bar{x} + t_{\alpha/2}\frac{s}{\sqrt{n}}$

8.76.

1. Hypotheses: H_o: $\mu = 90\%$; H_1: $\mu > 90\%$
2. Select test statistic: t = $(\bar{x} - 90)/(s / 2.4495)$
3. Rejection region: t > 2.015
4. Calculate the test statistic: t = (93.6 - 90)/(2.75 / 2.4495) = 3.207
5. α = .05 level test result: Reject the null hypothesis.
6. *P*-value = P(t > 3.207) is between 0.01 and 0.025.

8.77.

Mean = 77.833
Standard deviation = 33.512

1. Hypotheses: H_o: $\mu = 60$; H_1: $\mu \neq 60$
2. Select test statistic: t = $(\bar{x} - 60)/(s / 3.4641)$
3. Rejection region: | t | > 2.201
4. Calculate the test statistic: t = (77.833 - 60)/(33.512 / 3.4641) = 1.84
5. α = .05 level test result: Do not reject the null hypothesis.
6. *P*-value = 2 * P(t > 1.84) is between 0.05 and 0.1.

8.78.

We would not reject the test at .05 since the *P*-value = 0.093.

a.

T-Test of the Mean

Test of mu = 3.000 vs mu not = 3.000

Variable	N	Mean	StDev	SE Mean	T	P-Value
CARS	35	2.771	0.671	0.113	-2.02	0.051

T-Test of the Mean

Reject the null hypothesis at α = 0.10 but not at 0.50 since the *P*-value is 0.051.

b.

T-Test of the Mean

Test of mu = 10.500 vs mu not = 10.500

Variable	N	Mean	StDev	SE Mean	T	P-Value
Diesel	23	10.106	2.089	0.436	-0.91	0.38

Do not reject the null hypothesis since the P-value is not smaller than $\alpha = 0.05$.

8.79.

a.

Variable	N	Mean	StDev	SE Mean	90.0 % C.I.
NumCalls	75	89.60	17.32	2.00	(86.27, 92.93)

b.

Variable	N	Mean	StDev	SE Mean	95.0 % C.I.
Diesel	23	10.106	2.089	0.436	(9.202, 11.009)

Yes.

8.80.

a.

Confidence Intervals

Variable	N	Mean	StDev	SE Mean	95.0 % C.I.
C1	10	13.456	2.633	0.833	(11.572, 15.340)
C2	10	15.54	4.20	1.33	(12.53, 18.54)
C3	10	14.13	3.47	1.10	(11.65, 16.62)
C4	10	16.039	2.903	0.918	(13.961, 18.116)
C5	10	14.73	3.26	1.03	(12.40, 17.06)
C6	10	14.12	4.01	1.27	(11.26, 16.99)
C7	10	13.381	2.683	0.849	(11.461, 15.301)
C8	10	12.60	3.79	1.20	(9.89, 15.31)
C9	10	14.347	1.487	0.470	(13.283, 15.410)
C10	10	17.485	1.546	0.489	(16.379, 18.590)
C11	10	14.820	2.916	0.922	(12.733, 16.906)
C12	10	14.24	3.46	1.10	(11.76, 16.72)
C13	10	14.31	4.35	1.38	(11.20, 17.42)
C14	10	13.627	2.009	0.635	(12.190, 15.065)
C15	10	15.35	3.25	1.03	(13.02, 17.68)
C16	10	14.963	2.412	0.763	(13.238, 16.689)
C17	10	13.86	3.83	1.21	(11.12, 16.59)
C18	10	16.044	2.720	0.860	(14.098, 17.991)
C19	10	15.088	2.990	0.946	(12.948, 17.228)
C20	10	15.672	2.945	0.931	(13.564, 17.779)

Nineteen of the twenty confidence intervals cover the true mean of 15. (C10 does not.) This is consistent with the confidence level associated with these intervals which is 95%.

b.

Confidence Intervals

Variable	N	Mean	StDev	SE Mean	95.0 % C.I.
C1	5	13.89	2.88	1.29	(10.32, 17.46)
C2	5	16.64	2.73	1.22	(13.26, 20.03)
C3	5	14.034	1.666	0.745	(11.964, 16.103)
C4	5	16.71	4.86	2.17	(10.67, 22.75)
C5	5	14.585	1.852	0.828	(12.285, 16.885)
C6	5	13.367	0.965	0.431	(12.168, 14.565)
C7	5	15.17	3.71	1.66	(10.56, 19.78)
C8	5	16.22	2.63	1.17	(12.96, 19.48)
C9	5	13.72	2.44	1.09	(10.68, 16.76)
C10	5	15.911	1.760	0.787	(13.724, 18.097)
C11	5	15.194	1.644	0.735	(13.153, 17.236)
C12	5	15.19	3.29	1.47	(11.11, 19.28)
C13	5	11.438	2.137	0.956	(8.784, 14.092)
C14	5	14.691	2.107	0.942	(12.074, 17.307)
C15	5	14.68	2.71	1.21	(11.31, 18.04)
C16	5	15.628	1.946	0.870	(13.211, 18.045)
C17	5	14.02	3.19	1.43	(10.06, 17.99)
C18	5	14.83	4.35	1.95	(9.42, 20.24)
C19	5	12.48	3.22	1.44	(8.48, 16.48)
C20	5	15.64	3.93	1.76	(10.76, 20.52)

Eighteen of the twenty confidence intervals cover the true mean of 15. (C6 and C13 do not.) This is inconsistent with the 95% confidence level associated with these intervals since .95 is a long-run relative frequency.

8.81.

a.

Point estimate of μ = 1235 / 75 = 16.467

Estimated standard error = [(249 / 74) ^ 0.5] / (75 ^ 0.5) = 0.212

b.

Point estimate of μ = 88.6 / 48 = 1.846

Estimated standard error = [(307.2 / 47) ^ 0.5] / (48 ^ 0.5) = 0.369

8.82.

The sample size, n, must be increased by a factor of 3 for the standard error to be reduced to 1/3 of its original value.

8.83.

a. (1.96 * 8.5 / 0.6)^2 = 771

b. (2.17 * 103 / 12)^2 = 347

c. (1.44 * 0.14 / 0.03)^2 = 46

8.84.

a. Sample mean = 3022 / 65 = 46.49
s.d. = (2406 / 64) ^ 0.5 = 6.131
95% C.I.: 46.49 ± 1.96 * 6.131 / 8.0623 = 46.49 ± 1.4905 = (45.00 , 47.98)

b. Sample mean = 7225 / 50 = 144.5
s.d. = (408.2 / 49) ^ 0.5 = 2.886
90% C.I. = 144.4 ± 1.645 * 2.886 / 7.0712 = 144.4 ± 0.671 = (143.729 , 145.071)

8.85.

a. Can't tell
b. Yes
c. No

8.86.

a. Test statistic: $z = (\bar{x} - 21.6) / (s / \sqrt{n})$
Rejection region: $|z| > 1.96$

b. Test statistic: $z = (\bar{x} - 45.7) / (s / \sqrt{n})$
Rejection region: $z < -1.88$

8.87.

a. Test statistic: $z = (\bar{x} - 102) / (s / \sqrt{n})$
Rejection region: $|z| > 1.645$

b. Test statistic: $z = (\bar{x} - 6.4) / (s / \sqrt{n})$
Rejection region: $z > 1.34$

8.88.

a. Test statistic: z = (27.3 - 29) / (6.4 / 7.07) = -1.878
P-value = P(z > -1.878) = 0.970
Do not reject the null hypothesis. The evidence against the null hypothesis is extremely weak.

b. Test statistic: z = (27.3 - 31) / (6.4 / 7.07) = -4.09
P-value = P(z < -4.09) = 0
The evidence against the null hypothesis is very strong.

c. Test statistic: z = (27.3 - 29) / (6.4 / 7.07) = -1.878
P-value = 2 * P(|z| < -1.878) = 2 * 0.0301 = 0.060
The null hypothesis would be rejected at $\alpha = 0.1$ and not rejected at $\alpha = 0.05$. The evidence against the null hypothesis is moderately strong.

8.89. 90% C.I.: 15.91 ± 1.645 * 16.43 / 11.832 = 15.91 ± 2.284 = (13.626 , 18.194)

8.90. 95% C.I.: 89.60 ± 1.96 * 17.32 / 8.660 = 89.60 ± 3.92 = (85.68 , 93.52)

8.91.
Z-Test

Test of mu = 85.00 vs mu not = 85.00

Variable	N	Mean	StDev	SE Mean	Z	P-Value
Calls	75	89.60	17.32	2.00	2.30	0.021

The evidence against the null hypothesis is strong.

8.92.
a. Reject the null hypothesis of $\mu = 100$ since 100 is not in the confidence interval.
b. Do not reject the null hypothesis of $\mu = 110$ since 110 is in the confidence interval.

8.93.
a. Accept the null hypothesis of $\mu = 9$ since 9.0 is in the confidence interval.
b. Reject the null hypothesis of $\mu = 9.4$ since 9.4 is not in the confidence interval.

8.94.
Mean = 32
Standard deviation = 9.1378
D.f. = 4
90% C.I.: 32 ± 2.132 * 9.1378 / 2.236 = 32 ± 8.7128 = (23.29 , 40.71)

8.95. 95% C.I.: 8.72 ± 2.086 * 2.2 / 4.5826 = 8.72 ± 1.0014 = (7.72 , 9.72)

8.96. 95% C.I.: 41.6 ± 2.145 * 9.32 / 3.873 = 41.6 ± 5.1617 = (36.44 , 46.76)

8.97. Assumption: The data are from a normal population.
Mean = 32.1
Standard deviation = 8.9499
95% C.I.: 32.1 ± 2.262 * 8.9499 / 3.1623 = 32.1 ± 6.4019 = (25.70 , 38.50)

8.98. Assumption: The data are from a normal population.
Mean = 32.1
Standard deviation = 8.9499
Test statistic: t = (32.1 - 25) / (8.9499 / 3.1623) = 2.509
P-value = 2 * P(t > 2.509) is between 0.02 and 0.05.
At $\alpha = 0.05$, reject the null hypothesis but do not reject the null hypothesis at $\alpha = 0.10$.
The evidence against the null hypothesis is moderately strong.

8.99. Assumption: The data are from a normal population.
Confidence Intervals

Variable	N	Mean	StDev	SE Mean	95.0 % C.I.
C1	15	226.60	33.74	8.71	(207.91, 245.29)

8.100. Assumption: The data are from a normal population
Mean = 32.1
Standard deviation = 8.9499
Test statistic: $t = (32.1 - 205) / (8.9499 / 3.1623) = -61.091$
P-value = 2 * P(t > 61.091) = 0
At $\alpha = .05$, reject the null hypothesis that $\mu = 205$. The evidence against the null hypothesis is strong.

8.101.
H_o: $\mu = 1.5$ versus H_1: $\mu < 1.5$

8.102.
H_0: $\mu = 2.4$ versus H_1: $\mu > 2.4$
Test statistic: $t = (4.5 - 2.4) / (2.1 / 2.828) = 2.828$
Rejection region: $t > 1.895$
P-value = P(t > 2.828) is between 0.01 and 0.025.
Reject the null hypothesis. The evidence against the null hypothesis is strong.

11.1.

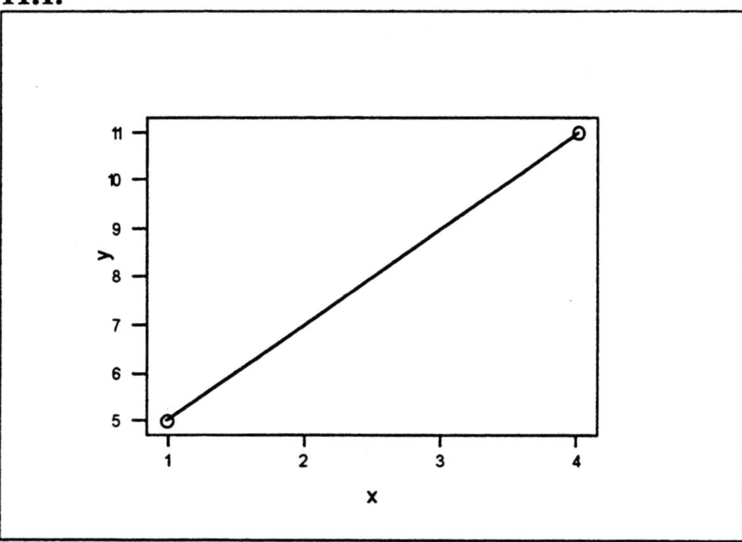

Y-intercept = 3; Slope = 2

11.2. $\beta_0 = 3$; $\beta_1 = 4$; $\sigma = 5$

11.3. $\beta_0 = 7$; $\beta_1 = -6$; $\sigma = 4$

11.4.

a. Mean = 11; Standard deviation = 1

b. Mean = 5; Standard deviation = 1

11.5.

a. Mean = 6; Standard deviation = 3

b. Mean = -6; Standard deviation = 3

11.6.

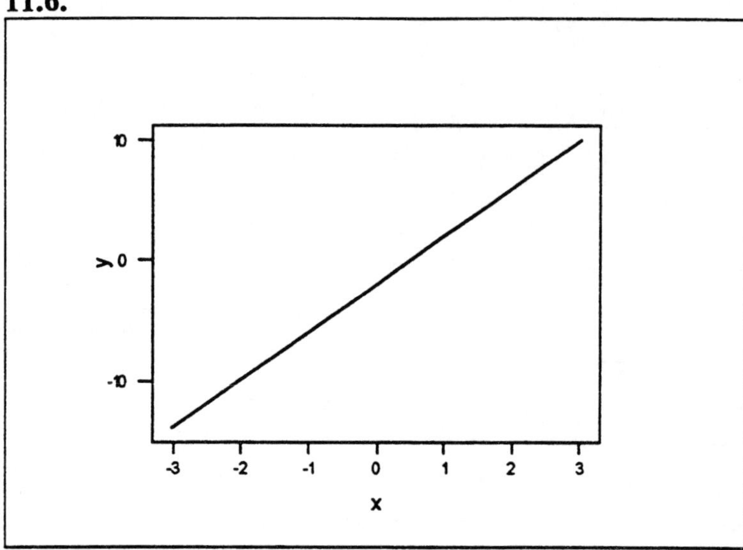

11.7.

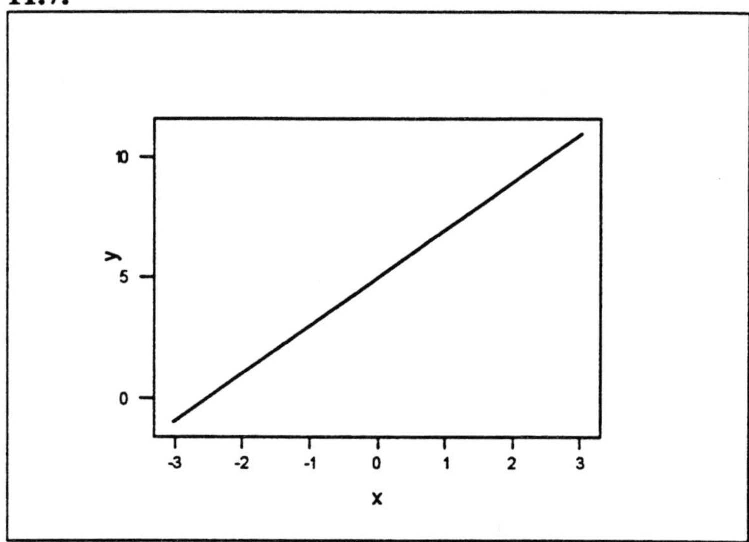

11.8.

a. When x = 3, Y = -5. When x = 6, Y = -8.

b. The response at x = 3 will not necessarily be larger than the response at x = 6. The responses are assumed to be normally distributed about their means with a common standard deviation of 2. Therefore, the response at x = 3 could be considerably below its mean while the response at x = 5 could be considerably above its mean.

11.9.

a. When x = 4, Y = 19. When x = 5, Y = 23.

b. With a standard deviation of 5, the response at $x = 5$ can, with high probability, range from 4 to 34, and the response at $x = 4$ can, with high probability, range from 8 to 38. Therefore the response at $x = 5$ will not always be larger than the response at $x = 4$.

11.10.

a. $Y = \beta_0 + \beta_1 x + \varepsilon$ where β_0 represents the fixed costs, β_1 represents the variable costs, x is the number of units, ε represents the unexpected variations or circumstances that effect the total production costs, and Y represents the total production cost.

b. β_0 represents the fixed costs and β_1 represents the variable costs.

c. One component variable that might be included in the error term ε is energy costs.

11.11.

a. $Y = \beta_0 + \beta_1 x + \varepsilon$ where β_0 represents the fixed costs such as location, β_1 represents the variable costs related to the number of square feet, ε represents the unexpected variations such as condition and type of construction, and Y represents the total cost.

b. β_1 represents the number of additional dollars of rent for each additional square foot of apartment.

c. One component variable that might be included in the error term ε is the condition of the apartment.

11.12.

a.

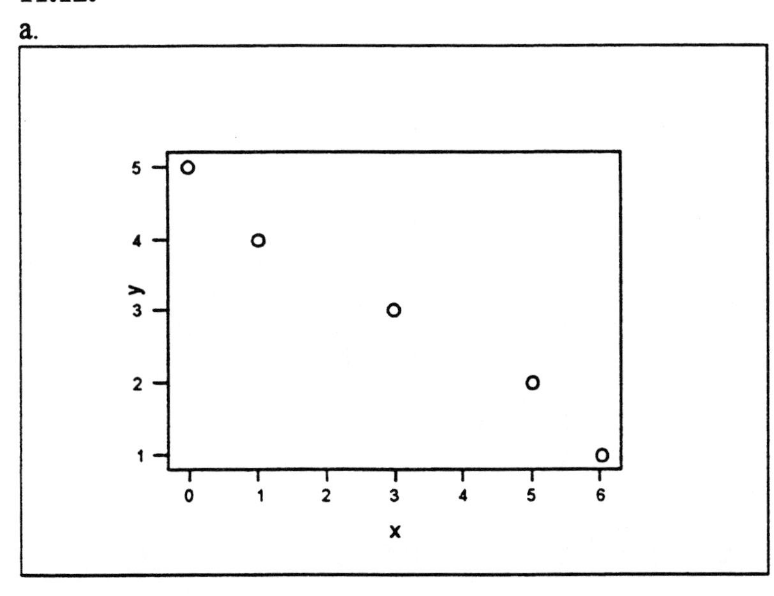

b.

Xmean =	3
Ymean =	3
Sxx =	26
Syy =	10
Sxy =	-16

c. $\hat{\beta}_0 = 4.85$; $\hat{\beta}_1 = -0.615$

d. The regression equation is $\hat{y} = 4.85 - 0.615\ x$

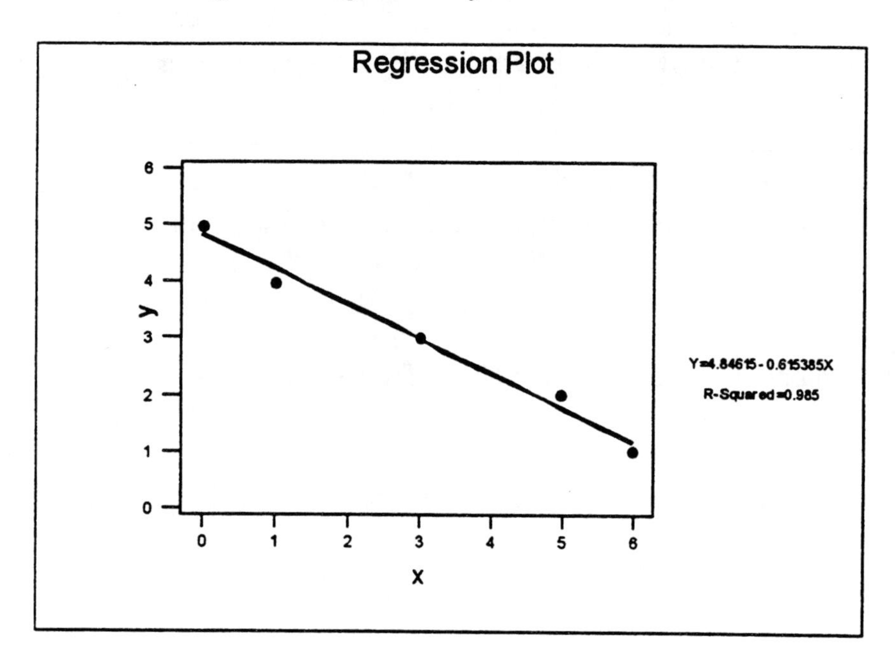

11.13.

a.

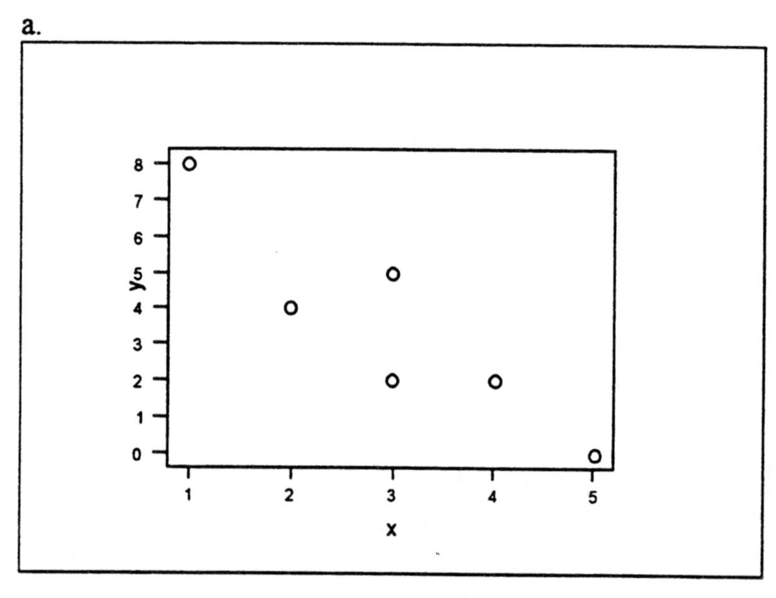

b.

Xmean =	3
Ymean =	3.5
Sxx =	10
Syy =	39.5
Sxy =	-18

c. $\hat{\beta}_0 = 8.9$; $\hat{\beta}_1 = -1.8$

d. The regression equation is $\hat{y} = 8.90 - 1.80$ x.

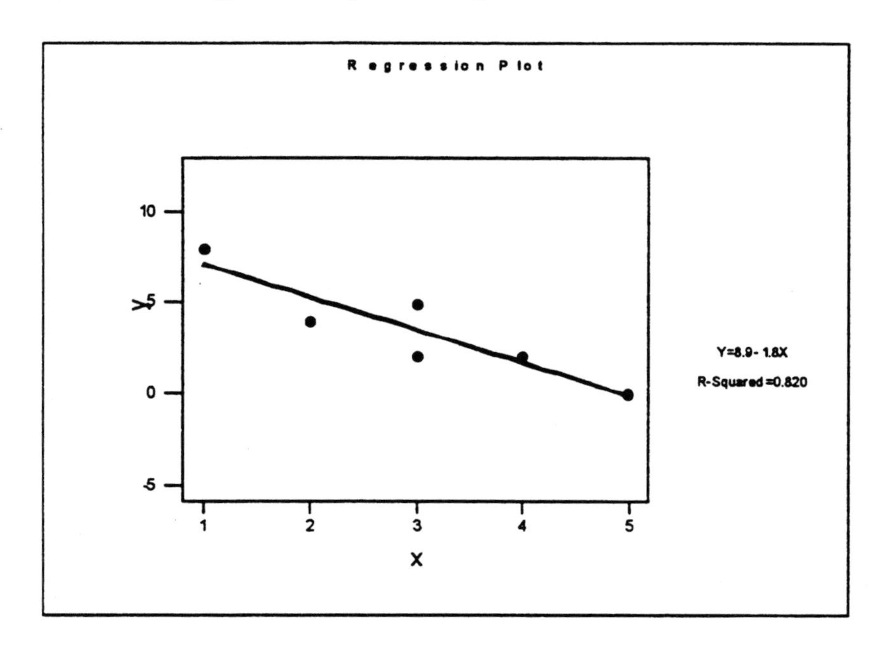

11.14.

a.

x	y	RESIDUALS
0	5	0.2
1	4	-0.4
6	1	-0.2
3	3	0.0
5	2	0.4

Sum of squares of RESIDUALS = 0.1539

b. SSE = 0.1539

c. $s^2 = 0.1539 / 3 = 0.0513$

11.15.

a.

x	y	RESIDUALS
1	8	0.9
2	4	-1.3
3	5	1.5
3	2	-1.5
4	2	0.3
5	0	0.1

Sum of squares of RESIDUALS = 7.1

b. SSE = 7.100

c. $s^2 = 7.1 / 4 = 1.775$

11.16.

a.

Xmean = 2
Ymean = 5
Sxx = 10
Syy = 30
Sxy = 16

b. $\hat{\beta}_0 = 1.8$; $\hat{\beta}_1 = 1.6$

c. The regression equation is $\hat{y} = 1.80 + 1.60\ x$

11.17.

a.

Xmean = 3.4
Ymean = 6
Sxx = 29.2
Syy = 26
Sxy = 25

b.

$\hat{\beta}_1 = 0.856$

$\hat{\beta}_0 = 3.089$

c.

$\hat{y} = 3.089 + 0.856x$

11.18.

a. $\hat{\beta}_1 = 2.76 / 10.2 = 0.271$

$\hat{\beta}_0 = 3.2 - 0.2706 * 5.4 = 1.739$

$\hat{y} = 1.739 + 0.271x$

b. SSE = 2.03 - 2.76^2/10.2 = 1.283

c. $s^2 = 1.2832 / 14 = 0.092$

11.19.

a. $\hat{\beta}_1 = 3.21 / 24.15 = 0.133$

$\hat{\beta}_0 = 5.1 - 0.1329 * 2.3 = 4.794$

b. SSE = 4.82 - 3.21^2 / 24.15 = 4.393

c. $s^2 = 4.3933 / 13 = 0.338$

11.20.

a. $SSE = S_{yy} - \beta_1 S_{xy} = S_{yy} - (S_{xy} / S_{xx}) * S_{xy} = S_{yy} - S^2_{xy} / S_{xx}$

b. $SSE = S_{yy} - \beta_1^2 S_{xx} = S_{yy} - (S_{xy} / S_{xx})^2 * S_{xx} = S_{yy} - S^2_{xy} / S_{xx}$

11.21.

$\bar{y} = \hat{\beta}_0 + \hat{\beta}_1\bar{x} = (\bar{y} - \hat{\beta}_1\bar{x}) + \hat{\beta}_1\bar{x}$

11.22.

a. $\hat{y}_i = \hat{\beta}_0 + \hat{\beta}_1 x_i = \bar{y} - \hat{\beta}_1\bar{x} + \hat{\beta}_1 x_i = \bar{y} + \hat{\beta}_1(x_i - \bar{x})$

b. $\hat{\varepsilon}_i = y_i - \hat{\beta}_0 - \hat{\beta}_1 x_i = y_i - \hat{y}_i = y_i - (\bar{y} + \hat{\beta}_1(x_i - \bar{x})) = (y_i - \bar{y}) - \hat{\beta}_1(x_i - \bar{x})$

c.

$$\sum \hat{\varepsilon}_i = \sum\left(y_i - \bar{y}\right) - \hat{\beta}_1 \sum (x_i - \bar{x}) = \sum y_i - \sum \bar{y} - \hat{\beta}_1\left(\sum\left(x_i - \bar{x}\right)\right) = n\bar{y} - n\bar{y} - \hat{\beta}_1\left(n\bar{x} - n\bar{x}\right) = 0$$

d. $$\sum \hat{\varepsilon}_i^2 = \sum\left[\left(y_i - \bar{y}\right)^2 - \hat{\beta}_1^2\left(x_i - \bar{x}\right)^2\right] = S_{yy} - \frac{S_{xy}^2}{S_{xx}^2}(S_{xx}) = S_{yy} - S_{xy}^2 / S_{xx}$$

11.23.

The regression equation is
Rent = 276 + 0.518 Sqft

Predictor	Coef	Stdev	t-ratio	p
Constant	275.5	115.3	2.39	0.031
Sqft	0.5177	0.1166	4.44	0.001

s = 69.83   R-sq = 58.5%   R-sq(adj) = 55.5%

Analysis of Variance

SOURCE	DF	SS	MS	F	p
Regression	1	96143	96143	19.71	0.001
Error	14	68273	4877		
Total	15	164416			

Unusual Observations

Obs.	Sqft	Rent	Fit	Stdev.Fit	Residual	St.Resid
2	900	595.0	741.5	19.6	-146.5	-2.19R

R denotes an obs. with a large st. resid.

b. 0.518 is the estimated mean rent for an additional square foot of space.

c. SSE = 68273
$s^2 = 68273 / 14 = 4876.643$

11.24.

a. Fitted line: Rent = 584 + 139 No.Bath
b. The mean increase in rent for an additional bathroom is 139.
c. $s^2 = 7334.210$

11.25.

a.

The regression equation is
y = 4.85 - 0.615 x

Predictor	Coef	Stdev	t-ratio	p
Constant	4.8462	0.1674	28.96	0.000
x	-0.61538	0.04441	. -13.86	0.001

s = 0.2265   R-sq = 98.5%   R-sq(adj) = 97.9%

Analysis of Variance

SOURCE	DF	SS	MS	F	p
Regression	1	9.8462	9.8462	192.00	0.001
Error	3	0.1538	0.0513		
Total	4	10.0000			

$\hat{\beta}_1 = -0.615$

$\hat{\beta}_0 = 4.85$

$s^2 = 0.1538 / 3 = 0.051$

b.

d.f. = 3

$t_{.025} = 3.182$

Rejection region: $|t| > 3.182$

Value of test statistic: t = - 0.615 / (0.2265 / (26^0.5)) = -13.845

Since the observed $|t| = 13.8450 > 3.182$, reject the null hypothesis at the 5% level and conclude a linear relation exists between x and y.

c. 90% C.I.: 4.85 ± 2.353 * 0.2265 * (1/5 + 3^2 / 26)^0.5 = 4.85 ± 2.353 * 0.1674 = 4.85 ± 0.3939 = (4.46 , 5.24)

d. When x* = 2.5, $\hat{y}$ = 4.85 - 0.615 *2.5 = 3.313.

90% C.I.: 3.3125 ± 2.353 * .2265 * (1/5 + (3.3125 - 3)^2 / 26)^0.5 = 3.3125 ± 2.353 * .1022 = $3.3125 \pm .2405$ = (3.07, 3.55)

11.26.

95% C.I.: -0.615 ± 3.182 * 0.2265 / (26^0.5) = -0.615 ± 3.182 * 0.0444 = -0.615 ± 0.1413 = (-0.76 , -0.47)

We are 95% confident that by increasing x by one unit, the mean value for y would decrease somewhere between 0.47 and 0.76 units.

11.27.

a.

The regression equation is

$\hat{y} = 0.400 + 0.700$ x

```
Predictor        Coef       Stdev    t-ratio       p
Constant       0.4000      0.2422       1.65   0.197
x             0.70000     0.07303       9.59   0.002

s = 0.2309      R-sq = 96.8%      R-sq(adj) = 95.8%

Analysis of Variance

SOURCE       DF         SS         MS        F       p
Regression    1     4.9000     4.9000    91.88   0.002
Error         3     0.1600     0.0533
Total         4     5.0600
```

$\hat{\beta}_0 = 0.4$; $\hat{\beta}_1 = 0.7$; $s^2 = .2309^2 = 0.053$

b. d.f. = 3
Rejection region: $|t| > 3.182$
Value of test statistic: (0.7 - 1) / (0.2309 / (10^0.5)) = -0.3 / 0.0730 = -4.109
Conclusion: Reject the null hypothesis that the slope is 1.

c. 90% C.I.: $0.7 \pm 2.353 * 0.2309 / (10^{0.5}) = 0.7 \pm 0.1718 = (0.53 , 0.87)$

d. When x = 3.5, $\hat{y} = 0.400 + 0.700 * 3.5 = 2.85$.
95% C.I.: $2.85 \pm 3.182 * 0.2309 * (1/5 + (3.5 - 3)^2 / 10)^{0.5} = 2.85 \pm 0.3459$
= (2.50 , 3.20)

11.28.

a.

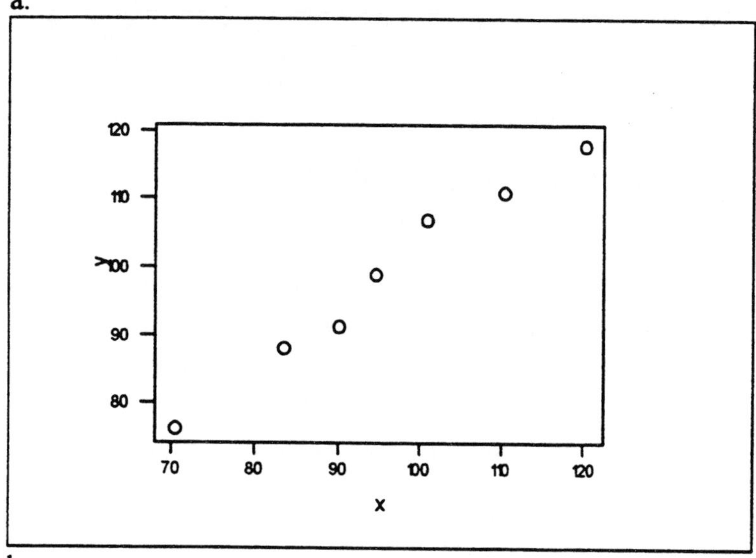

b.
The regression equation is
$\hat{y} = 15.5 + 0.869$ x

Predictor	Coef	Stdev	t-ratio	p
Constant	15.462	5.529	2.80	0.038
x	0.86942	0.05707	15.23	0.000

s = 2.314   R-sq = 97.9%   R-sq(adj) = 97.5%

Analysis of Variance

SOURCE	DF	SS	MS	F	p
Regression	1	1242.9	1242.9	232.08	0.000
Error	5	26.8	5.4		
Total	6	1269.7			

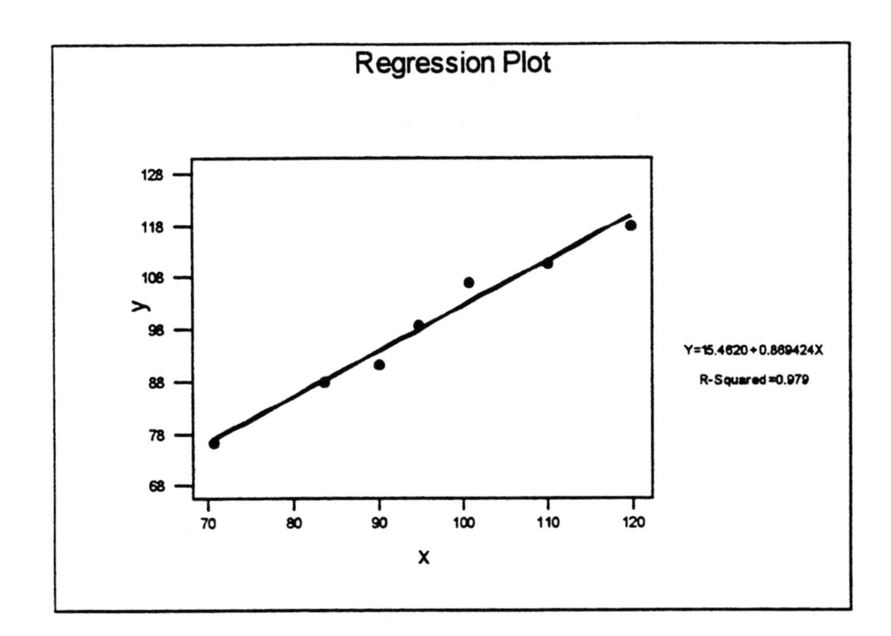

c. 95% C.I.: 0.869 ± 2.571 * 2.314 / (1644.317)^0.5 = 0.869 ± 0.1467
= (0.72 , 1.02)

d. It is dangerous to extrapolate this equation to x = $140,000. Relationship of y with x may change.

11.29.

a. When x = $90.0, $\hat{y}$ = 15.5 + 0.869 * 90,000 = 93.710
95% C.I.: (91.313, 96.108)

b. 95% prediction interval for the selling price of a home assessed at $90,000:
(87.295, 100.125)

11.30.

a.

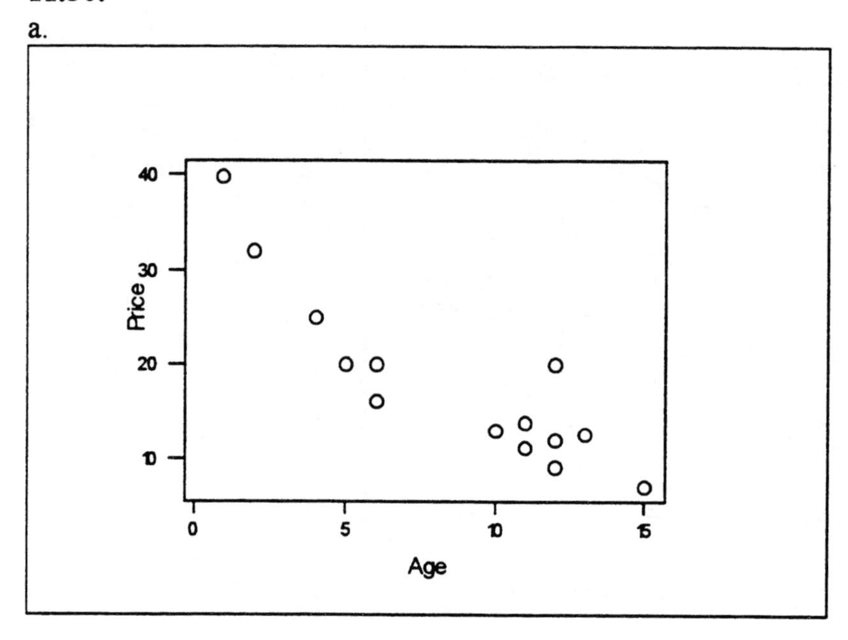

b.

The regression equation is
Price = 33.7 - 1.86 Age

Predictor	Coef	Stdev	t-ratio	p
Constant	33.726	2.588	13.03	0.000
Age	-1.8621	0.2651	-7.02	0.000

s = 4.343 R-sq = 79.1% R-sq(adj) = 77.5%

Analysis of Variance

SOURCE	DF	SS	MS	F	p
Regression	1	930.63	930.63	49.34	0.000
Error	13	245.19	18.86		
Total	14	1175.82			

Unusual Observations

Obs.	Age	Price	Fit	Stdev.Fit	Residual	St.Resid
1	1.0	39.90	31.86	2.35	8.04	2.20R
11	12.0	20.00	11.38	1.41	8.62	2.10R

R denotes an obs. with a large st. resid.

c. When Age = 19, Price = 33.7 - 1.86 * 19 = -1.64.
There is danger in predicting a negative price for this Corvette. Extrapolation of prices outside the original range of the data may give incorrect values, such as this negative price.

11.31.

a. Slope = -0.271; Y-intercept = 18.0
Equation of the fitted line: $\hat{y}$ = 18.0 - 0.271x

b. s = 10.61
When n and S_{xx} are both large, the estimated standard deviation of the prediction is essentially s. Hence, under these conditions s is the (estimated) standard error of the prediction (estimate).

c. Value of test statistic: t = -0.271/(10.61 / (3780.58)^0.5) = -1.571
Rejection region: |t| > 1.761
Do not reject the null hypothesis.

11.32.

The regression equation is
Profits = 25.0 - 0.713 Employee

Predictor	Coef	Stdev	t-ratio	p
Constant	25.013	5.679	4.40	0.001
Employee	-0.7125	0.2912	-2.45	0.029

```
s = 9.839          R-sq = 31.5%          R-sq(adj) = 26.3%
```

Analysis of Variance

SOURCE	DF	SS	MS	F	p
Regression	1	579.40	579.40	5.99	0.029
Error	13	1258.40	96.80		
Total	14	1837.80			

Value of test statistic: t = -0.713 / (9.839 / (1141.18)^0.5) = -2.448
Rejection region: |t| > 1.761
Reject the null hypothesis.
The conclusion in 11.31(c) changes to reject H_o. This shows the large influence that a single observation can have when the number of observations is small. The Dunn and Bradstreet observation is an outlier. It may or may not make sense to throw it out depending on the objectives of the study.

11.33.

a. Equation of the fitted least squares line: $\hat{y}$= 1.65 + 0.002 x

b. 90% C.I.: 0.00202 ± 1.697 /(1.274 / (15599892)^0.5) = 0.00202 ± 5261.060
= (-5261.058 , 5261.062)

c. When x = 948, $\hat{y}$ = 1.65 + 0.00202 * 948 = 3.565
95% C.I.: 3.565 ± 2.042 * 1.274 * (36 / 35 + (948 - 954.325)^2 / 16626804)^0.5
= 3.565 ± 2.6387 = (0.93 , 6.20)

11.34.

a. Value of test statistic: t = 0.02
Rejection region: |t| > 2.064
Do not reject the null hypothesis since the test statistic is not in the rejection region.

b. Value of test statistic: t = 14.71 - 0.968 / 14.71 = 14.644
Rejection region: |t| > 2.064
Reject the null hypothesis since the test statistic is in the rejection region.

c. The results in parts a and b suggest that the regression equation might have a y-intercept = 0.

d. **Regression Analysis**

```
The regression equation is
Ln(Act) = 0.969 Ln(Est)

Predictor        Coef         Stdev     t-ratio        p
Noconstant
Ln(Est)       0.96903       0.03207       30.21     0.000

s = 0.4038
Analysis of Variance

SOURCE        DF          SS           MS          F          p
Regression     1      148.83       148.83     912.79      0.000
Error         25        4.08         0.16
Total         26      152.91
```

A straight-line model with a y-intercept = 0 seems to be adequate for these data. The sum of the residuals is 0.016858, however, which is not as close to zero as the sum of the residuals for the original regression equation.

11.35.

a. $\ln(\widehat{\text{Predicted actual cost}}) = 0.003 + 0.968 * \ln(10) = 0.003 + 0.968 * 2.3026 = 2.232$

Antilog(2.232) = 9.318

b. 90% C.I. for lny: (2.025 , 2.302)

90% C.I. for y when x = 10: (7.576 , 9.994)

11.36.

a. $r^2 = 0.378 / 2.01 = 0.188$

b. r = 0.434

11.37.

Source	Sum of Squares	d.f.	Mean Square	F-ratio
Regression	0.378	1	0.378	2.779
Error	1.632	12	0.136	
Total	2.010	13		

11.38.

a. r^2 = (-160)^2 / (92 * 457) = 0.609

b. r = -0.780

11.39.

Source	Sum of Squares	d.f.	Mean Square	F-ratio
Regression	278.261	1	278.261	48.259
Error	178.739	31	5.766	
Total	457	32		

11.40.

r^2 = 8.563 / 16.5 = 0.519
r = 0.720

11.41.

a. r^2 = 9.846 / 10 = 0.985

b. r = -0.992

c. Value of test statistic: 0.992 * (3)^0.5 / (1 - 0.992^2)^0.5 = 13.611
Rejection region: |t| > 3.182
Reject the null hypothesis.

11.42.

a. r^2 = 1242.937 / 1269.714 = 0.979

b. r = 0.989

c. Value of test statistic: t = 0.9894 * (5^0.5) / (1 - 0.979)^0.5 = 15.267
Rejection region: |t| > 2.571
Reject the null hypothesis.

11.43.

a. r^2 = 0.791

b. F= 49.34
Rejection region: F > 4.67
Reject the null hypothesis of a zero slope.
The regression is significant at α close to 0.

c. $t = -7.02$; $t^2 = 49.3 = F$

11.44.

a. $r^2 = 15\%$
SSR is a minor portion of SST; therefore, the straight-line model is not a good fit to the data.

b. Sample correlation coefficient = -0.387

c. The *P*-value related to the F-ratio is 0.138, indicating that the regression is not significant at the 10% level.

11.45.

a. $r^2 = 0.528$ This means that 52.8% of the variability in *y* is explained by the fitted straight-line.

b. r = 0.727

c. The *P*-value related to the F-ratio is essentially 0, indicating that the regression is significant at the 1% level.

d. To test the null hypothesis of a zero correlation, compute:
t = 0.727 * (35)^0.5 / (1-0.528)^0.5 = 6.260.
Rejection region: $|t| > 2.750$
Reject the null hypothesis.

11.46. The adjusted r-squared value is 1 - (9/8) * (1 - 0.799) = 0.774.

11.47.

$$\left(\frac{r^2}{1-r^2}\right)(n-2) = \frac{\frac{SSR}{SST}}{1-\frac{SSR}{SST}}(n-2) = \frac{SSR}{SST-SSR}(n-2) = \frac{SSR}{SSE}(n-2) = \frac{MSR}{MSE} = F$$

11.48.

$$(1-r^2)S_{yy} = \left(1-\frac{SSR}{SST}\right)S_{yy} = \left(\frac{SST-SSR}{SST}\right)S_{yy} = \left(\frac{SS_{yy}-\frac{S^2{}_{xy}}{S_{xx}}}{S_{yy}}\right)S_{yy} = SS_{yy} - \frac{S^2{}_{xy}}{S_{xx}} = SSE$$

$$\hat{\beta}_1{}^2 S_{xx} = \left(\frac{S_{xy}}{S_{xx}}\right)^2 S_{xx} = \frac{S_{xy}{}^2}{S_{xx}}$$

11.49.
The regression equation is
$\ln \hat{y} = 2.54 + 0.288$ Year

Predictor	Coef	Stdev	t-ratio	p
Constant	2.54170	0.09735	26.11	0.000
Year	0.28837	0.01569	18.38	0.000

s = 0.1425 R-sq = 97.7% R-sq(adj) = 97.4%

Analysis of Variance

SOURCE	DF	SS	MS	F	p
Regression	1	6.8605	6.8605	337.83	0.000
Error	8	0.1625	0.0203		
Total	9	7.0230			

a. $\gamma = 0.3$ corresponds to $\beta_1 = \ln(\gamma + 1) = 0.262$
Value of test statistic: $t = (0.288 - 0.220) / 0.016 = 4.25$
Rejection region: $t > 1.397$
Reject the null hypothesis. The growth rate is greater than 0.3.

b. 90% C.I. for β_1: $0.288 \pm 1.860 * 0.016 = 0.288 \pm 0.0298 = (0.258, 0.318)$
90% C.I. for $\gamma = (1.29, 1.37)$

11.50. When x = 11, $\ln \hat{y} = 2.54 + 0.288 * 11 = 5.708$

Fit	Stdev.Fit	95.0% C.I.	95.0% P.I.
5.7138	0.0974	(5.4892, 5.9383)	(5.3157, 6.1119)

Taking antilogs gives: (203.51 , 451.20)

11.51.

a.

Regression Analysis

```
The regression equation is
Yt = 1.27 - 0.217 Yt-1

46 cases used 2 cases contain missing values
```

Predictor	Coef	Stdev	t-ratio	p
Constant	1.2670	0.5491	2.31	0.026
Yt-1	-0.2167	0.1375	-1.58	0.122

s = 3.564   R-sq = 5.3%   R-sq(adj) = 3.2%

Analysis of Variance

SOURCE	DF	SS	MS	F	p
Regression	1	31.53	31.53	2.48	0.122
Error	44	559.03	12.71		
Total	45	590.56			

Unusual Observations

Obs.	Yt-1	Yt	Fit	Stdev.Fit	Residual	St.Resid
20	1.5	8.470	0.933	0.528	7.537	2.14R
26	0.9	-7.130	1.083	0.527	-8.213	-2.33R
30	-0.9	8.800	1.460	0.596	7.340	2.09R
48	10.6	-4.490	-1.025	1.399	-3.465	-1.06 X

R denotes an obs. with a large st. resid.
X denotes an obs. whose X value gives it large influence.

b. We cannot reject H_o: $\beta_1 = 0$ at the 10% level. Also, $r^2 = 0.053$. (Very little of the variation in Yt is explained by Yt-1.) Both of these suggest that the model $y_t = \beta_0 + \varepsilon_t$ is adequate for these data. That is,successive S&P 500 monthly returns are essentially independent.

c.

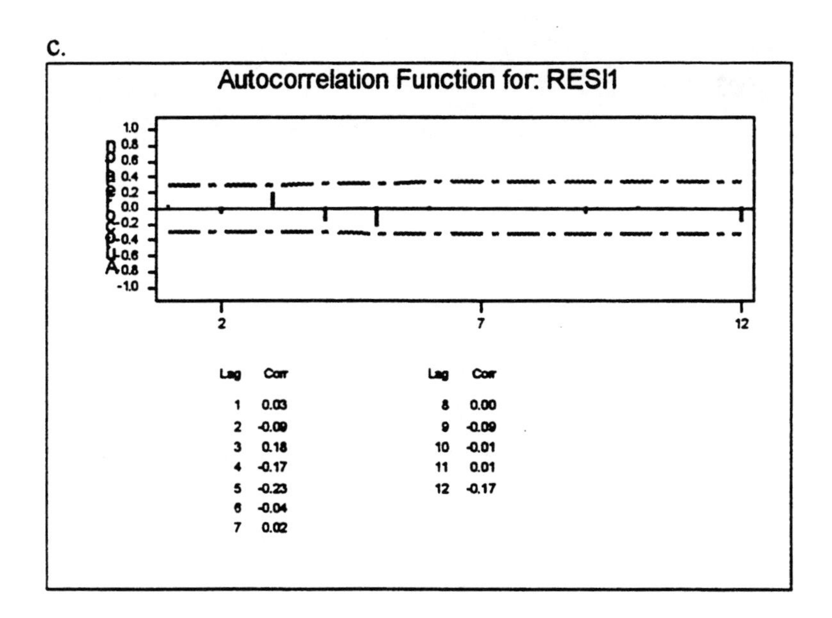

Lag	Corr	Lag	Corr
1	0.03	8	0.00
2	-0.09	9	-0.09
3	0.18	10	-0.01
4	-0.17	11	0.01
5	-0.23	12	-0.17
6	-0.04		
7	0.02		

The assumption of independent errors is warranted since the autocorrelation coefficients are between ± 2 standard errors (± 2/(46^.5) = 0.295) of zero.

11.52.

a.

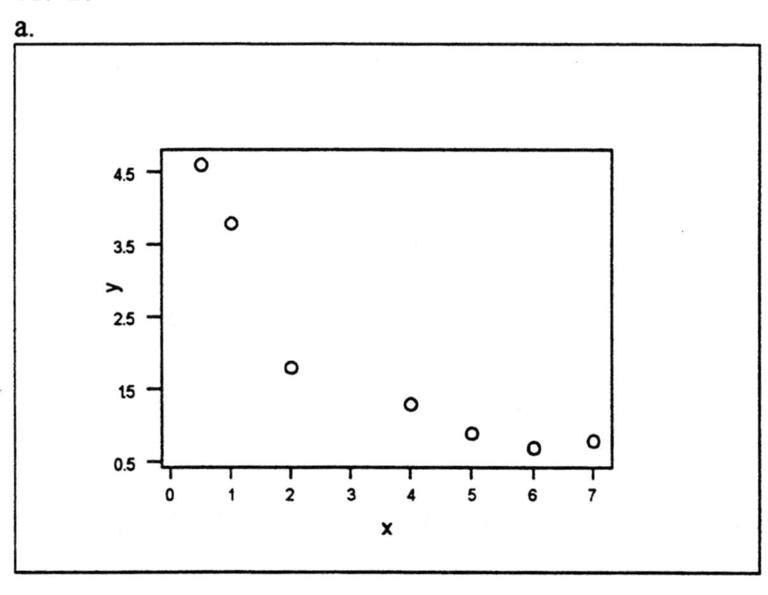

b.

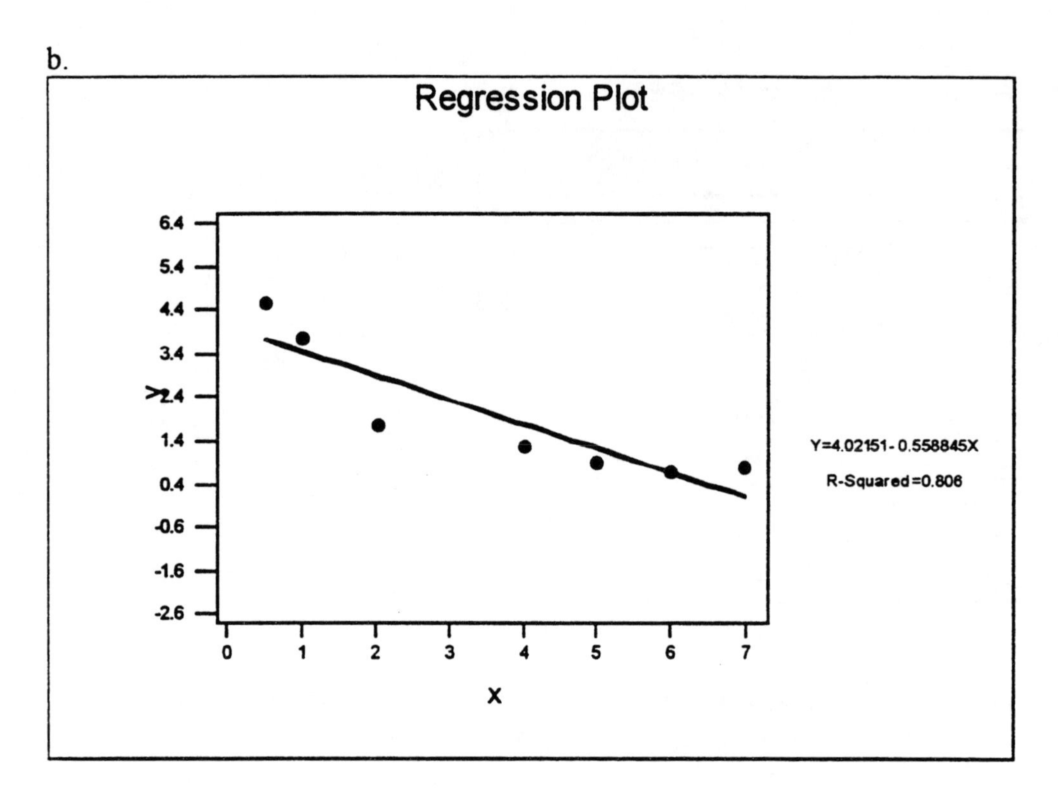

c. The regression equation is

$\hat{y} = 4.02 - 0.559\, x$

Predictor	Coef	Stdev	t-ratio	p
Constant	4.0215	0.5315	7.57	0.001
x	-0.5588	0.1227	-4.55	0.006

s = 0.7602 R-sq = 80.6% R-sq(adj) = 76.7%

Analysis of Variance

SOURCE	DF	SS	MS	F	p
Regression	1	11.979	11.979	20.73	0.006
Error	5	2.889	0.578		
Total	6	14.869			

80.6% of the variability in the response y is explained by the fitted straight line.

d. The residuals are: 0.8579 0.3373 -1.1038 -0.4861 -0.3273 0.0316 0.6904

Plotting the residuals against the fitted values gives:

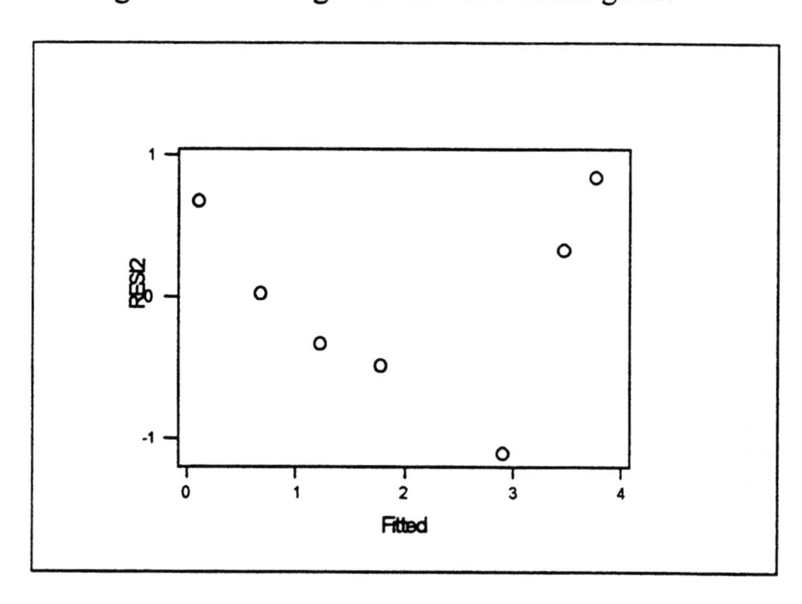

The residual plot is bowed suggesting a curvilinear pattern not captured by a straight-line model. A straight-line model with a transformed response variable might provide a better fit.

11.53.

a.

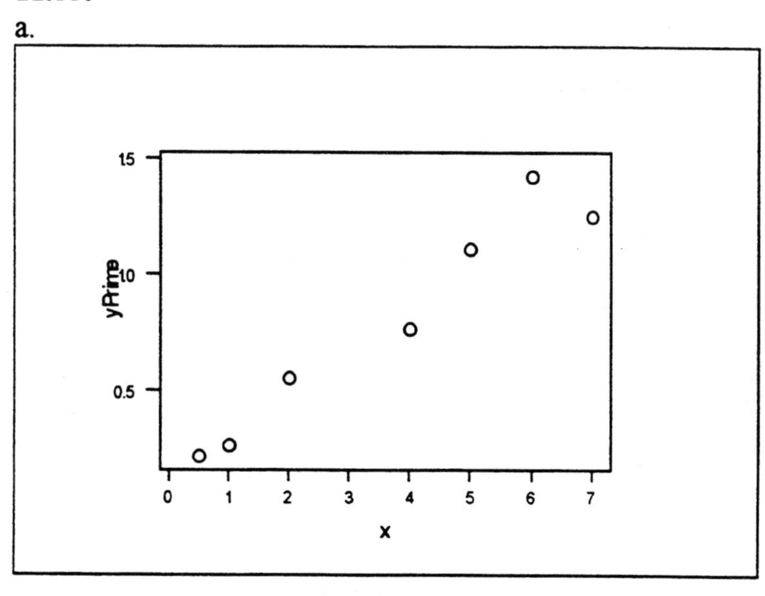

b.
The regression equation is
$\hat{y}$ Prime = 0.129 + 0.184 x

Predictor	Coef	Stdev	t-ratio	p
Constant	0.12926	0.09121	1.42	0.216
x	0.18393	0.02106	8.73	0.000

s = 0.1305   R-sq = 93.8%   R-sq(adj) = 92.6%

Analysis of Variance

SOURCE	DF	SS	MS	F	p
Regression	1	1.2976	1.2976	76.25	0.000
Error	5	0.0851	0.0170		
Total	6	1.3827			

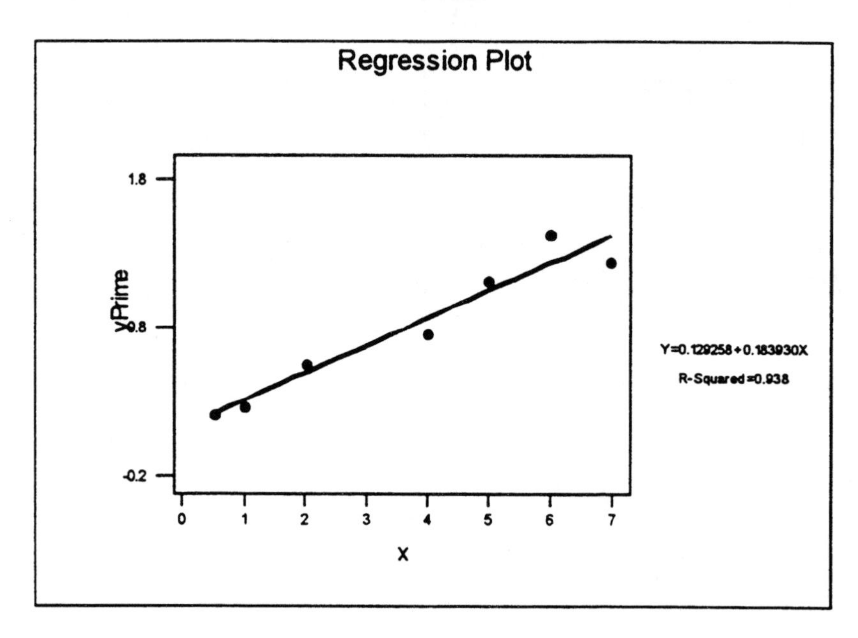

c. r^2 = 93.8% of the variability in the response y is explained by the fitted straight line. This transformation is a better fit than the original fitted-line equation.

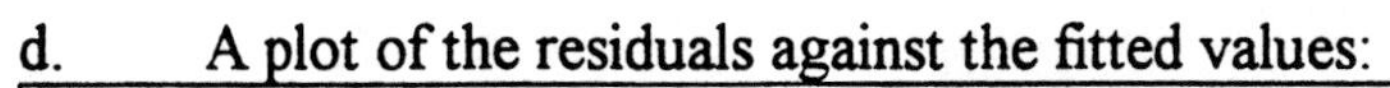

d. A plot of the residuals against the fitted values:

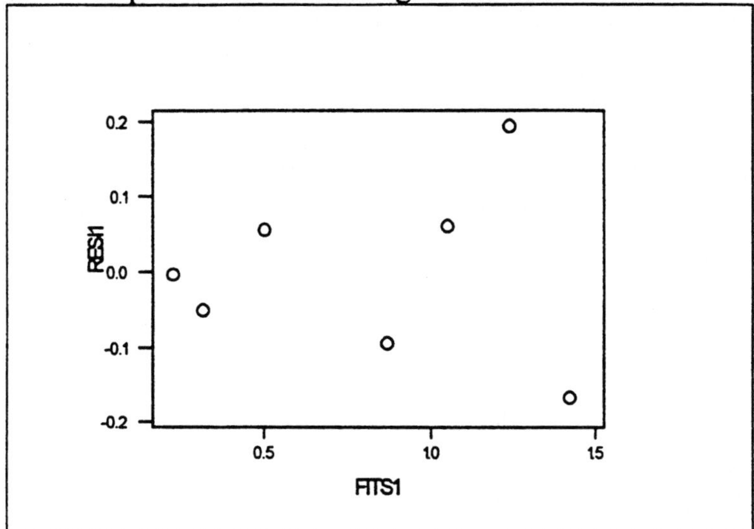

The transformed data appear to be consistent with a straight line model since the residuals are scattered in a horizontal band about zero.

11.54.

a.

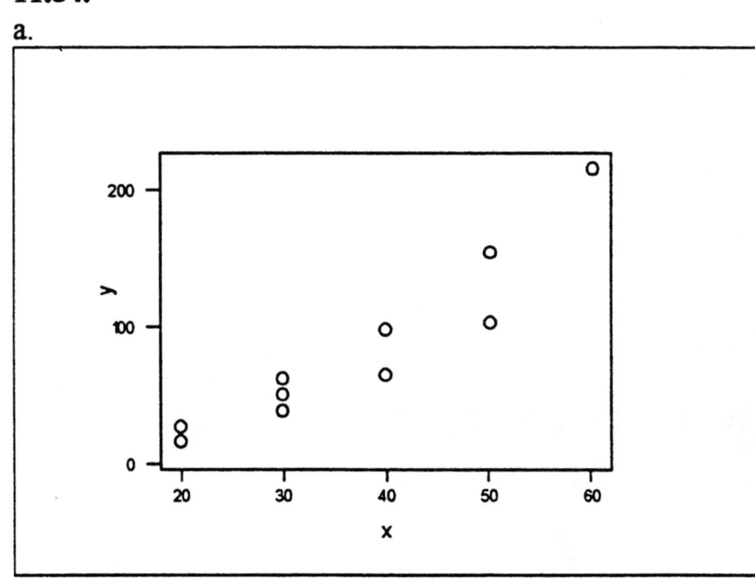

b.
The regression equation is
$\hat{y} = -78.3 + 4.38\ x$

Predictor	Coef	Stdev	t-ratio	p
Constant	-78.33	22.50	-3.48	0.008
x	4.3820	0.5753	7.62	0.000

s = 23.08   R-sq = 87.9%   R-sq(adj) = 86.4%

Analysis of Variance

SOURCE	DF	SS	MS	F	p
Regression	1	30915	30915	58.02	0.000
Error	8	4263	533		
Total	9	35178			

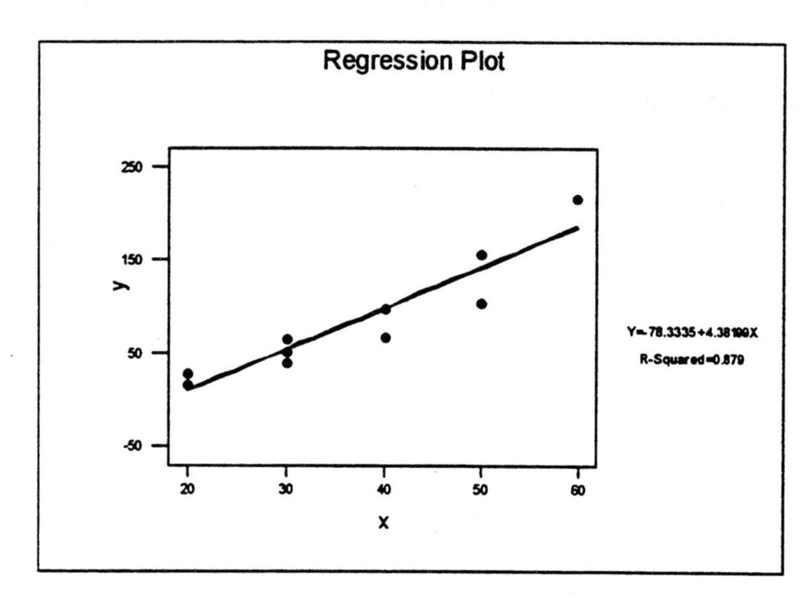

c. A graph of the residuals against the explanatory value:

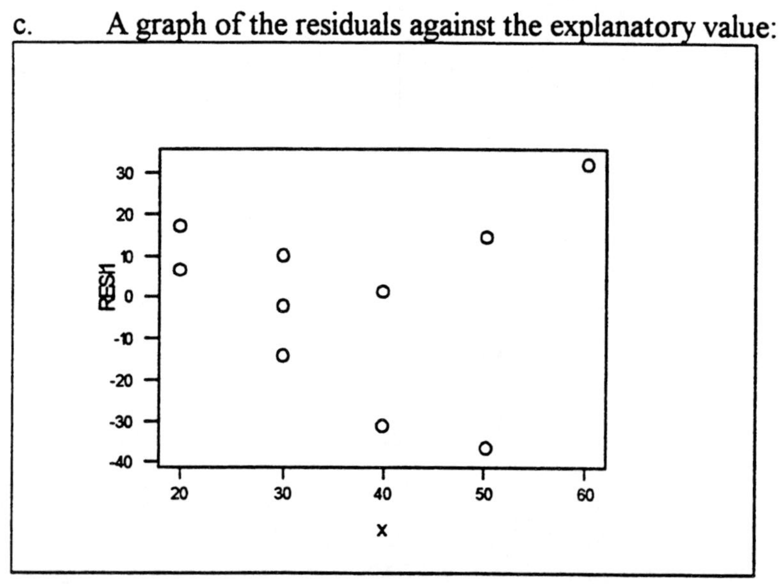

Slight bowing in the residual plot suggests that a curvilinear model or a straight-line model with a transformed response variable might provide a better fit.

11.55.

a.

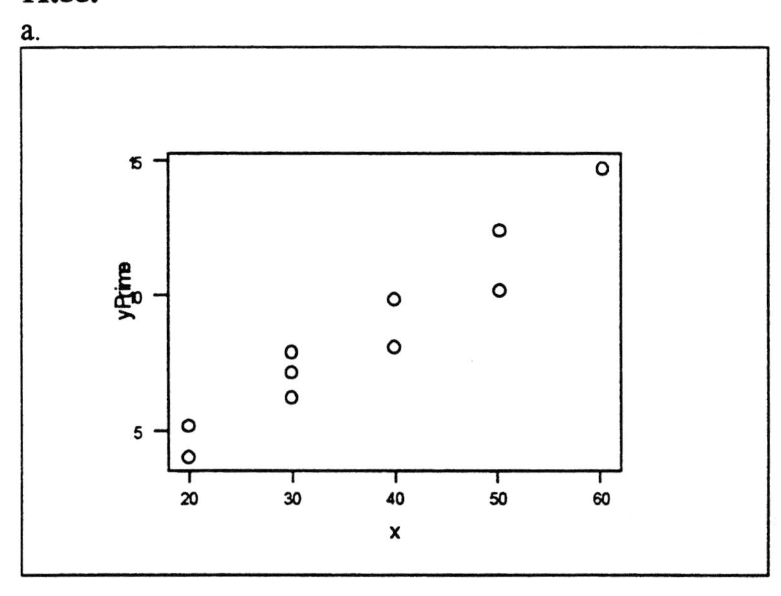

b.

The regression equation is

$\hat{y}$Prime = - 0.167 + 0.237 x

Predictor	Coef	Stdev	t-ratio	p
Constant	-0.1665	0.9323	-0.18	0.863
x	0.23703	0.02383	9.95	0.000

s = 0.9563 R-sq = 92.5% R-sq(adj) = 91.6%

Analysis of Variance

SOURCE	DF	SS	MS	F	p
Regression	1	90.456	90.456	98.91	0.000
Error	8	7.316	0.915		
Total	9	97.773			

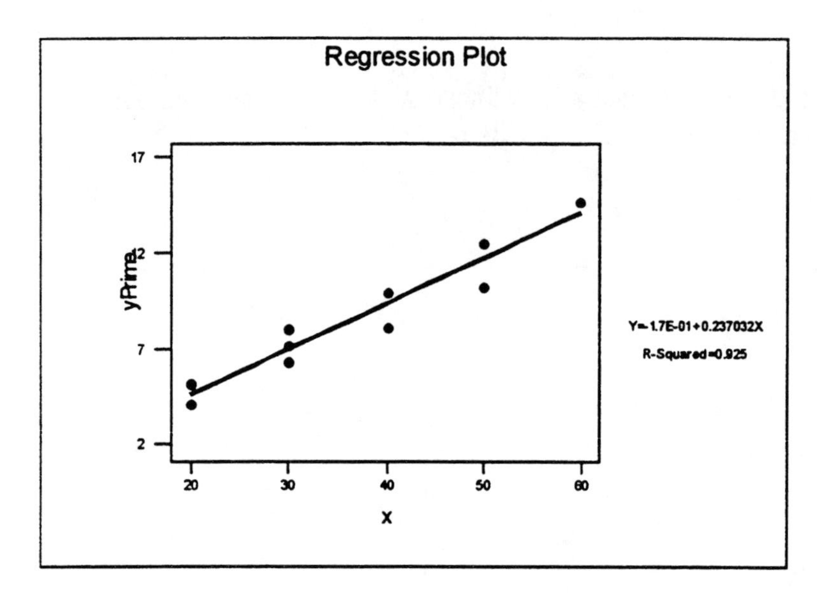

c. $r^2 = 92.5\%$ of the variability is explained by the fitted line equation. This is an adequate fit.

d. A plot of the residuals against the explanatory variable:

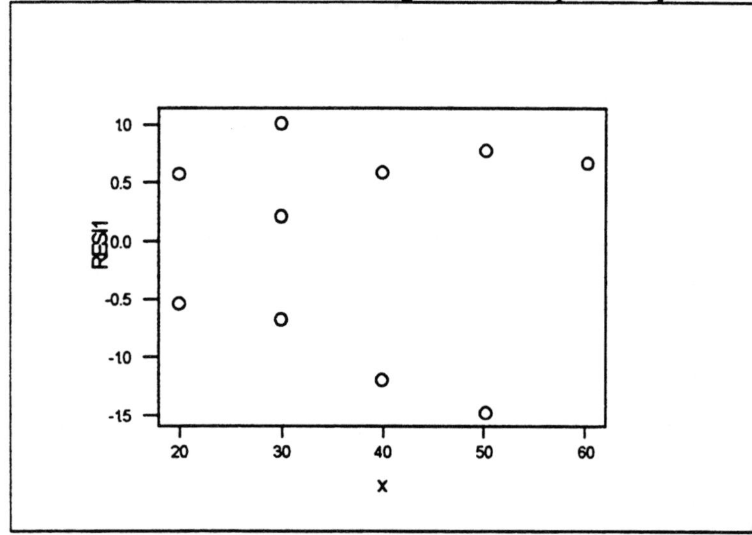

The transformed data appear to be consistent with a straight-line model since the residuals are scattered in a horizontal band about zero.

11.56.

a. We must choose a value for β to minimize $S(\beta) =$

$$\sum(y_i - \beta x_i)^2 = \sum y_i^2 - 2\beta\sum x_i y_i + \beta^2\sum x_i^2 = \sum y_i^2 + \left(\sum x_i^2\right)\left(\beta - \frac{\sum x_i y_i}{\sum x_i^2}\right)^2 - \frac{\left(\sum x_i y_i\right)^2}{\sum x_i^2}$$

All terms on the right are positive and only the middle term involves β. We can make the middle term zero (and minimize S(β)) by picking $\beta = \hat{\beta} = \dfrac{\sum x_i y_i}{\sum x_i^2}$.

b. $$s^2 = \frac{SSE}{n-1} = \frac{\sum_{i=1}^{n}\left(y_i - \hat{\beta}x_i\right)^2}{n-1}$$

11.57. $\hat{\beta} = \dfrac{\sum x_i y_i}{\sum x_i^2} = \sum\left(\dfrac{x_i}{\sum x_i^2}\right) y_i = \sum w_i y_i$ where $w_i = x_i / \sum x_i^2$. Since the y_i 's are independent, and each y_i has variance σ^2,

$$\text{Var}\left(\hat{\beta}\right) = \sum w_i^2 Var(y_i) = \sigma^2 \sum w_i^2 = \sigma^2 \sum \frac{x_i^2}{\left(\sum x_i^2\right)^2} = \sigma^2 \frac{\sum x_i^2}{\left(\sum x_i^2\right)^2} = \frac{\sigma^2}{\sum x_i^2}.$$

11.58. 90% C.I. for β: 0.992 ± 1.714 * 0.01059= 0.992 ± 0.01815 = (0.97 , 1.01)

11.59.
Value of test statistic: t = (0.992 - 1) / 0.01059 = -0.755
Rejection region: |t| > 1.714
Do not reject the null hypothesis. There is no reason to believe that the slope is different from 1.

11.60.
a.

```
Regression Analysis

The regression equation is
ExcRateY = 41.9 + 0.694 x

24 cases used 1 cases contain missing values

Predictor        Coef        Stdev     t-ratio        p
Constant        41.91        16.49        2.54     0.019
x              0.6945       0.1176        5.91     0.000

s = 6.542        R-sq = 61.3%      R-sq(adj) = 59.6%

Analysis of Variance

SOURCE        DF          SS          MS         F        p
Regression     1      1493.5      1493.5     34.89    0.000
Error         22       941.6        42.8
Total         23      2435.1

Unusual Observations
Obs.         x   ExcRateY         Fit  Stdev.Fit    Residual    St.Resid
  2        168     155.77      158.56       3.57       -2.79      -0.51 X
X denotes an obs. whose X value gives it large influence.
```

Test for the significance of the regression:
F = 34.89
d.f. = 23
Rejection region: F ≥ 2.94
Reject the null hypothesis. The regression is significant.

b. r^2 = 61.3% of the variability in the response y is explained by the fitted line. The linear relation is moderately strong.

c. Test: H_o: $\beta_0 = 0$ versus $\beta_0 \neq 0$.
Value of test statistic: t = 2.54
Rejection region: |t| > 2.069
Reject the null hypothesis. The y-intercept is not zero.

d.

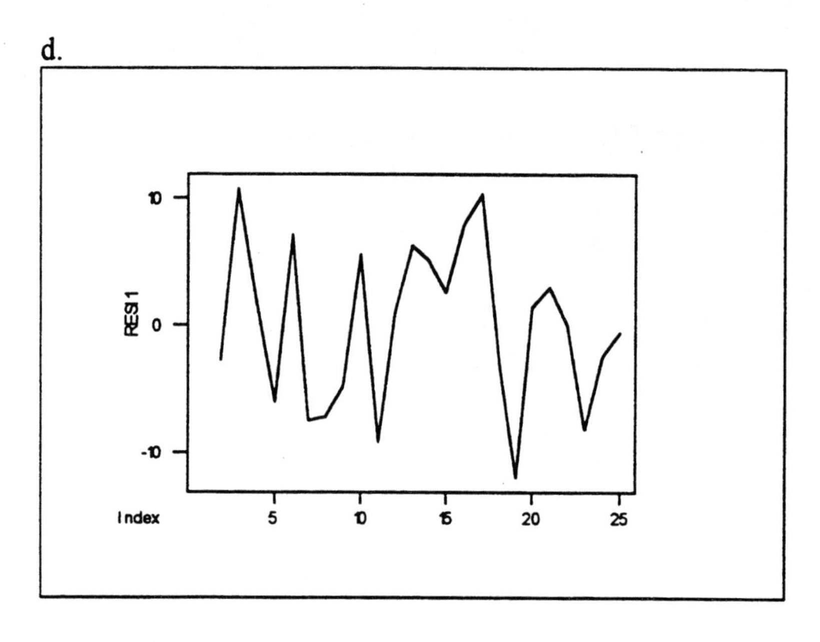

Autocorrelation Function

```
ACF of RESI1

          -1.0 -0.8 -0.6 -0.4 -0.2  0.0  0.2  0.4  0.6  0.8  1.0
            +----+----+----+----+----+----+----+----+----+----+
  1    0.056                         XX
  2   -0.203                     XXXXXX
  3    0.088                         XXX
  4    0.161                         XXXXX
  5   -0.312                  XXXXXXXXX
  6   -0.251                    XXXXXXX
```

±2/(25)^0.5 = ±0.4

Since all the autocorrelations are within ±0.4 of zero, independence errors appear to be a reasonable assumption.

11.61.

a.

Regression Analysis

```
The regression equation is
ExRateY = 0.975 x

24 cases used 1 cases contain missing values

Predictor          Coef          Stdev     t-ratio         p
Noconstant
x              0.974663     0.005272      184.88     0.000

s = 62.03
Analysis of Variance

SOURCE         DF          SS           MS          F        p
Regression      1   131503864    131503864   34180.11    0.000
Error          23       88490         3847
Total          24   131592352

Unusual Observations
Obs.          x     ExRateY         Fit  Stdev.Fit     Residual     St.Resid
  9        2462      2249.4      2400.1       13.0       -150.7       -2.48R
 10        2477      2281.0      2414.6       13.1       -133.6       -2.20R

R denotes an obs. with a large st. resid.
```

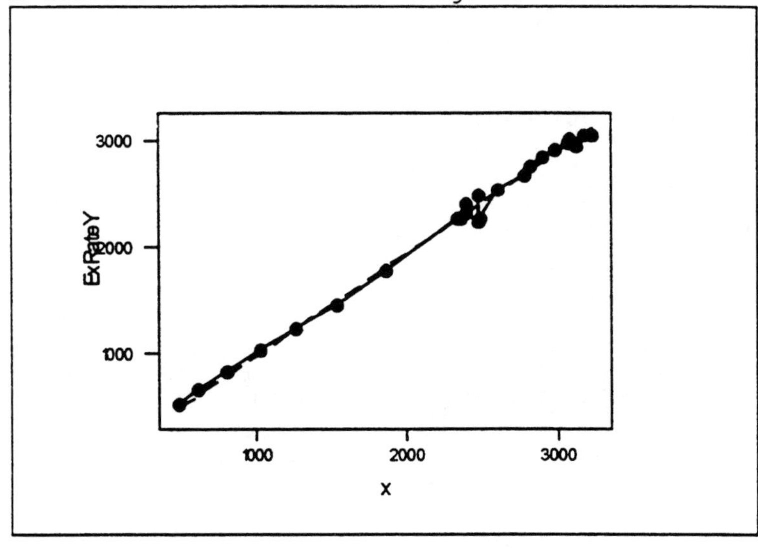

b. 95% C.I. for β: $0.975 \pm 2.069 * 0.00527 = 0.975 \pm 0.011 = (0.96, 0.99)$
$\beta = 1$ is not in the confidence interval as the theory would suggest. Mexico attempted to manage its economy and froze its exchange rate during quarters 10-12. This action disturbed the economic relationship.

c. The linear relation between Mexico and the U.S. is stronger than the relation between Japan and the United States. The regression analysis below shows an r^2 value of 99.5% for Mexico/U.S. compared to 61.3% for the Japan/U.S. regression analysis.

Regression Analysis

```
The regression equation is
ExRateY = 41.6 + 0.958 x

24 cases used 1 cases contain missing values

Predictor        Coef        Stdev     t-ratio       p
Constant        41.63        36.33        1.15    0.264
x             0.95840      0.01513       63.36    0.000

s = 61.61       R-sq = 99.5%     R-sq(adj) = 99.4%
```

d. An unmanaged exchange rate would fluctuate more, causing the relationship to appear weaker.

11.62. a.
Regression Analysis

```
The regression equation is
ExRateY = 1.01 Lag1

23 cases used 2 cases contain missing values

Predictor        Coef        Stdev     t-ratio       p
Noconstant
Lag1          1.01029      0.01397       72.30    0.000

s = 158.2
Analysis of Variance

SOURCE        DF          SS          MS         F          p
Regression     1   130769224   130769224   5226.59      0.000
Error         22      550439       25020
Total         23   131319664

Unusual Observations
Obs.      Lag1     ExRateY         Fit  Stdev.Fit    Residual     St.Resid
  9       1856      2249.4      1875.2       25.9       374.2        2.40R

R denotes an obs. with a large st. resid.
```

For the second quarter of 1992, the estimate of the exchange rate is 3243.13.

b. This is a strong linear relation for prediction.

c.

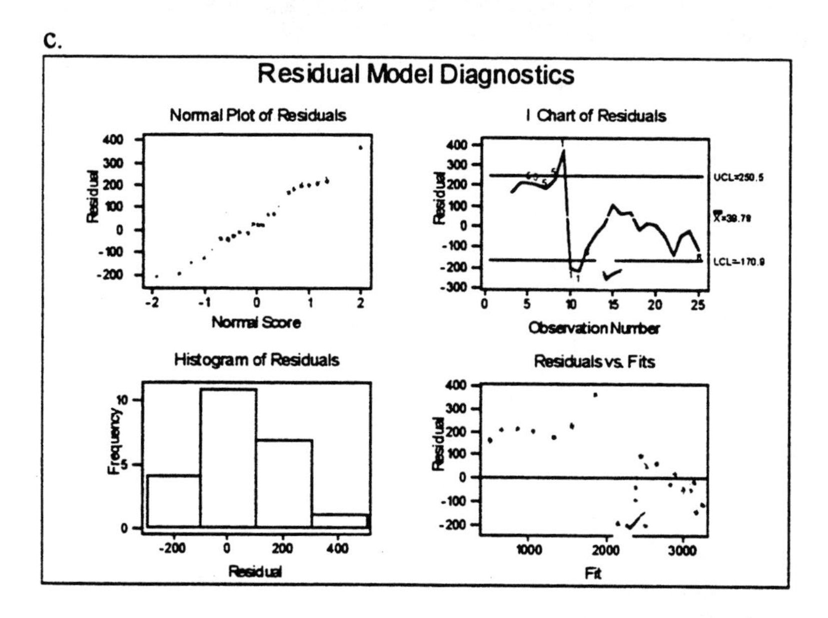

Refer to the I-chart of the residuals. The large negative residuals at quarters 10 and 11 makes interpreting the time-sequence of residuals difficult. Apart from residuals 10 and 11, the remaining residuals appear to vary randomly in a relatively narrow band about zero.

d.

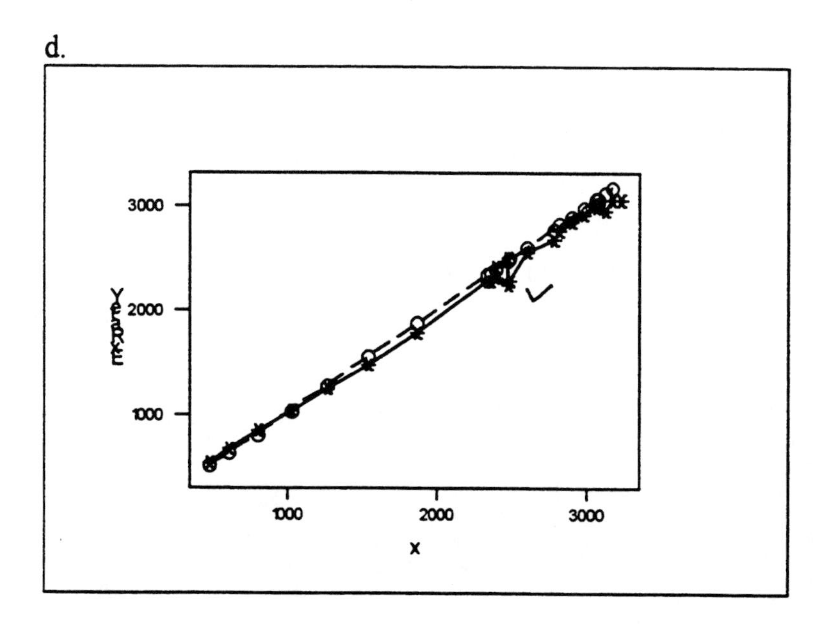

When Mexico froze the exchange rate (quarters 10-12), this created an unusual prediction for quarters 10 and 11. The predicted y is much greater than the actual (managed) y for this x.

✓ Corresponds to observations # 10 and 11 which are highly unusual. Residual plots and residual autocorrelations are greatly affected by these two large residuals. Residual autocorrelations are not informative in this case.

11.63.

a.

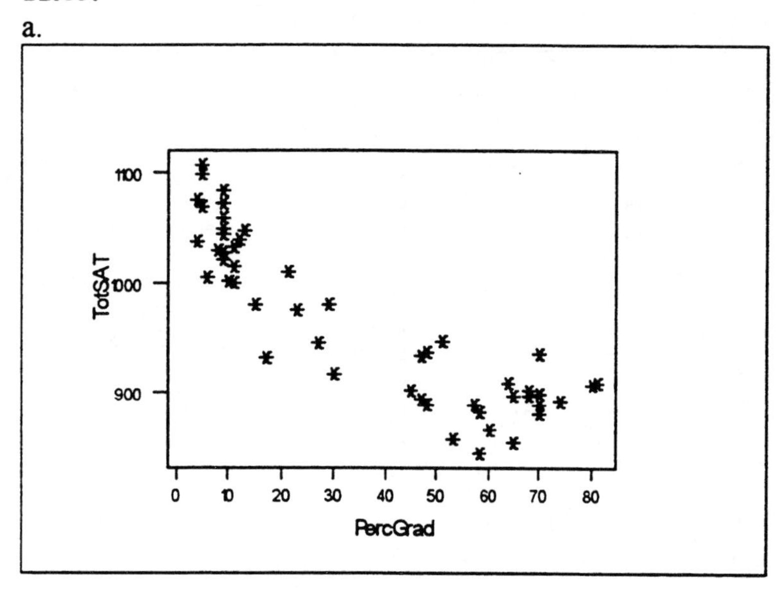

b.

Regression Analysis

```
The regression equation is
TotSAT = 1053 - 2.51 PercGrad

Predictor        Coef         Stdev      t-ratio         p
Constant      1053.34          8.41       125.27     0.000
PercGrad      -2.5147        0.1899       -13.24     0.000

s = 35.73        R-sq = 78.2%       R-sq(adj) = 77.7%

Analysis of Variance

SOURCE        DF          SS          MS          F         p
Regression     1      223887      223887     175.38     0.000
Error         49       62554        1277
Total         50      286441

Unusual Observations
Obs. PercGrad     TotSAT         Fit  Stdev.Fit    Residual    St.Resid
 49       17.0     932.00     1010.59       6.12      -78.59      -2.23R
```

R denotes an obs. with a large st. resid.

r^2 = 78.2% of the variability in the response y is explained by the fitted straight line. This is a fairly strong linear relation.

c.
Value of test statistic for significant regression: F = 175.38
Rejection region: F > 4.05
Reject the null hypothesis of H_0: β_1 =0. The regression is significance.

d.

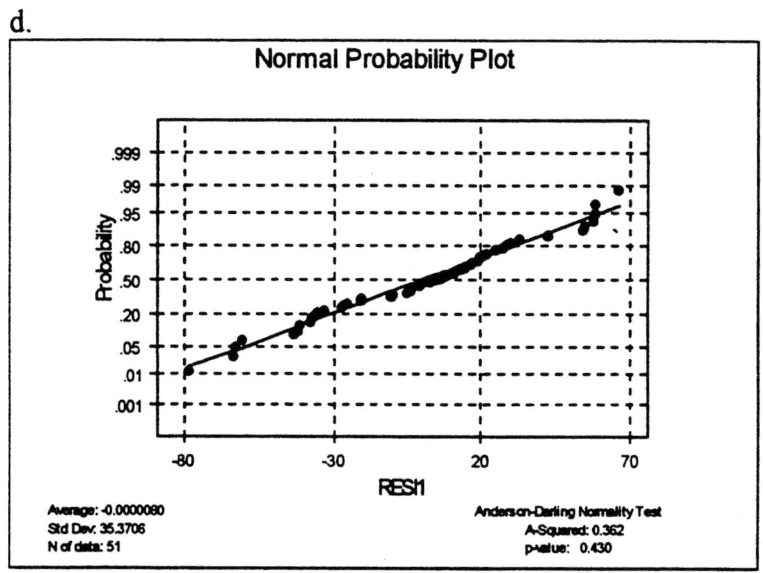

Since this graph is nearly a straight line, the normal assumption for the errors (responses) seems reasonable.

11.64.

a. When there are 35% graduates, Total SAT = 1053 - 2.51 * 35 = 965.32

```
     Fit  Stdev.Fit         95.0% C.I.            95.0% P.I.
  965.32       5.00   (  955.26,   975.38)   (  892.80, 1037.84)
```

b. When there are 98% graduates, Total SAT = 1053 - 2.51 * 98 = 806.90

```
     Fit  Stdev.Fit         95.0% C.I.            95.0% P.I.
  806.90      12.86   (  781.04,   832.75)   (  730.57,   883.23) X
X  denotes a row with X values away from the center
```

c. When fewer students take the SAT test, mainly the best students are taking the test, which raises the overall average for that state. If at least 50% of the students in each state took the test, the relationship between SAT score and percent graduates taking the test might disappear or be nearly horizontal.

11.65.

a.

The regression equation is
OpExBsbl = 18.9 + 1.30 PlCtBsbl

Predictor	Coef	Stdev	t-ratio	p
Constant	18.883	4.138	4.56	0.000
PlCtBsbl	1.3016	0.1528	8.52	0.000

s = 5.382 R-sq = 75.1% R-sq(adj) = 74.1%

Analysis of Variance

SOURCE	DF	SS	MS	F	p
Regression	1	2101.7	2101.7	72.56	0.000
Error	24	695.2	29.0		
Total	25	2796.9			

Unusual Observations

Obs.	PlCtBsbl	OpExBsbl	Fit	Stdev.Fit	Residual	St.Resid
23	18.0	60.00	42.31	1.64	17.69	3.45R

R denotes an obs. with a large st. resid.

Fitted straight line: OpExBsbl = 18.9 + 1.30 PlCtBsbl

r^2 = 75.1% of the variability of the response y is explained by the fitted straight line. This is a fairly strong linear relation.

b. Value of test statistic for H_0: $\rho = 0$: t = 0.867 * (24)^0.5 / (1 - 0.751)^0.5 = 8.512
Rejection region: |t| > 2.797
Reject the null hypothesis. The population correlation is not zero.

c. We cannot conclude that operating costs are roughly twice player costs. The estimated intercept, 18.9, represents fixed operating costs. The estimated slope coefficient, 1.3, indicates that, after fixed costs, mean operating expenses increase slightly more than increases in player costs.

d. When player costs are $30.5 million:

Fit	Stdev.Fit	95.0% C.I.	95.0% P.I.
58.58	1.24	(56.01, 61.15)	(47.18, 69.98)

e.

Unusual Observations

Obs.	PlCtBsbl	OpExBsbl	Fit	Stdev.Fit	Residual	St.Resid
7	18.0	60.00	42.31	1.64	17.69	3.45R

R denotes an obs. with a large st. resid

11.66.

a.

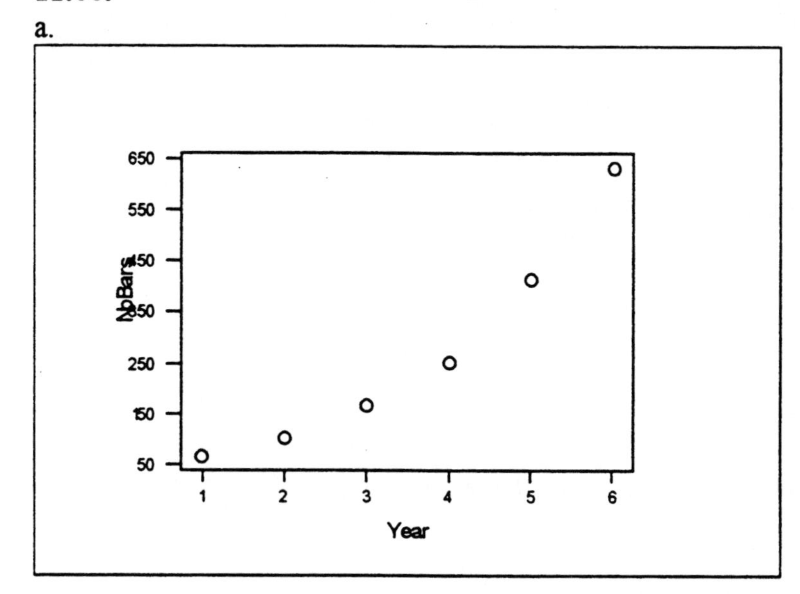

This plot is an upward curve.

b.

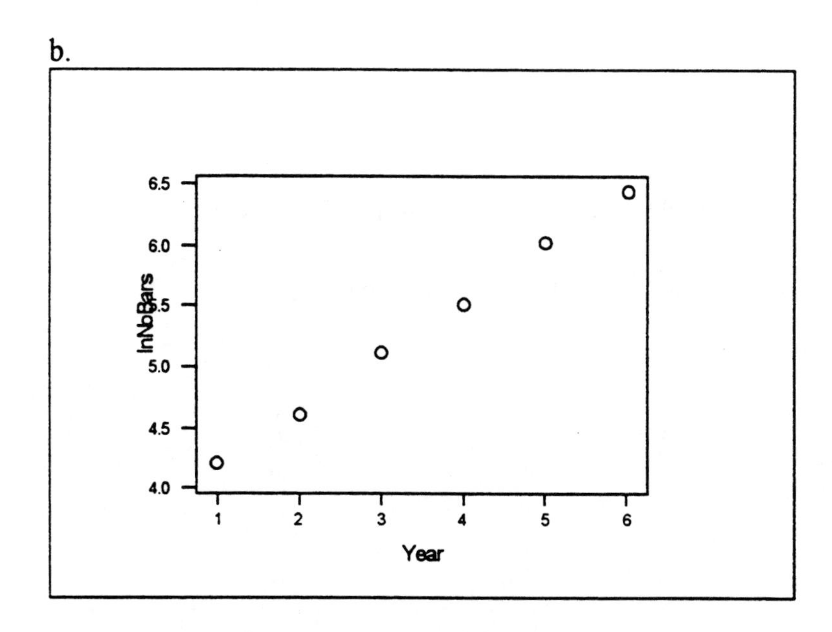

This plot is consistent with linear growth.

c.

The regression equation is
lnNoBars = 3.73 + 0.455 Year

Predictor	Coef	Stdev	t-ratio	p
Constant	3.73054	0.02878	129.64	0.000
Year	0.454746	0.007389	61.54	0.000

s = 0.03091 R-sq = 99.9% R-sq(adj) = 99.9%

Analysis of Variance

SOURCE	DF	SS	MS	F	p
Regression	1	3.6189	3.6189	3787.55	0.000
Error	4	0.0038	0.0010		
Total	5	3.6227			

r^2 = 99.9% of the variability in the response y is explained by the fitted straight line. This is an extremely strong relation.

d. Value of the test statistic for the significance of the linear regression:
F-ratio = 3787.55
Rejection region: F > 7.71
Reject the null hypothesis.

11.67.

a.

95% C.I. for β_1 = 0.455 ± 2.776 * 0.007389 = 0.375 ± .0205 = (0.434 , 0.476)
95% C.I. for γ = (0.543 , 0.609)
0.5 is not a plausible value for γ since it is not in the confidence interval.

b. lnCoBars = 3.73 + 0.455 * 7 = 6.915
Prediction interval:

Fit	Stdev.Fit	95.0% C.I.	95.0% P.I.
6.8751	0.0282	(6.7967, 6.9535)	(6.7589, 6.9913)

Number of coffee bars: (861.69 , 1087.13)

c. lnCoBars = 3.73 + 0.455 * 11 = 8.735
Coffee Bars = 6217
I do not have much faith in this prediction since it is an extrapolation.

11.68.

a.

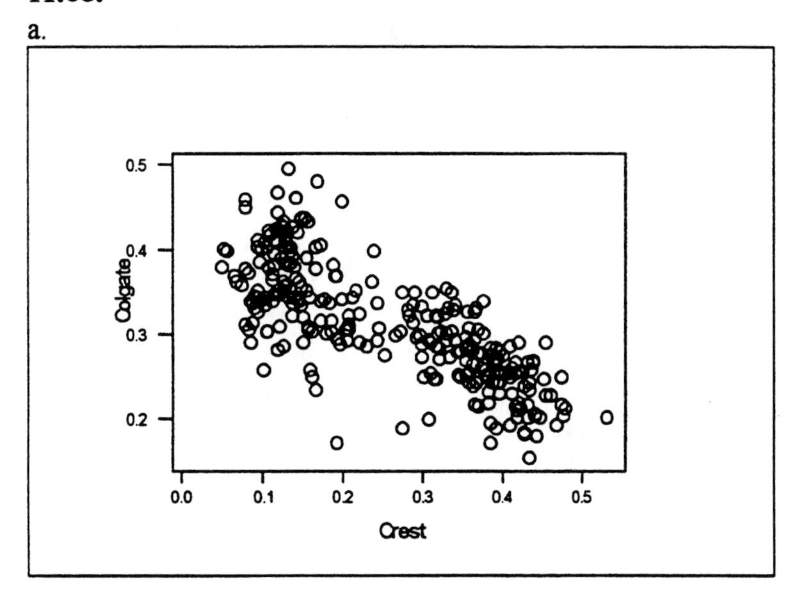

b.

The regression equation is
Colgate = 0.419 - 0.416 Crest

Predictor	Coef	Stdev	t-ratio	p
Constant	0.418850	0.006010	69.69	0.000
Crest	-0.41623	0.02110	-19.72	0.000

s = 0.04443 R-sq = 58.7% R-sq(adj) = 58.5%

Analysis of Variance

SOURCE	DF	SS	MS	F	p
Regression	1	0.76785	0.76785	389.00	0.000
Error	274	0.54085	0.00197		
Total	275	1.30870			

Unusual Observations

Obs.	Crest	Colgate	Fit	Stdev.Fit	Residual	St.Resid
2	0.166	0.48200	0.34976	0.00327	0.13224	2.98R
12	0.131	0.49700	0.36432	0.00374	0.13268	3.00R
17	0.140	0.46400	0.36058	0.00361	0.10342	2.34R
28	0.086	0.29000	0.38305	0.00446	-0.09305	-2.11R
43	0.165	0.23500	0.35017	0.00328	-0.11517	-2.60R
44	0.160	0.25000	0.35225	0.00334	-0.10225	-2.31R
47	0.100	0.25700	0.37723	0.00423	-0.12023	-2.72R
76	0.159	0.25900	0.35267	0.00336	-0.09367	-2.11R
78	0.197	0.45800	0.33685	0.00294	0.12115	2.73R
102	0.191	0.17200	0.33935	0.00300	-0.16735	-3.78R
123	0.119	0.47000	0.36932	0.00392	0.10068	2.28R
203	0.307	0.19900	0.29107	0.00289	-0.09207	-2.08R
239	0.274	0.19000	0.30480	0.00270	-0.11480	-2.59R

R denotes an obs. with a large st. resid.

s = 0.04443 is an estimate of σ.

c. The sign of the slope coefficient is negative, signifying that as Crest's market share goes up, Colgate's market share goes down.

95% C.I. for β_1: $-0.416 \pm 1.96 * 0.02110 = -0.416 \pm 0.041356 = (-0.46, -0.37)$

d. Time series plot of the residuals:

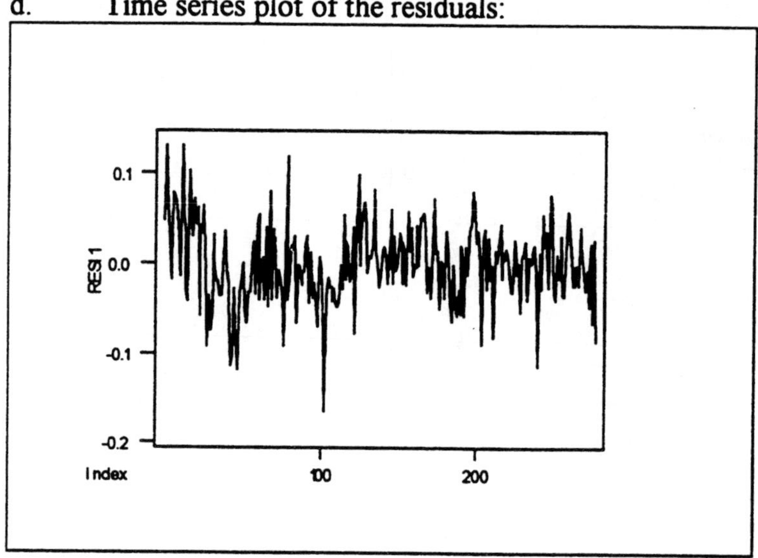

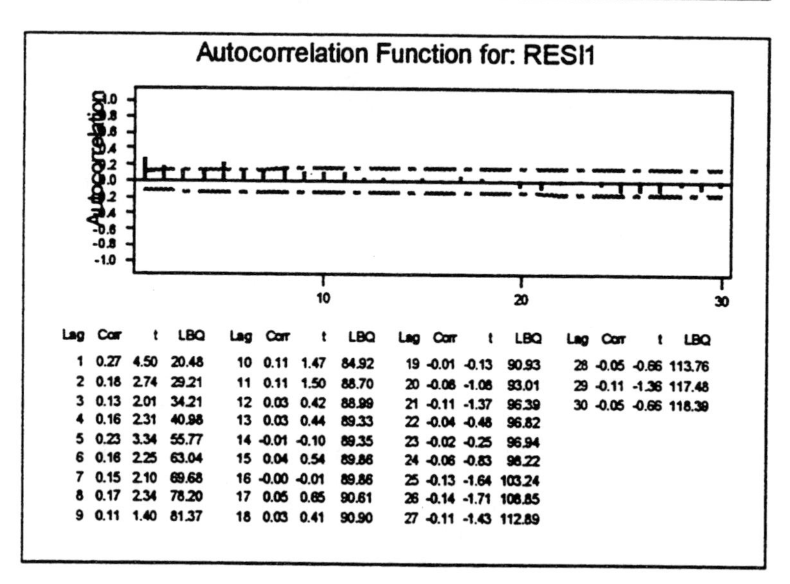

Lag	Corr	t	LBQ	Lag	Corr	t	LBQ	Lag	Corr	t	LBQ	Lag	Corr	t	LBQ
1	0.27	4.50	20.48	10	0.11	1.47	84.92	19	-0.01	-0.13	90.93	28	-0.05	-0.66	113.76
2	0.18	2.74	29.21	11	0.11	1.50	88.70	20	-0.08	-1.08	93.01	29	-0.11	-1.36	117.48
3	0.13	2.01	34.21	12	0.03	0.42	88.99	21	-0.11	-1.37	96.39	30	-0.05	-0.66	118.39
4	0.16	2.31	40.98	13	0.03	0.44	89.33	22	-0.04	-0.48	96.82				
5	0.23	3.34	55.77	14	-0.01	-0.10	89.35	23	-0.02	-0.25	96.94				
6	0.16	2.25	63.04	15	0.04	0.54	89.86	24	-0.06	-0.83	98.22				
7	0.15	2.10	69.68	16	-0.00	-0.01	89.86	25	-0.13	-1.64	103.24				
8	0.17	2.34	78.20	17	0.05	0.65	90.61	26	-0.14	-1.71	108.85				
9	0.11	1.40	81.37	18	0.03	0.41	90.90	27	-0.11	-1.43	112.89				

ACF of RESI1

```
          -1.0 -0.8 -0.6 -0.4 -0.2  0.0  0.2  0.4  0.6  0.8  1.0
            +----+----+----+----+----+----+----+----+----+----+
  1   0.282                          XXXXXXXX
  2   0.182                          XXXXXX
  3   0.135                          XXXX
  4   0.153                          XXXXX
  5   0.228                          XXXXXXX
  6   0.167                          XXXXX
  7   0.159                          XXXXX
```

```
  8   0.178                       XXXXX
  9   0.108                       XXXX
 10   0.110                       XXXX
 11   0.126                       XXXX
 12   0.037                       XX
 13   0.038                       XX
 14  -0.010                       X
 15   0.038                       XX
 16   0.009                       X
 17   0.056                       XX
 18   0.034                       XX
 19  -0.003                       X
 20  -0.079                     XXX
 21  -0.100                     XXX
 22  -0.043                      XX
 23  -0.015                       X
 24  -0.061                     XXX
 25  -0.121                    XXXX
 26  -0.132                    XXXX
 27  -0.123                    XXXX
 28  -0.056                      XX
 29  -0.116                    XXXX
 30  -0.058                      XX
```

The independence assumption for the errors is not justified as shown by the autocorrelations for lags 1 - 11 that are uniformly larger than ± 2/n^.5 = 0.12.

12.1.

a. $\hat{y}$ = 4.21+ 8 *11.37+ 30 *(-.51) = 79.9

b. $\hat{y}$ = 4.21+ 8 *11.37+ 50 *(-.51) = 69.7

c. $\hat{y}$ = 4.21+ 3 *11.37+ 50 *(-.51) = 12.8

12.2.

a. When x_1 is increased by 1 unit, E(Y) is increased by 1 unit.
b. When x_2 is decreased by 1 unit, E(Y) is increased by 4 units.
c. When x_1 and x_2 are both increased by 1 unit, E(Y) is decreased by 3 units.

12.3.

a. E(Y) = -1+ 15 *(-2)+ 10 *3 = -1
b. E(Y) = -1+ 12 *(-2)+ 11 *3 = 8
c. E(Y) = -1+ 12 *(-2)+ 18 *3 = 29

12.4.

E(Y) = -1+ 5 *(-2)+ 12 *3= 25
P(X > 20) = P(Z > -1.25) = 0.894

12.5.

a. s = (46.25 / (20 - 2 - 1))^.5 = 1.649
b. R^2 = 236.70 / (236.70 + 46.25) = 0.837
83.7% of the variability in the response y is explained by the fitted multiple regression equation.

12.6.

a. 95% C.I. for β_0: 4.21 ± 2.110 * 2.26 = 4.21 ± 4.7686 = (-0.56 , 8.98)
95% C.I. for β_2: -.51 ± 2.110 * .098 = -.51 ± .2068 = (-0.72 , -0.30)

b. Value of test statistic: t = 11.37 / 1.08 = 10.528
Rejection region: t > 1.740
Reject the null hypothesis and conclude that β_1 > 12.

12.7.

a. **Regression analysis:**

The regression equation is
Newsprnt = - 43492 + 3853 Papers + 4508 LnFamily

Predictor	Coef	Stdev	t-ratio	p
Constant	-43492	12448	-3.49	0.004
Papers	3853	1526	2.53	0.027
LnFamily	4508	1431	3.15	0.008

s = 2211 R-sq = 74.3% R-sq(adj) = 70.0%

Analysis of Variance

SOURCE	DF	SS	MS	F	p
Regression	2	169208400	84604200	17.31	0.000
Error	12	58652444	4887704		
Total	14	227860848			

SOURCE	DF	SEQ SS
Papers	1	120708064
LnFamily	1	48500344

Unusual Observations

Obs.	Papers	Newsprnt	Fit	Stdev.Fit	Residual	St.Resid
8	1.00	1399	5443	1347	-4044	-2.31R
10	2.00	13907	9297	1162	4610	2.45R

R denotes an obs. with a large st. resid.

b. F = 17.31 and its associated *P*-value of 0 indicate that there is strong evidence for rejecting the null hypothesis H_o: $\beta_1 = \beta_2 = 0$. The regression is significant.

c. The independent variables are related. They do not act "independently" on the response. Dropping the x variable LnRetSal changes the magnitudes of the effects of the remaining variables on the response.

d. 74.3% of the variability in the response y is explained by the fitted multiple regression equation.

12.8.

a. **Regression analysis:**

The regression equation is
$y = -2.27 + 0.0475\, x - 0.000045\, xSq$

Predictor	Coef	Stdev	t-ratio	p
Constant	-2.2707	0.4299	-5.28	0.000
x	0.047513	0.003387	14.03	0.000
xSq	-0.00004482	0.00000622	-7.21	0.000

s = 0.1534   R-sq = 99.6%   R-sq(adj) = 99.5%

Analysis of Variance

SOURCE	DF	SS	MS	F	p
Regression	2	42.155	21.078	895.24	0.000
Error	7	0.165	0.024		
Total	9	42.320			

SOURCE	DF	SEQ SS
x	1	40.932
xSq	1	1.223

b. 99.6% of the variability in the response y is explained by the fitted multiple regression equation.

c. The second order term is required since its t-value is significant. It has a P-value of 0.

d. The residual plots indicate the regression assumptions are reasonable.

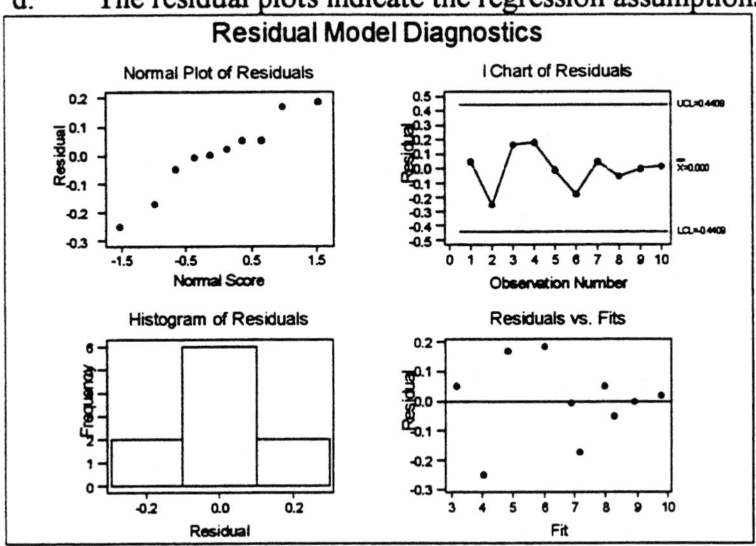

12.9.

a. **Regression analysis:**

The regression equation is

95Price = - 5426 + 90.1 HP + 2.76 Weight

Predictor	Coef	Stdev	t-ratio	p
Constant	-5426	4460	-1.22	0.240
HP	90.11	48.33	1.86	0.080
Weight	2.765	2.896	0.95	0.353

s = 2998 R-sq = 61.6% R-sq(adj) = 57.1%

Analysis of Variance

SOURCE	DF	SS	MS	F	p
Regression	2	245598768	122799384	13.66	0.000
Error	17	152829808	8989989		
Total	19	398428576			

SOURCE	DF	SEQ SS
HP	1	237407248
Weight	1	8191518

Unusual Observations

Obs.	HP	95Price	Fit	Stdev.Fit	Residual	St.Resid
19	200	31934	22987	1757	8947	3.68R

R denotes an obs. with a large st. resid.

F = 13.66 and its associated *P*-value of 0 indicate that the regression is significant. The t-ration for Weight indicates this variable may be dropped (do not reject $H_0{:}\beta_2 = 0$) if the variable HP is in the model.

b. 61.6% of the variability in the response of y is explained by the fitted multiple regression equation.

c.

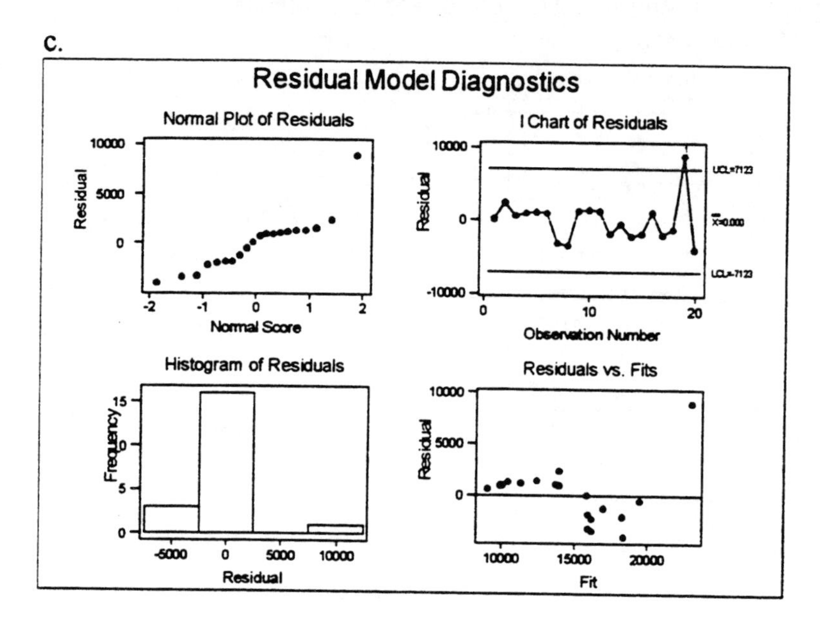

The residual plots identify one observation as an outlier. Observation 19 has a large 95Price relative to the other cars in the sample. Without this observation, the regression assumptions are reasonable.

12.10.

a. **Regression analysis:**

The regression equation is
MPG = 48.4 - 0.0235 HP - 0.00505 Weight

Predictor	Coef	Stdev	t-ratio	p
Constant	48.447	3.431	14.12	0.000
HP	-0.02348	0.03718	-0.63	0.536
Weight	-0.005048	0.002228	-2.27	0.037

s = 2.307 R-sq = 63.3% R-sq(adj) = 59.0%

Analysis of Variance

SOURCE	DF	SS	MS	F	p
Regression	2	156.081	78.040	14.66	0.000
Error	17	90.469	5.322		
Total	19	246.550			

SOURCE	DF	SEQ SS
HP	1	128.773
Weight	1	27.308

```
Unusual Observations
Obs.        HP        MPG         Fit   Stdev.Fit    Residual      St.Resid
  6        102     40.000      34.790       0.927       5.210         2.47R
  9        105     30.000      34.295       0.849      -4.295        -2.00R

R denotes an obs. with a large st. resid.
```

F = 14.66 and its associated *P*-value of zero indicate that the regression is significant. The t-value, -0.63, for HP indicates that if the variable Weight is in the regression function, the variable HP can be dropped (do not reject H_o: $\beta_1 = 0$).

b. The estimate of the cofficient β_1 is - 0.0235 indicating that as the HP increases, there is a small decrease in the mean MPG.

95% C.I: for β_1: $-0.0235 \pm 2.110 * 0.03718 = -0.0235 \pm 0.0784$
$= (-0.10 , 0.05)$
Notice this C.I. includes $\beta_1 = 0$.

c.

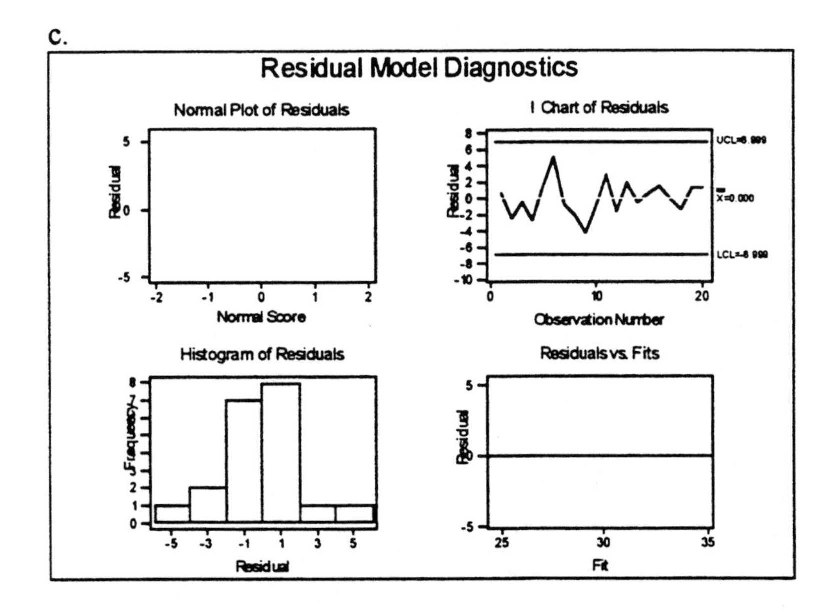

The residual plots indicate the regression assumptions are reasonable.

12.11.

a. Fitted regression function:

LnComp = 5.36 + 0.263 LnSales + 0.0157 Exper - 0.429 Educate - 0.0308 PctOwn

$\hat{\beta}_3 = -0.429$ indicates that the mean log compensation decreases by 0.429 units for each unit increase in the level of education.

b. LnComp = 5.36 + 0.263 *ln(7818) + 0.0157 *6 - 0.429 *1 - 0.0308 *0.004
= 7.3828

Compensation in millions of dollars: $1.608 or $1,608,000.

c. 46.8% of the variability in log compensation is explained by the fitted regression function.

d. The *P*-value associated with the F statistic is 0 indicating that the regression is significant.

12.12.
The normal scores plot is a fairly straight line indicating the normality assumption is adequate. The I Chart scores indicates no apparent pattern in the residuals. The independence assumption is reasonable. The histogram is symmetric. The plot of the residuals versus the fits shows the residuals distributed relatively uniformly in a horizontal band about zero. There is no reason to doubt the constant variance assumption. To summarize, there is no reason to doubt the regression model assumptions.

12.13.
a.

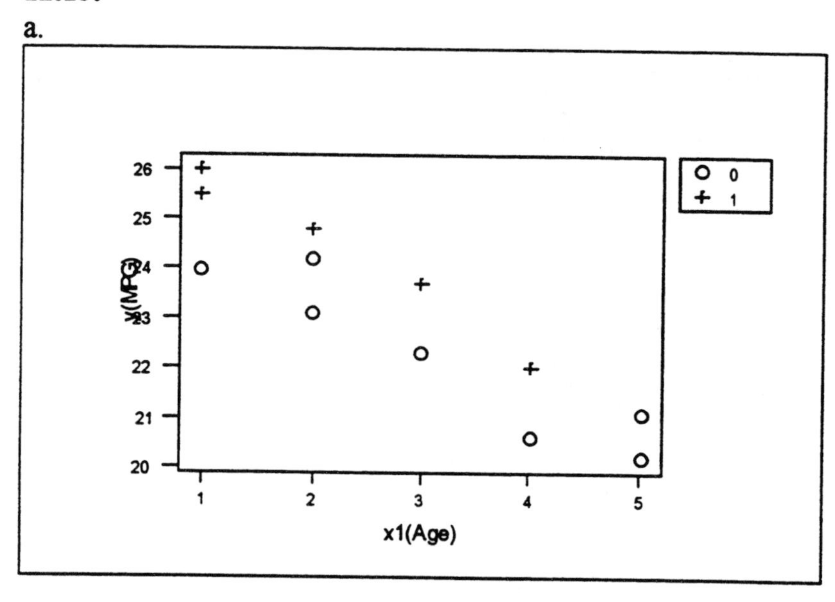

b.
Regression analysis:
The regression equation is
y(MPG) = 25.5 - 1.04 x1(Age) + 1.21 x2(Gend)

Predictor	Coef	Stdev	t-ratio	p
Constant	25.4815	0.4162	61.22	0.000
x1(Age)	-1.0396	0.1156	-9.00	0.000
x2(Gend)	1.2055	0.3332	3.62	0.006

s = 0.5378   R-sq = 93.5%   R-sq(adj) = 92.0%

```
Analysis of Variance

SOURCE          DF          SS          MS         F        p
Regression       2      37.339      18.670     64.55    0.000
Error            9       2.603       0.289
Total           11      39.942

SOURCE          DF      SEQ SS
x1(Age)          1      33.554
x2(Gend)         1       3.786
```

Gender appears to make a difference in MPG. Female drivers increase mean MPG by $\hat{\beta}_2$ = 1.21 miles per gallon for all ages of taxis considered.

c.

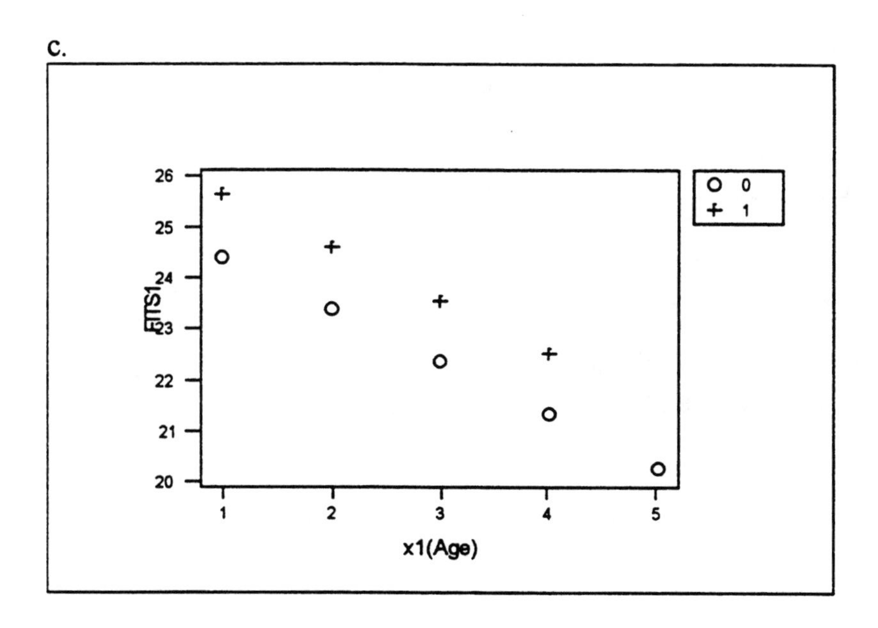

Male "line": $\hat{y} = 25.5 - 1.04x_1$

Female "line": $\hat{y} = 26.71 - 1.04x_1$

d.

Regression Analysis:

The regression equation is

y(MPG) = 26.4 - 1.18 x1(Age)

```
Predictor          Coef         Stdev     t-ratio         p
Constant        26.3598        0.5025       52.46     0.000
x1(Age)         -1.1763        0.1623       -7.25     0.000

s = 0.7993        R-sq = 84.0%      R-sq(adj) = 82.4%
```

Analysis of Variance

SOURCE	DF	SS	MS	F	p
Regression	1	33.554	33.554	52.52	0.000
Error	10	6.389	0.639		
Total	11	39.942			

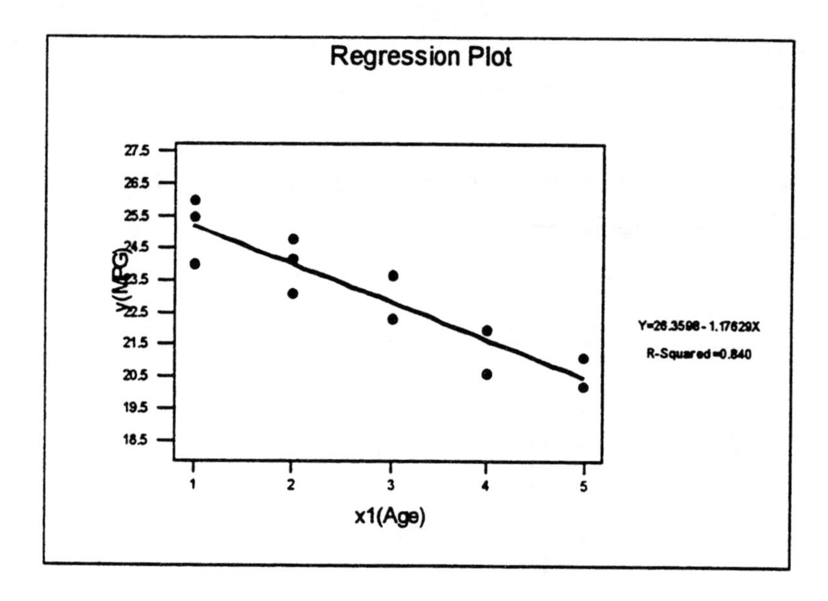

Gender is important. If gender is ignored, the fitted straight line tends to overestimate MPG for male drivers and underestimate MPG for female drivers.

12.14.

a.

Regression analysis:

The regression equation is

$\hat{y}$ = - 43.2 + 0.372 x1 + 0.352 x2 + 19.1 x3

Predictor	Coef	Stdev	t-ratio	p	VIF
Constant	-43.15	31.67	-1.36	0.192	
x1	0.3716	0.3397	1.09	0.290	1.5
x2	0.3515	0.2917	1.21	0.246	1.4
x3	19.12	11.04	1.73	0.103	1.5

s = 13.91 R-sq = 49.8% R-sq(adj) = 40.4%

Analysis of Variance

SOURCE	DF	SS	MS	F	p
Regression	3	3071.1	1023.7	5.29	0.010
Error	16	3096.7	193.5		
Total	19	6167.8			

SOURCE	DF	SEQ SS
x1	1	1821.3
x2	1	669.3
x3	1	580.5

```
Unusual Observations
Obs.        x1          y         Fit   Stdev.Fit    Residual     St.Resid
 20         95      57.00       84.43        4.73      -27.43       -2.10R
```

R denotes an obs. with a large st. resid.

The P-value for the F-ratio is 0.01. The regression is significant.

b. $\hat{y} = -43.2 + 0.372 * 86 + 0.352 * 77 + 19.1 * 3.4 = 80.836$

c. The variables x1 and x2 are each not significant if the other predictor variables remain in the regression function since their P-values are 0.290 and 0.246 respectively. If x1 and x2 are in the model, x3 is not significant at significance levels $\alpha \leq 0.10$.

The VIF's are 1.5, 1.4, and 1.5. The VIF values, the relatively small t-values, and the significant F statistic (P-value = 0.01) together imply that multicollinearity might be a problem. The x's are linearly related.

12.15.

a. The mean leverage is 0.20. None of the observations are high leverage points since their values are not greater than 3(0.20) = 0.60.

b.

```
Unusual Observations
Obs.        x1          y         Fit   Stdev.Fit    Residual     St.Resid
 20         95      57.00       84.43        4.73      -27.43       -2.10R
```

R denotes an obs. with a large st. resid.

The fitted model over-predicted the response for this observation since the residual, $y - \hat{y}$, is negative. Given the first two exam scores and the current gpa, this student's final exam score should have been higher according to the fitted regression function.

12.16.

Response is y

```
                Adj.                          x x x
Vars   R-sq     R-sq      C-p           s     1 2 3

   1   37.7     34.2      3.9      14.610         X
   1   30.1     26.3      6.3      15.472       X
   2   46.0     39.7      3.2      13.992       X X
   2   45.2     38.8      3.5      14.096     X   X
   3   49.8     40.4      4.0      13.912     X X X
```

Regression analysis:
The regression equation is
y = - 32.6 + 0.451 x2 + 23.3 x3

Predictor	Coef	Stdev	t-ratio	p
Constant	-32.62	30.34	-1.07	0.297
x2	0.4514	0.2787	1.62	0.124
x3	23.31	10.42	2.24	0.039

s = 13.99 R-sq = 46.0% R-sq(adj) = 39.7%

Analysis of Variance

SOURCE	DF	SS	MS	F	p
Regression	2	2839.5	1419.8	7.25	0.005
Error	17	3328.3	195.8		
Total	19	6167.8			

SOURCE	DF	SEQ SS
x2	1	1859.0
x3	1	980.5

Unusual Observations

Obs.	x2	y	Fit	Stdev.Fit	Residual	St.Resid
1	85.0	91.00	68.69	8.72	22.31	2.04R

R denotes an obs. with a large st. resid.

Either of the two variable models is about as good as the three variable model in terms of R^2 and the Cp statistic. Notice, however, that the coefficient of x_2 (exam 2) is not significantly different from 0 at the 10% level provided x_3 (GPA) is in the model. This suggests a $p = 1$ predictor variable model with x_3 may be adequate for these data.

Stepwise Regression

```
 F-to-Enter:        3.50    F-to-Remove:        3.50

 Response is     y      on  3 predictors, with N =   20

    Step          1
Constant     -26.24

x3             31.4
T-Ratio        3.30

S              14.6
R-Sq          37.71
```

The stepwise procedure with F-to-enter = F-to-remove = 3.5 chooses a $p = 1$ variable model with predictor variable x_3 (GPA). The result is consistent with the all possible regression procedure.

12.17.

a.

Regression analysis:

The regression equation is

95Price = - 12943 + 93.8 HP + 3.55 Weight + 155 MPG

Predictor	Coef	Stdev	t-ratio	p	VIF
Constant	-12943	16282	-0.79	0.438	
HP	93.76	50.04	1.87	0.079	3.8
Weight	3.548	3.382	1.05	0.310	4.8
MPG	155.2	322.6	0.48	0.637	2.7

s = 3069 R-sq = 62.2% R-sq(adj) = 55.1%

Analysis of Variance

SOURCE	DF	SS	MS	F	p
Regression	3	247776880	82592296	8.77	0.001
Error	16	150651696	9415731		
Total	19	398428576			

SOURCE	DF	SEQ SS
HP	1	237407248
Weight	1	8191518
MPG	1	2178106

Unusual Observations

Obs.	HP	95Price	Fit	Stdev.Fit	Residual	St.Resid
19	200	31934	23176	1841	8758	3.57R

R denotes an obs. with a large st. resid.

Multicollinearity is a problem since the VIF values of 3.8, 2.8, and 2.7 are large. The separate effects of horsepower, weight, and miles per gallon on selling price cannot be estimated. There is redundant information among the predictor variables.

b.

Stepwise Regression

F-to-Enter: 4.00 F-to-Remove: 4.00

Response is 95Price on 3 predictors, with N = 20

Step	1
Constant	-2771
HP	129
T-Ratio	5.15
S	2991
R-Sq	59.59

Regression Analysis:

```
The regression equation is
95Price = - 2771 + 129 HP

Predictor          Coef         Stdev      t-ratio          p
Constant          -2771          3478        -0.80      0.436
HP               129.48         25.13         5.15      0.000

s = 2991          R-sq = 59.6%       R-sq(adj) = 57.3%

Analysis of Variance

SOURCE        DF           SS           MS           F          p
Regression     1    237407248    237407248       26.54      0.000
Error         18    161021328      8945629
Total         19    398428576

Unusual Observations
Obs.       HP     95Price          Fit   Stdev.Fit     Residual      St.Resid
 19       200       31934        23125        1747         8809        3.63RX

X denotes an obs. whose X value gives it large influence.
```

This observation has a large 95Price value.

This choice seems reasonable. R^2 for the selected one variable model is about as large as R^2 for the three variable model and multicollinearity among predictor variables has been eliminated.

12.18.

Stepwise Regression

F-to-Enter: 3.50 F-to-Remove: 3.50

```
Response is   LnComp   on   9 predictors, with N =    50

    Step         1         2         3         4
Constant     7.608     5.525     5.686     6.154

Educate      -0.49     -0.47     -0.50     -0.59
T-Ratio      -3.68     -3.94     -4.31     -5.23

LnSales                0.263     0.255     0.215
T-Ratio                 3.53      3.52      3.12

PctOwn                          -0.025    -0.088
T-Ratio                          -1.90     -3.45

Valuate                                  0.00158
T-Ratio                                     2.83

S            0.566     0.509     0.495     0.461
R-Sq         22.02     38.38     42.84     51.46
```

Regression Analysis

```
The regression equation is
LnComp = 6.15 - 0.593 Educate + 0.215 LnSales + 0.00158 Valuate - 0.0884
PctOwn

Predictor        Coef         Stdev     t-ratio        p
Constant       6.1543        0.5922       10.39    0.000
Educate       -0.5934        0.1135       -5.23    0.000
LnSales       0.21523       0.06906        3.12    0.003
Valuate     0.0015828     0.0005600        2.83    0.007
PctOwn       -0.08839       0.02560       -3.45    0.001

s = 0.4615       R-sq = 51.5%      R-sq(adj) = 47.1%

Analysis of Variance

SOURCE        DF          SS          MS         F        p
Regression     4     10.1603      2.5401     11.93    0.000
Error         45      9.5837      0.2130
Total         49     19.7439

SOURCE        DF      SEQ SS
Educate        1      4.3481
LnSales        1      3.2295
Valuate        1      0.0442
PctOwn         1      2.5384

Unusual Observations
Obs.  Educate      LnComp        Fit  Stdev.Fit   Residual    St.Resid
 26      2.00      7.7160     6.7784     0.0947     0.9376       2.08R
 31      2.00      6.5338     6.3744     0.4410     0.1594       1.17 X
 33      0.00      6.3969     6.2574     0.3759     0.1395       0.52 X

R denotes an obs. with a large st. resid.
X denotes an obs. whose X value gives it large influence.
```

The stepwise procedure (with F-to-enter = F-to-remove = 3.5) selects a $p = 4$ variable model with predictor variables Educate, LnSales, PctOwn, and Valuate. This result is consistent with the all possible regression procedure. The best $p = 3$ variable model uses the predictor variables above with the exception of Valuate.

12.19. The series is stationary in the mean.

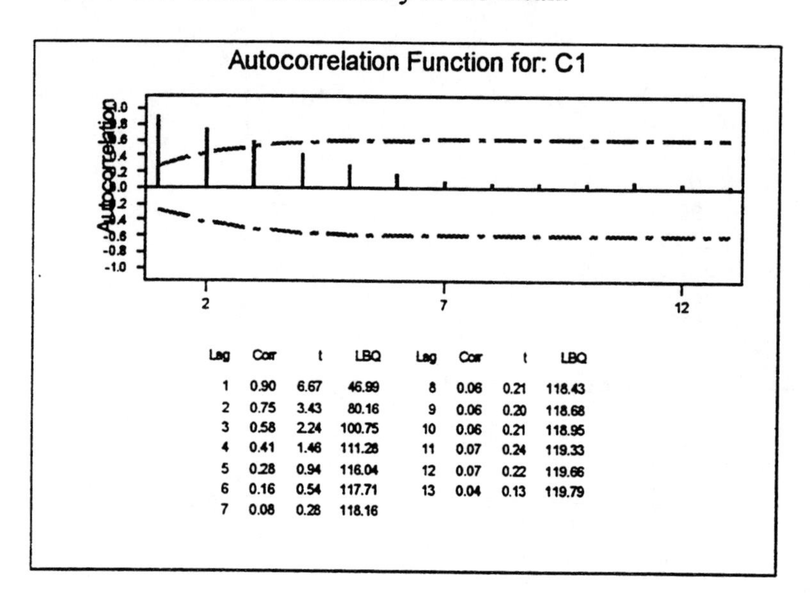

Lag	Corr	t	LBQ	Lag	Corr	t	LBQ
1	0.90	6.67	46.99	8	0.06	0.21	118.43
2	0.75	3.43	80.16	9	0.06	0.20	118.68
3	0.58	2.24	100.75	10	0.06	0.21	118.95
4	0.41	1.46	111.28	11	0.07	0.24	119.33
5	0.28	0.94	116.04	12	0.07	0.22	119.66
6	0.16	0.54	117.71	13	0.04	0.13	119.79
7	0.08	0.28	118.16				

The autocorrelations at small lags are large but die out fairly rapidly; the inventory series is stationary in the mean.

Autocorrelation for the series of first differences:

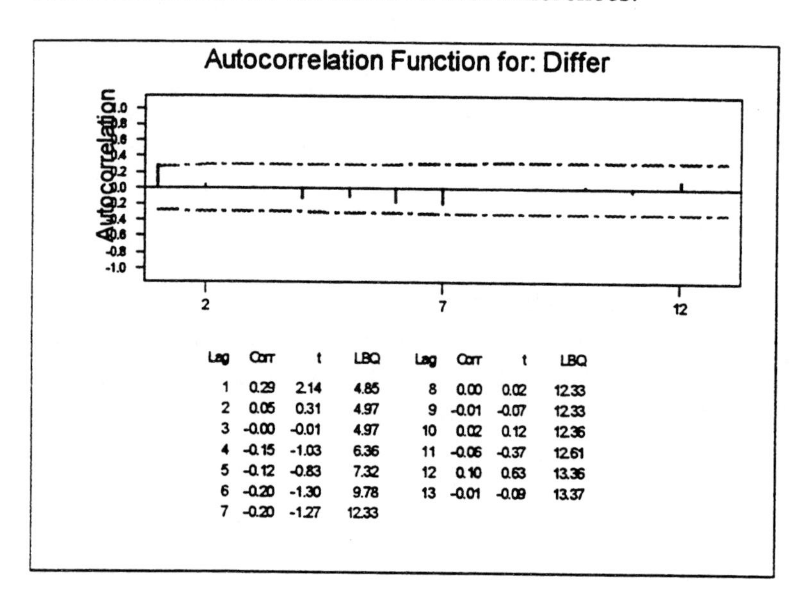

Lag	Corr	t	LBQ	Lag	Corr	t	LBQ
1	0.29	2.14	4.85	8	0.00	0.02	12.33
2	0.05	0.31	4.97	9	-0.01	-0.07	12.33
3	-0.00	-0.01	4.97	10	0.02	0.12	12.36
4	-0.15	-1.03	6.36	11	-0.06	-0.37	12.61
5	-0.12	-0.83	7.32	12	0.10	0.63	13.36
6	-0.20	-1.30	9.78	13	-0.01	-0.09	13.37
7	-0.20	-1.27	12.33				

The autocorrelations for the series of first differences are all fairly small relative to twice their standard errors. The series is also stationary. (The differences of a stationary series will be stationary.)

12.20.

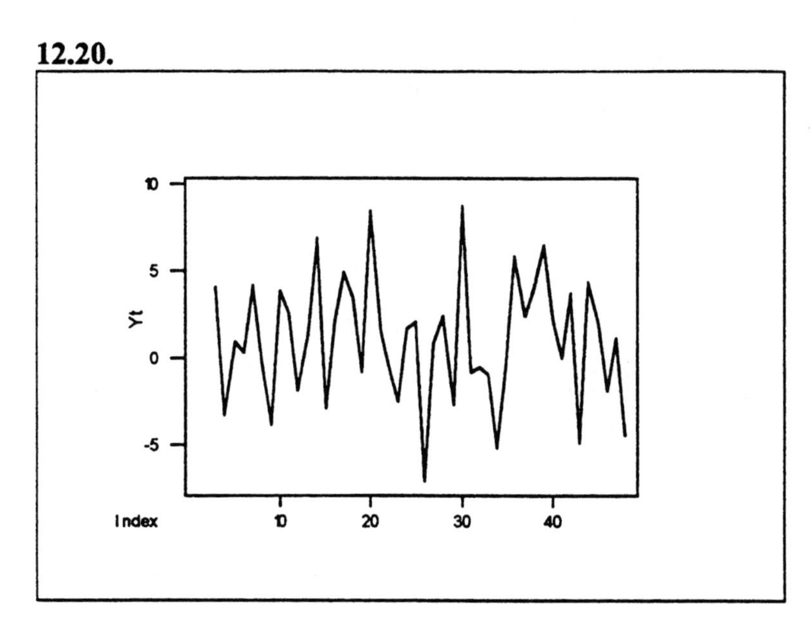

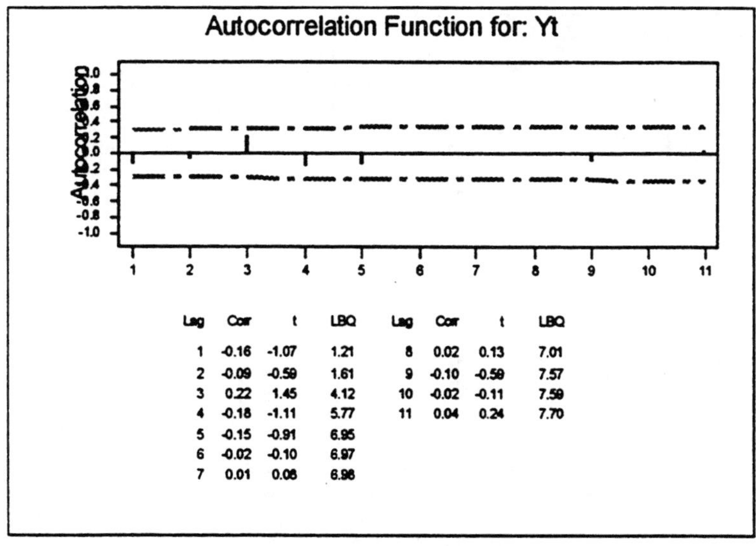

Lag	Corr	t	LBQ	Lag	Corr	t	LBQ
1	-0.16	-1.07	1.21	8	0.02	0.13	7.01
2	-0.09	-0.59	1.61	9	-0.10	-0.59	7.57
3	0.22	1.45	4.12	10	-0.02	-0.11	7.59
4	-0.18	-1.11	5.77	11	0.04	0.24	7.70
5	-0.15	-0.91	6.95				
6	-0.02	-0.10	6.97				
7	0.01	0.08	6.98				

This series is fairly stationary in the mean since the autocorrelations are all small and the time series plot shows the series varies about a constant level that is roughly zero.

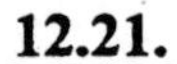

12.21.

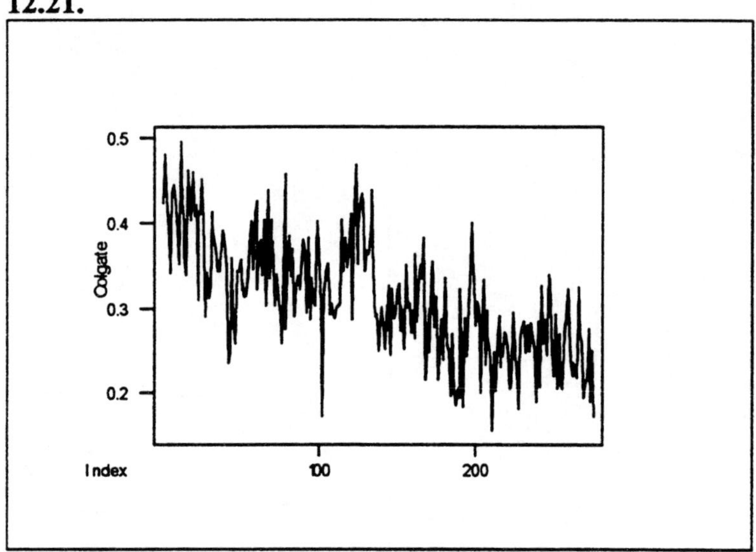

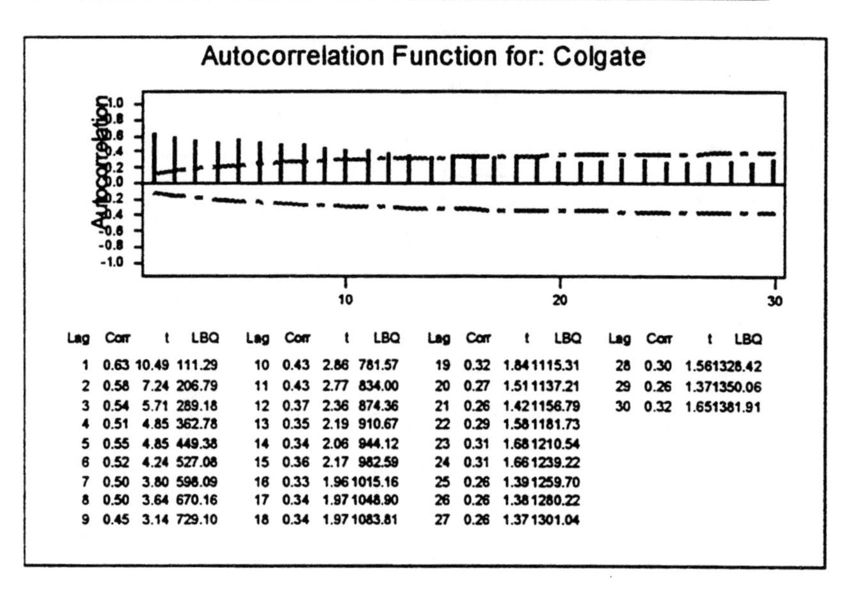

Lag	Corr	t	LBQ	Lag	Corr	t	LBQ	Lag	Corr	t	LBQ	Lag	Corr	t	LBQ
1	0.63	10.49	111.29	10	0.43	2.86	781.57	19	0.32	1.84	1115.31	28	0.30	1.56	1328.42
2	0.58	7.24	206.79	11	0.43	2.77	834.00	20	0.27	1.51	1137.21	29	0.26	1.37	1350.06
3	0.54	5.71	289.18	12	0.37	2.36	874.36	21	0.26	1.42	1156.79	30	0.32	1.65	1381.91
4	0.51	4.85	362.78	13	0.35	2.19	910.67	22	0.29	1.58	1181.73				
5	0.55	4.85	449.38	14	0.34	2.06	944.12	23	0.31	1.68	1210.54				
6	0.52	4.24	527.08	15	0.36	2.17	982.59	24	0.31	1.66	1239.22				
7	0.50	3.80	598.09	16	0.33	1.96	1015.16	25	0.26	1.39	1259.70				
8	0.50	3.64	670.16	17	0.34	1.97	1048.90	26	0.26	1.38	1280.22				
9	0.45	3.14	729.10	18	0.34	1.97	1083.81	27	0.26	1.37	1301.04				

This series is nonstationary in the mean since the level of the series seems to change and the autocorrelations do not die out rapidly.

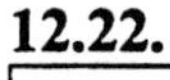

12.22.

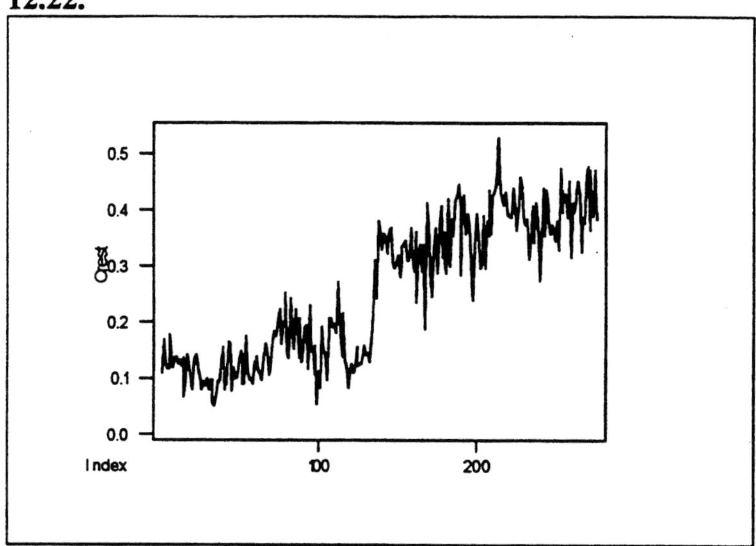

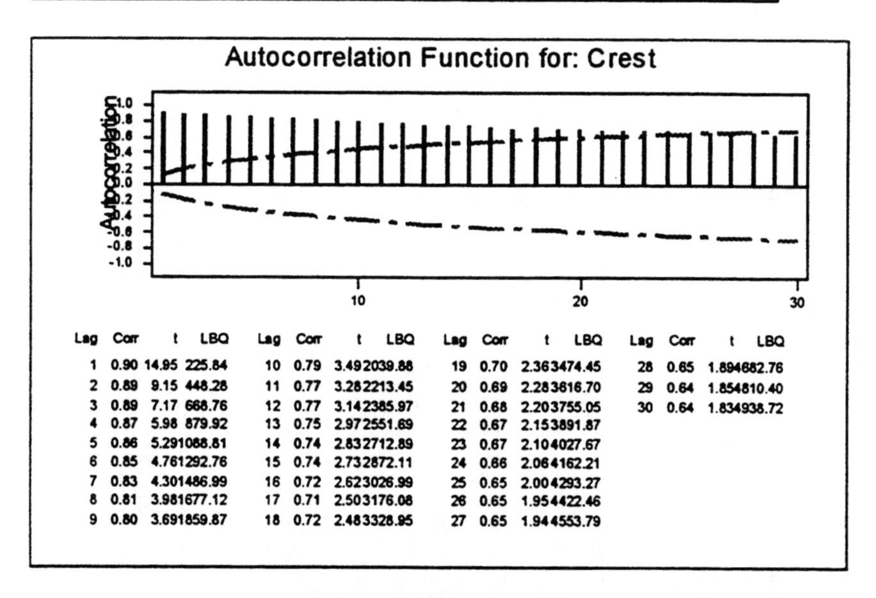

Lag	Corr	t	LBQ	Lag	Corr	t	LBQ	Lag	Corr	t	LBQ	Lag	Corr	t	LBQ
1	0.90	14.95	225.84	10	0.79	3.49	2039.88	19	0.70	2.36	3474.45	28	0.65	1.89	4682.76
2	0.89	9.15	448.28	11	0.77	3.28	2213.45	20	0.69	2.28	3616.70	29	0.64	1.85	4810.40
3	0.89	7.17	668.76	12	0.77	3.14	2385.97	21	0.68	2.20	3755.05	30	0.64	1.83	4938.72
4	0.87	5.98	879.92	13	0.75	2.97	2551.69	22	0.67	2.15	3891.87				
5	0.86	5.29	1088.81	14	0.74	2.83	2712.89	23	0.67	2.10	4027.67				
6	0.85	4.76	1292.76	15	0.74	2.73	2872.11	24	0.66	2.06	4162.21				
7	0.83	4.30	1486.99	16	0.72	2.62	3026.99	25	0.65	2.00	4293.27				
8	0.81	3.98	1677.12	17	0.71	2.50	3176.08	26	0.65	1.95	4422.46				
9	0.80	3.69	1859.87	18	0.72	2.48	3328.95	27	0.65	1.94	4553.79				

This series is nonstationary in the mean since the first plot shows the level of the series seems to change and the autocorrelation does not die out rapidly.

12.23. Answers will vary.

12.24.
Time Series Decomposition

```
Data      Employm
Length    180.000
NMissing 2.00000

Trend Line Equation

Yt = 58.2807 + 5.72E-03*t
```

Seasonal Indices

Period	Index
1	-5.49934
2	-5.37017
3	-5.08267
4	-4.41601
5	-1.27395
6	7.40483
7	11.8527
8	11.3486
9	1.60275
10	-1.86601
11	-3.75559
12	-4.94517

Accuracy of Model

MAPE: 2.20305
MAD: 1.31641
MSD: 3.07319

Row	Employm	Trend	Seasonal	Detrend	Deseason	Model	RESI1
1	56.3	58.2864	-4.9452	-1.9864	61.2452	53.3412	2.95878
2	55.7	58.2921	-5.4993	-2.5921	61.1993	52.7928	2.90723
3	55.8	58.2978	-5.3702	-2.4978	61.1702	52.9277	2.87234
4	56.3	58.3036	-5.0827	-2.0036	61.3827	53.2209	3.07912
5	57.2	58.3093	-4.4160	-1.1093	61.6160	53.8933	3.30673
6	59.1	58.3150	-1.2739	0.7850	60.3739	57.0410	2.05895
7	71.5	58.3207	7.4048	13.1793	64.0952	65.7255	5.77445
8	72.2	58.3264	11.8527	13.8736	60.3473	70.1792	2.02081
9	72.7	58.3322	11.3486	14.3678	61.3514	69.6807	3.01926
10	61.5	58.3379	1.6027	3.1621	59.8973	59.9406	1.55938
11	57.4	58.3436	-1.8660	-0.9436	59.2660	56.4776	0.92241
12	56.9	58.3493	-3.7556	-1.4493	60.6556	54.5937	2.30627
13	55.3	58.3550	-4.9452	-3.0550	60.2452	53.4099	1.89013
14	54.9	58.3608	-5.4993	-3.4608	60.3993	52.8614	2.03857
15	54.9	58.3665	-5.3702	-3.4665	60.2702	52.9963	1.90369
16	54.9	58.3722	-5.0827	-3.4722	59.9827	53.2895	1.61047
17	54.6	58.3779	-4.4160	-3.7779	59.0160	53.9619	0.63808
18	57.7	58.3836	-1.2739	-0.6836	58.9739	57.1097	0.59030
19	68.2	58.3894	7.4048	9.8106	60.7952	65.7942	2.40580
20	70.6	58.3951	11.8527	12.2049	58.7473	70.2478	0.35217
.							
.							
.							
158	54.2	59.1846	-5.4993	-4.9846	59.6993	53.6852	0.51476
159	54.6	59.1903	-5.3702	-4.5903	59.9702	53.8201	0.77987
160	54.3	59.1960	-5.0827	-4.8960	59.3827	54.1134	0.18665
161	54.8	59.2017	-4.4160	-4.4017	59.2160	54.7857	0.01426
162	58.1	59.2075	-1.2739	-1.1075	59.3739	57.9335	0.16648
163	68.1	59.2132	7.4048	8.8868	60.6952	66.6180	1.48198
164	73.3	59.2189	11.8527	14.0811	61.4473	71.0717	2.22835
165	75.5	59.2246	11.3486	16.2754	64.1514	70.5732	4.92679
166	66.4	59.2304	1.6027	7.1697	64.7973	60.8331	5.56691
167	60.5	59.2361	-1.8660	1.2639	62.3660	57.3701	3.12993
168	57.7	59.2418	-3.7556	-1.5418	61.4556	55.4862	2.21380
169	55.8	59.2475	-4.9452	-3.4475	60.7452	54.3023	1.49766
170	54.7	59.2532	-5.4993	-4.5532	60.1993	53.7539	0.94610

171	55.0	59.2590	-5.3702	-4.2590	60.3702	53.8888	1.11121
172	55.6	59.2647	-5.0827	-3.6647	60.6827	54.1820	1.41800
173	56.4	59.2704	-4.4160	-2.8704	60.8160	54.8544	1.54561
174	60.6	59.2761	-1.2739	1.3239	61.8739	58.0022	2.59783
175	70.8	59.2818	7.4048	11.5182	63.3952	66.6867	4.11333
176	76.4	59.2876	11.8527	17.1124	64.5473	71.1403	5.25970
177	74.8	59.2933	11.3486	15.5067	63.4514	70.6419	4.15814
178	62.2	59.2990	1.6027	2.9010	60.5973	60.9017	1.29825
179	*	59.3047	-1.8660	*	*	57.4387	*
180	*	59.3104	-3.7556	*	*	55.5549	*

Forecasts

Row	Period	Forecast
1	181	54.3710
2	182	53.8225
3	183	53.9574
4	184	54.2507
5	185	54.9230
6	186	58.0708
7	187	66.7553
8	188	71.2090
9	189	70.7105
10	190	60.9704
11	191	57.5074
12	192	55.6235

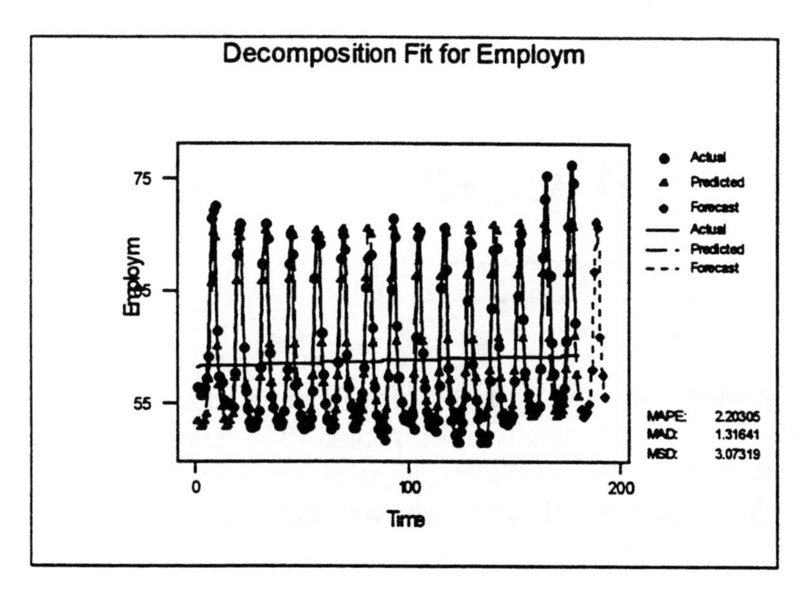

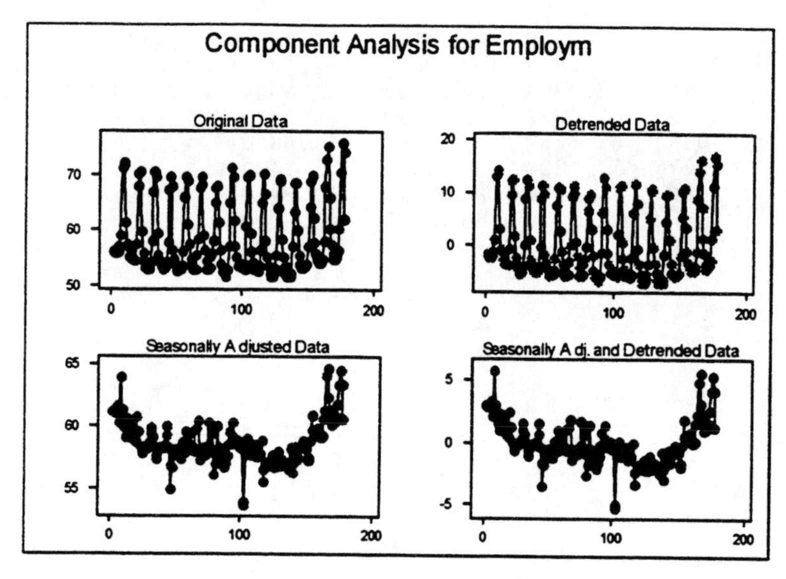

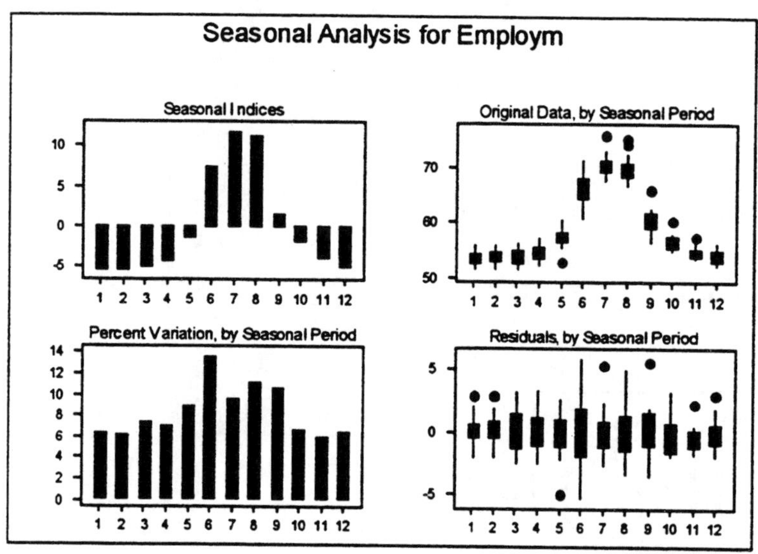

The seasonal indices are positive and relatively large during the summer months. The largest variation in the series also occurs during the summer months. The residuals have the most variation in the summer months. Clearly, the "extra" employment in food and kindred products occurs during the summer and it can be relatively variable.

12.25.

a. $\hat{y}_{n+1} = \beta_0 + \beta_1 y_n + \beta_2 y_{n-1} + \beta_3 y_{n-2}$

b. $\hat{y}_{n+1} = \beta_0 + y_n + \beta_1 (y_n - y_{n-1})$

c. $\hat{y}_{n+1} = \beta_0 + y_{n-3} + \beta_1 (y_n - y_{n-4})$

12.26.

For $l = 2$:

a. $\hat{y}_{n+2} = \beta_0 + \beta_1 \hat{y}_{n+1} + \beta_2 y_n + \beta_3 y_{n-1}$

b. $\hat{y}_{n+2} = \beta_0 + \hat{y}_{n+1} + \beta_1(\hat{y}_{n+1} - y_n)$

c. $\hat{y}_{n+2} = \beta_0 + y_{n-2} + \beta_1(\hat{y}_{n+1} - y_{n-3})$

For $l = 3$:

a. $\hat{y}_{n+3} = \beta_0 + \beta_1 \hat{y}_{n+2} + \beta_2 \hat{y}_{n+1} + \beta_3 y_n$

b. $\hat{y}_{n+3} = \beta_0 + \hat{y}_{n+2} + \beta_1(\hat{y}_{n+2} - \hat{y}_{n+1})$

c. $\hat{y}_{n+3} = \beta_0 + y_{n-1} + \beta_1(\hat{y}_{n+2} - y_{n-2})$

12.27. $\hat{y}_{n+l} = \beta_0 + \beta_1 \hat{y}_{n+l-1} + \beta_2 \hat{y}_{n+l-2}$,for $l = 3, 4, \ldots$

12.28. $\hat{y}_{n+2} = \beta_0 + \beta_1 \hat{y}_{n+1} = \beta_0 + \beta_1(\beta_0 + \beta_1 y_n) = \beta_0(1 + \beta_1) + \beta_1^2 y_n$

In general, $\hat{y}_{n+l} = \beta_0(1 + \beta_1 + \ldots + \beta_1^{\,l}) + \beta_1^{\,l} y_n$.

12.29. 95% C.I. for y_{63} : $5.9 \pm 2 * 0.3 = 5.9 \pm 0.6 = (5.3, 6.5)$

Element #:	Value
$y61$	5.5
$y62$	5.8
$\hat{y}63$	5.9
$\hat{y}64$	5.81
$\hat{y}65$	5.595
$\hat{y}66$	5.3445
$\hat{y}67$	5.14075

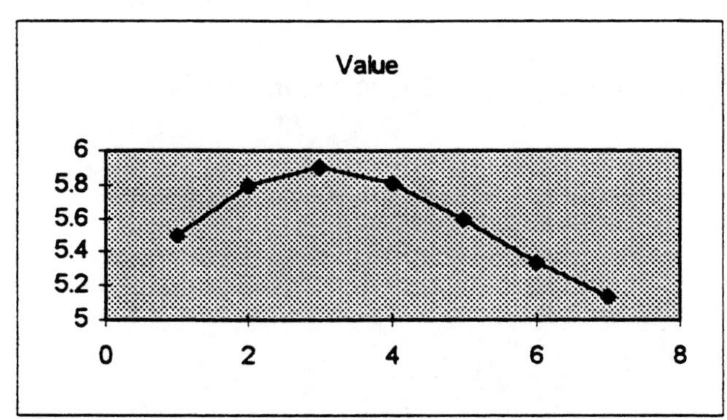

The forecasts tend to the series mean as the lead time increases.

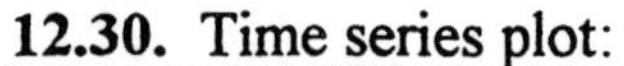

12.30. Time series plot:

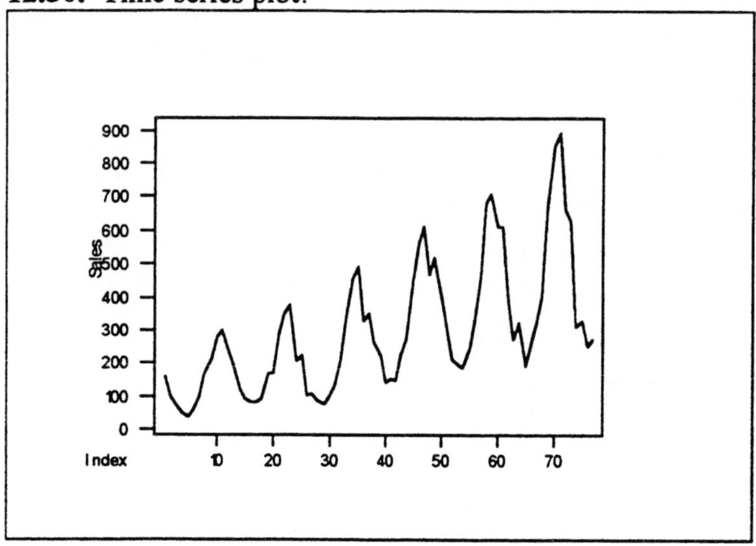

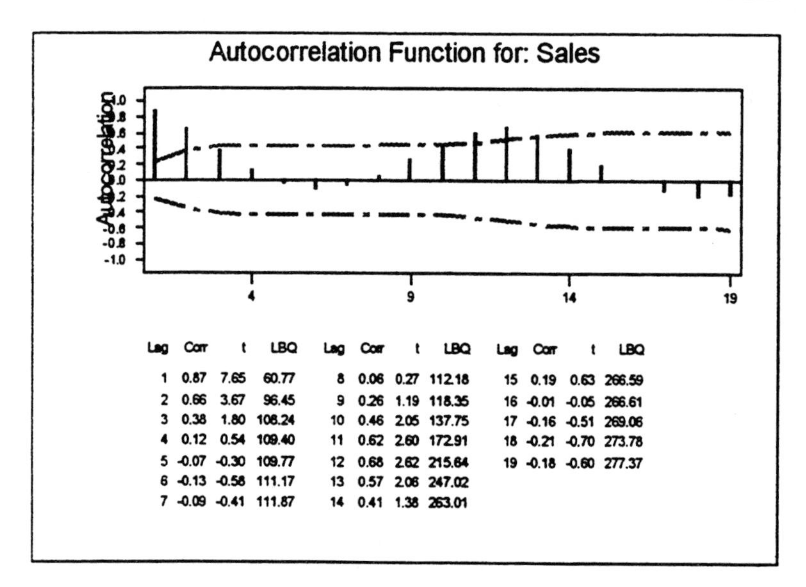

Lag	Corr	t	LBQ	Lag	Corr	t	LBQ	Lag	Corr	t	LBQ
1	0.87	7.65	60.77	8	0.06	0.27	112.18	15	0.19	0.63	266.59
2	0.66	3.67	96.45	9	0.26	1.19	118.35	16	-0.01	-0.05	266.61
3	0.38	1.80	108.24	10	0.46	2.05	137.75	17	-0.16	-0.51	269.06
4	0.12	0.54	109.40	11	0.62	2.60	172.91	18	-0.21	-0.70	273.78
5	-0.07	-0.30	109.77	12	0.68	2.62	215.64	19	-0.18	-0.60	277.37
6	-0.13	-0.58	111.17	13	0.57	2.06	247.02				
7	-0.09	-0.41	111.87	14	0.41	1.38	263.01				

This series is nonstationary since the first plot shows an upward drift in the series and the autocorrelations do not die out rapidly. This is a seasonal series as indicated by the large autocorrelations around $k = 12$, the seasonal lag. Notice that the variability of the series increases with the level. A transformation like the logarithmic transformation will stabilize the variability.

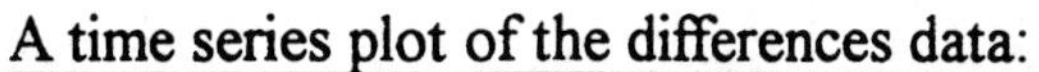
A time series plot of the differences data:

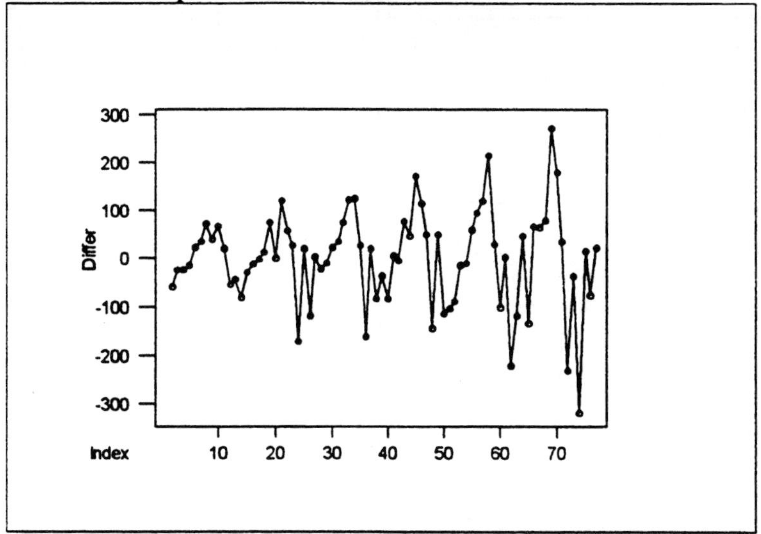

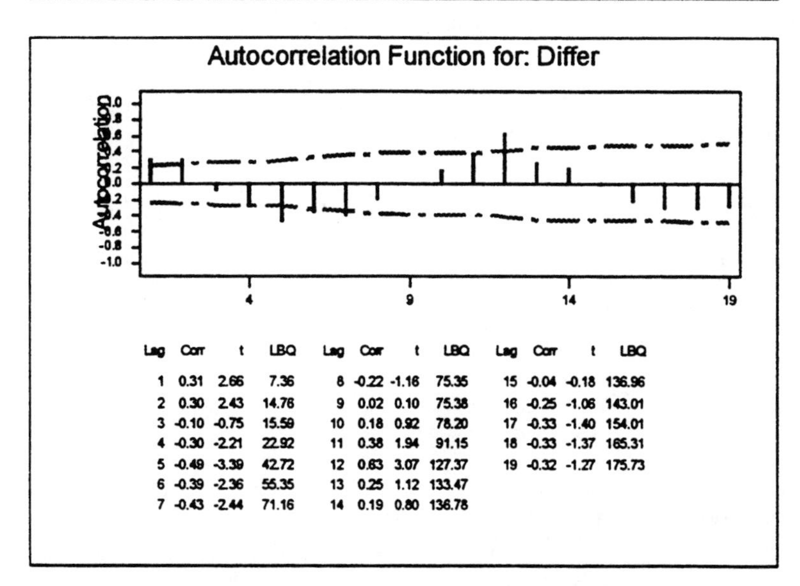

Lag	Corr	t	LBQ	Lag	Corr	t	LBQ	Lag	Corr	t	LBQ
1	0.31	2.66	7.36	8	-0.22	-1.16	75.35	15	-0.04	-0.18	136.96
2	0.30	2.43	14.76	9	0.02	0.10	75.38	16	-0.25	-1.06	143.01
3	-0.10	-0.75	15.59	10	0.18	0.92	78.20	17	-0.33	-1.40	154.01
4	-0.30	-2.21	22.92	11	0.38	1.94	91.15	18	-0.33	-1.37	165.31
5	-0.49	-3.39	42.72	12	0.63	3.07	127.37	19	-0.32	-1.27	175.73
6	-0.39	-2.36	55.35	13	0.25	1.12	133.47				
7	-0.43	-2.44	71.16	14	0.19	0.80	136.78				

The first difference series appears to be stationary with a seasonal component. It is a good practice to look at autocorrelations of the seasonal differences, $y_t - y_{t-12}$, in situations like this one.

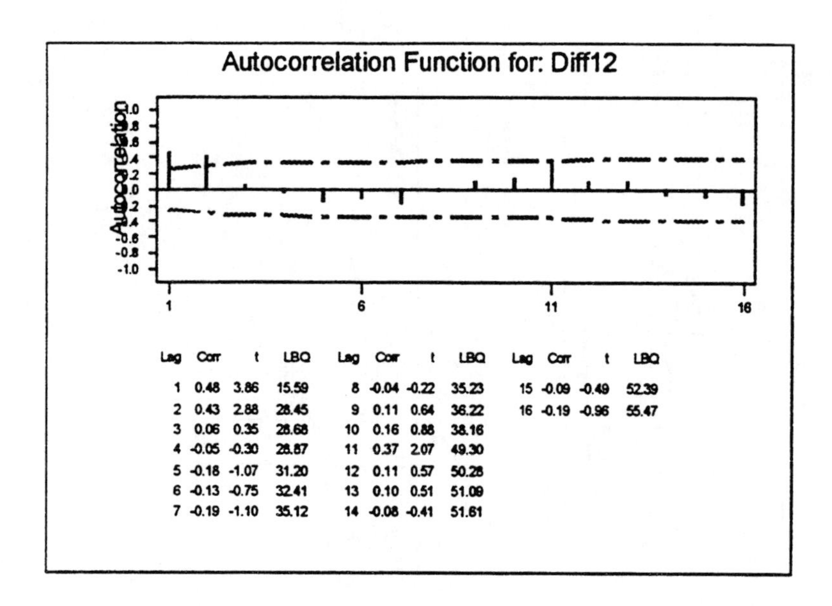

Lag	Corr	t	LBQ
1	0.48	3.86	15.59
2	0.43	2.88	28.45
3	0.06	0.35	28.68
4	-0.05	-0.30	28.87
5	-0.18	-1.07	31.20
6	-0.13	-0.75	32.41
7	-0.19	-1.10	35.12
8	-0.04	-0.22	35.23
9	0.11	0.64	36.22
10	0.16	0.88	38.16
11	0.37	2.07	49.30
12	0.11	0.57	50.28
13	0.10	0.51	51.09
14	-0.08	-0.41	51.61
15	-0.09	-0.49	52.39
16	-0.19	-0.96	55.47

This series might be represented by an autoregressive model in the seasonal differences.

12.31.

ARIMA Model (Fit AR(1) to seasonal difference for sales.)

```
Estimates at each iteration
Iteration          SSE      Parameters
    0           219622     0.100    56.818
    1           197706     0.250    47.374
    2           186568     0.400    37.915
    3           185040     0.476    33.104
    4           185036     0.480    32.817
    5           185036     0.480    32.799
Relative change in each estimate less than  0.0010

Final Estimates of Parameters
Type        Estimate      St. Dev.   t-ratio
AR   1        0.4800       0.1106       4.34
Constant     32.799         6.721       4.88

Differencing: 0 regular, 1 seasonal of order 12
No. of obs.:  Original series 77, after differencing 65
Residuals:    SS =  184984   (backforecasts excluded)
              MS =    2936   DF = 63

Modified Box-Pierce (Ljung-Box) chisquare statistic
Lag                 12           24             36             48
Chisquare    33.1(DF=11)   42.4(DF=23)   49.5(DF=35)   71.2(DF=47)
```

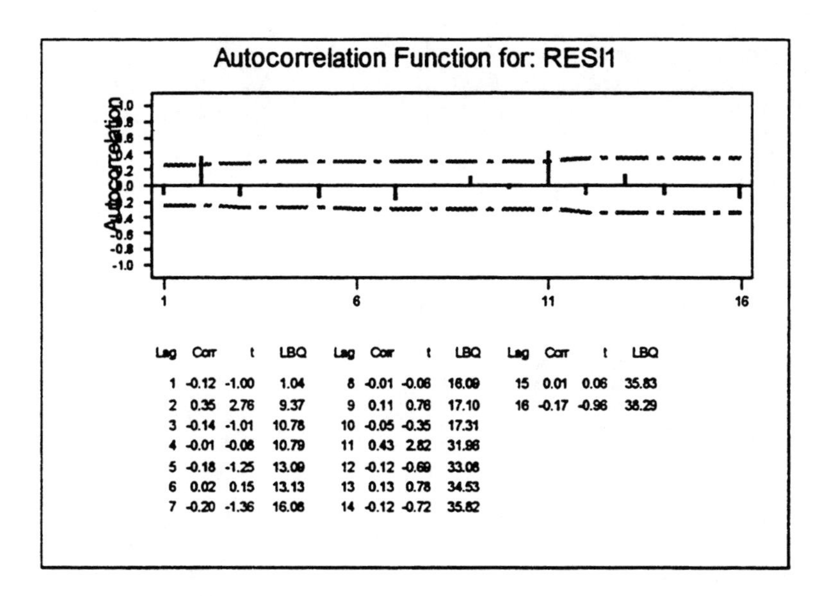

Lag	Corr	t	LBQ	Lag	Corr	t	LBQ	Lag	Corr	t	LBQ
1	-0.12	-1.00	1.04	8	-0.01	-0.06	16.09	15	0.01	0.06	35.83
2	0.35	2.76	9.37	9	0.11	0.78	17.10	16	-0.17	-0.96	38.29
3	-0.14	-1.01	10.78	10	-0.05	-0.35	17.31				
4	-0.01	-0.06	10.79	11	0.43	2.82	31.86				
5	-0.18	-1.25	13.09	12	-0.12	-0.69	33.06				
6	0.02	0.15	13.13	13	0.13	0.78	34.53				
7	-0.20	-1.36	16.06	14	-0.12	-0.72	35.82				

Some significant residual autocorrelation is remaining. This implies the model can be improved; however, the AR(1) model fit to the seasonally differenced data does a pretty good job of explaining the variation in the monthly sales of Company X.

12.32.

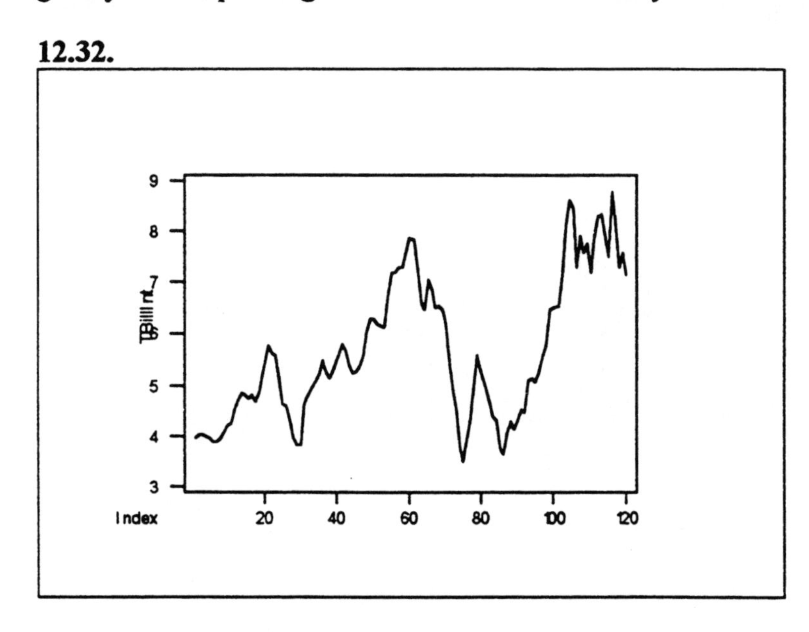

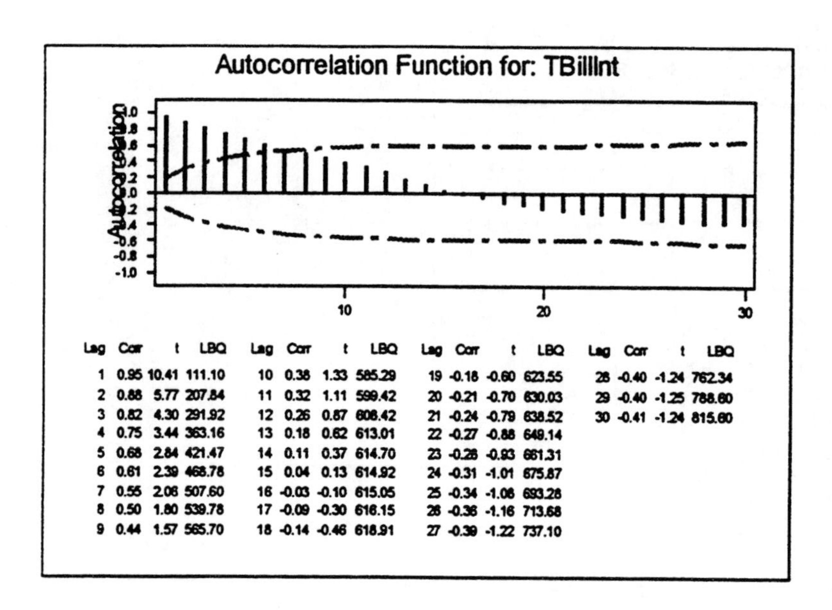

Lag	Corr	t	LBQ	Lag	Corr	t	LBQ	Lag	Corr	t	LBQ	Lag	Corr	t	LBQ
1	0.95	10.41	111.10	10	0.38	1.33	585.29	19	-0.18	-0.60	623.55	28	-0.40	-1.24	762.34
2	0.88	5.77	207.84	11	0.32	1.11	599.42	20	-0.21	-0.70	630.03	29	-0.40	-1.25	788.60
3	0.82	4.30	291.92	12	0.26	0.87	608.42	21	-0.24	-0.79	638.52	30	-0.41	-1.24	815.60
4	0.75	3.44	363.16	13	0.18	0.62	613.01	22	-0.27	-0.88	649.14				
5	0.68	2.84	421.47	14	0.11	0.37	614.70	23	-0.28	-0.93	661.31				
6	0.61	2.39	468.78	15	0.04	0.13	614.92	24	-0.31	-1.01	675.87				
7	0.55	2.06	507.60	16	-0.03	-0.10	615.05	25	-0.34	-1.06	693.28				
8	0.50	1.80	539.78	17	-0.09	-0.30	616.15	26	-0.36	-1.16	713.68				
9	0.44	1.57	565.70	18	-0.14	-0.46	618.91	27	-0.39	-1.22	737.10				

These data may be nonstationary in the mean since the autocorrelations do not die out rapidly. These data are non-seasonal.

For the differenced series:

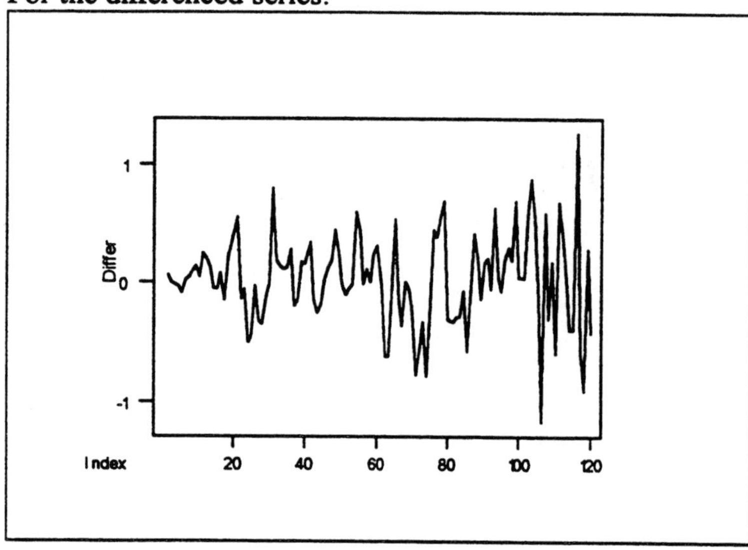

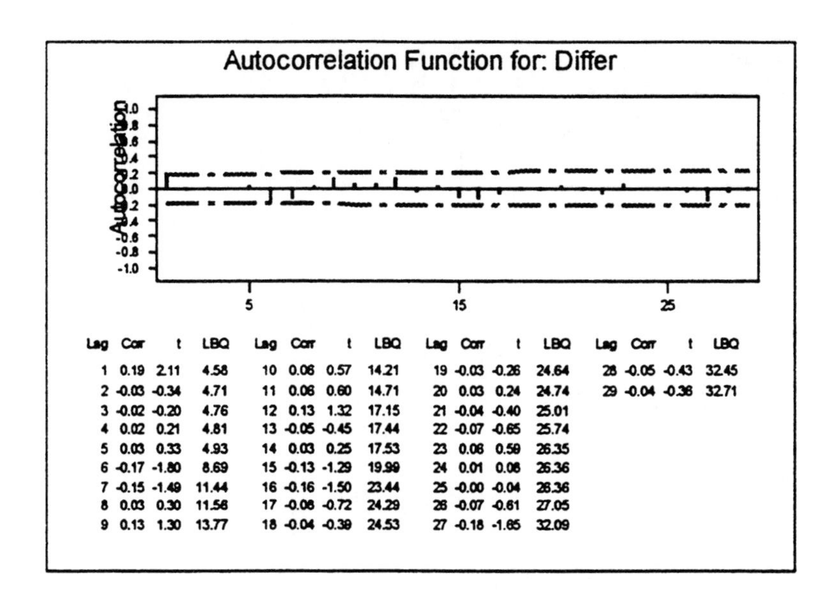

Lag	Corr	t	LBQ	Lag	Corr	t	LBQ	Lag	Corr	t	LBQ	Lag	Corr	t	LBQ
1	0.19	2.11	4.58	10	0.06	0.57	14.21	19	-0.03	-0.26	24.64	28	-0.05	-0.43	32.45
2	-0.03	-0.34	4.71	11	0.06	0.60	14.71	20	0.03	0.24	24.74	29	-0.04	-0.36	32.71
3	-0.02	-0.20	4.76	12	0.13	1.32	17.15	21	-0.04	-0.40	25.01				
4	0.02	0.21	4.81	13	-0.05	-0.45	17.44	22	-0.07	-0.65	25.74				
5	0.03	0.33	4.93	14	0.03	0.25	17.53	23	0.06	0.58	26.35				
6	-0.17	-1.80	8.69	15	-0.13	-1.29	19.99	24	0.01	0.06	26.36				
7	-0.15	-1.49	11.44	16	-0.16	-1.50	23.44	25	-0.00	-0.04	26.36				
8	0.03	0.30	11.56	17	-0.06	-0.72	24.29	26	-0.07	-0.61	27.05				
9	0.13	1.30	13.77	18	-0.04	-0.39	24.53	27	-0.18	-1.65	32.09				

The difference series appears to be stationary in the mean and is not seasonal.

12.33.

ARIMA Model (Fit AR(1) to differences in TBillInt.)

```
Estimates at each iteration
Iteration          SSE       Parameters
    0          18.1962      0.100     0.114
    1          17.0797      0.191     0.034
    2          17.0613      0.196     0.022
    3          17.0612      0.196     0.021
    4          17.0612      0.196     0.021
    5          17.0612      0.196     0.021
Relative change in each estimate less than  0.0010

Final Estimates of Parameters
Type          Estimate      St. Dev.   t-ratio
AR    1        0.1961        0.0912       2.15
Constant     0.02107       0.03501       0.60

Differencing: 1 regular difference
No. of obs.:  Original series 120, after differencing 119
Residuals:    SS = 17.0612  (backforecasts excluded)
              MS =   0.1458  DF = 117

Modified Box-Pierce (Ljung-Box) chisquare statistic
Lag                      12             24              36              48
Chisquare     11.5(DF=11)    19.7(DF=23)    35.2(DF=35)    57.9(DF=47)
```

The fitted model is $\hat{w}_t = 0.021 + 0.196w_{t-1}$ where $w_t = y_t - y_{t-1}$. Since $\beta_0 = 0.021$ is not significantly different from 0, drop this term and then we have, in terms of the original observations, $\hat{y}_t - y_{t-1} = 0.196(y_{t-1} - y_{t-2})$ or $\hat{y}_t = 1.196y_{t-1} - 0.196y_{t-2}$.

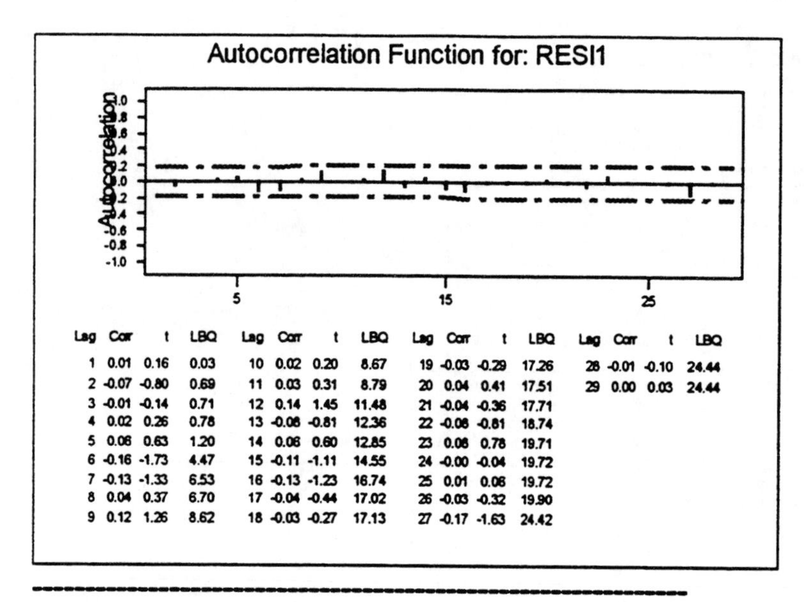

Lag	Corr	t	LBQ	Lag	Corr	t	LBQ	Lag	Corr	t	LBQ	Lag	Corr	t	LBQ
1	0.01	0.16	0.03	10	0.02	0.20	8.67	19	-0.03	-0.29	17.26	28	-0.01	-0.10	24.44
2	-0.07	-0.80	0.69	11	0.03	0.31	8.79	20	0.04	0.41	17.51	29	0.00	0.03	24.44
3	-0.01	-0.14	0.71	12	0.14	1.45	11.48	21	-0.04	-0.36	17.71				
4	0.02	0.26	0.78	13	-0.08	-0.81	12.36	22	-0.08	-0.81	18.74				
5	0.06	0.63	1.20	14	0.06	0.60	12.85	23	0.08	0.78	19.71				
6	-0.16	-1.73	4.47	15	-0.11	-1.11	14.55	24	-0.00	-0.04	19.72				
7	-0.13	-1.33	6.53	16	-0.13	-1.23	16.74	25	0.01	0.08	19.72				
8	0.04	0.37	6.70	17	-0.04	-0.44	17.02	26	-0.03	-0.32	19.90				
9	0.12	1.26	8.62	18	-0.03	-0.27	17.13	27	-0.17	-1.63	24.42				

ARIMA Model (Fit AR(2) model to **original series** TBillInt.)

```
Estimates at each iteration
Iteration        SSE       Parameters
    0          156.215     0.100      0.100      4.601
    1          119.559     0.250      0.058      3.980
    2           89.607     0.400      0.011      3.388
    3           65.261     0.550     -0.038      2.809
    4           46.057     0.700     -0.089      2.238
    5           31.733     0.850     -0.140      1.670
    6           22.128     1.000     -0.191      1.097
    7           17.274     1.150     -0.237      0.501
    8           16.785     1.193     -0.241      0.276
    9           16.768     1.197     -0.239      0.238
   10           16.767     1.199     -0.239      0.227
   11           16.767     1.199     -0.239      0.225
Relative change in each estimate less than  0.0010

Final Estimates of Parameters
Type         Estimate      St. Dev.  t-ratio
AR   1        1.1990       0.0907     13.22
AR   2       -0.2391       0.0910     -2.63
Constant     0.22546      0.03449      6.54
Mean          5.6150       0.8590

No. of obs.:  120
Residuals:    SS =  16.6062  (backforecasts excluded)
              MS =   0.1419  DF = 117

Modified Box-Pierce (Ljung-Box) chisquare statistic
Lag                  12           24            36             48
Chisquare    11.1(DF=10)   18.7(DF=22)   35.0(DF=34)    54.2(DF=46)
```

The fitted model is $\hat{y}_t = 0.225 + 1.199 y_{t-1} - 0.239 y_{t-2}$ or

$\hat{y}_t - 5.615 = 1.199(y_{t-1} - 5.615) - 0.239(y_{t-2} - 5.615)$ where $5.615 = \bar{y}$ is the sample mean. This stationary model is nearly the same as the non-stationary model involving the series of differences. Notice there is no mean in the model involving the differences since a non-stationary series has no mean. In this case, it is difficult to select one model over the other on the basis of the fit. The selection may have to be made on the basis of the nature of the forecasts.

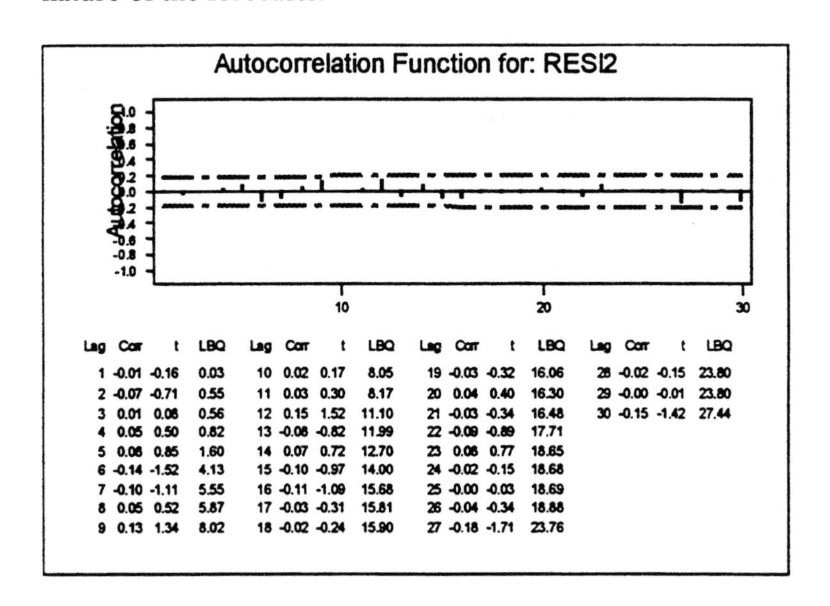

Lag	Corr	t	LBQ	Lag	Corr	t	LBQ	Lag	Corr	t	LBQ	Lag	Corr	t	LBQ
1	-0.01	-0.16	0.03	10	0.02	0.17	8.05	19	-0.03	-0.32	16.06	28	-0.02	-0.15	23.80
2	-0.07	-0.71	0.55	11	0.03	0.30	8.17	20	0.04	0.40	16.30	29	-0.00	-0.01	23.80
3	0.01	0.08	0.56	12	0.15	1.52	11.10	21	-0.03	-0.34	16.48	30	-0.15	-1.42	27.44
4	0.05	0.50	0.82	13	-0.08	-0.82	11.99	22	-0.09	-0.89	17.71				
5	0.08	0.85	1.60	14	0.07	0.72	12.70	23	0.08	0.77	18.85				
6	-0.14	-1.52	4.13	15	-0.10	-0.97	14.00	24	-0.02	-0.15	18.68				
7	-0.10	-1.11	5.55	16	-0.11	-1.09	15.68	25	-0.00	-0.03	18.69				
8	0.05	0.52	5.87	17	-0.03	-0.31	15.81	26	-0.04	-0.34	18.88				
9	0.13	1.34	8.02	18	-0.02	-0.24	15.90	27	-0.18	-1.71	23.76				

12.34.

ARIMA Model

```
Estimates at each iteration
Iteration          SSE      Parameters
    0          1.12105      0.100      0.133
    1          0.35505      0.122      0.048
    2          0.35088      0.121      0.042
    3          0.35087      0.121      0.042
    4          0.35087      0.121      0.042
Relative change in each estimate less than  0.0010

Final Estimates of Parameters
Type          Estimate      St. Dev.   t-ratio
AR   1          0.1210        0.1036      1.17
Constant    0.041675      0.006302      6.61

Differencing: 0 regular, 1 seasonal of order 12
No. of obs.:  Original series 107, after differencing 95
Residuals:    SS = 0.350871  (backforecasts excluded)
              MS = 0.003773  DF = 93

Modified Box-Pierce (Ljung-Box) chisquare statistic
Lag                    12              24              36              48
Chisquare     21.3(DF=11)     38.5(DF=23)     62.9(DF=35)     82.2(DF=47)
```

The AR(1) parameter is not significant. That is, we cannot reject H_o: β_1=0 at any reasonable significance level. However, we shall retain this parameter for forecasting.

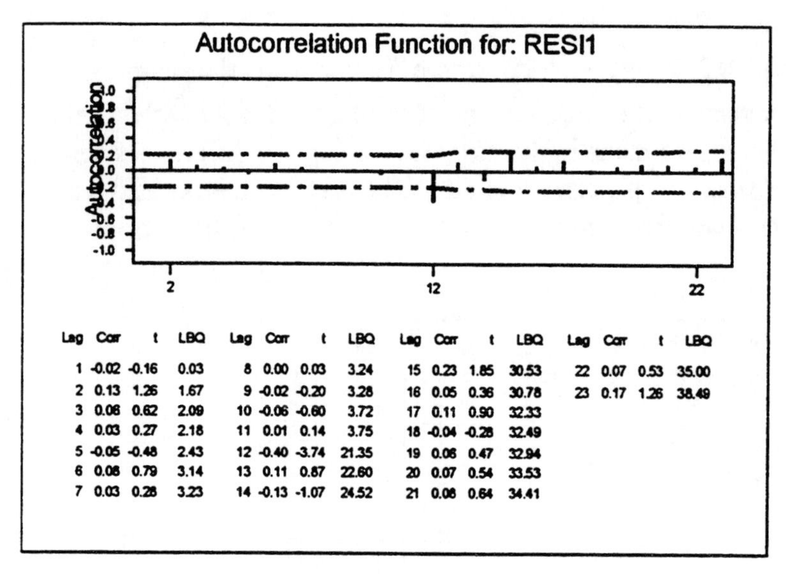

Lag	Corr	t	LBQ	Lag	Corr	t	LBQ	Lag	Corr	t	LBQ	Lag	Corr	t	LBQ
1	-0.02	-0.16	0.03	8	0.00	0.03	3.24	15	0.23	1.85	30.53	22	0.07	0.53	35.00
2	0.13	1.26	1.67	9	-0.02	-0.20	3.28	16	0.05	0.36	30.78	23	0.17	1.26	38.49
3	0.08	0.62	2.09	10	-0.06	-0.60	3.72	17	0.11	0.90	32.33				
4	0.03	0.27	2.18	11	0.01	0.14	3.75	18	-0.04	-0.28	32.49				
5	-0.05	-0.48	2.43	12	-0.40	-3.74	21.35	19	0.08	0.47	32.94				
6	0.08	0.79	3.14	13	0.11	0.87	22.60	20	0.07	0.54	33.53				
7	0.03	0.28	3.23	14	-0.13	-1.07	24.52	21	0.08	0.64	34.41				

lnForecast	Forecast
6.38320	580.403
6.40099	590.788
6.38258	580.044
6.34637	559.480
6.23362	499.996
6.31461	542.042
6.35003	561.523
6.33713	554.346
6.43933	613.806
6.42073	602.530
6.22743	496.918
6.43260	609.706

We prefer the model-based forecasts. The patterns of forecasts produced by the two methods are very similar.

12.35.

a.

Yt	EstYt	Resid
31	29.8	1.20000
34	35.1	-1.10000
40	38.2	1.80000
42	40.1	1.90000
39	40.5	-1.50000
34	31.2	2.80000
29	27.5	1.50000
20	22.4	-2.40000
17	19.5	-2.50000
14	15.1	-1.10000
15	16.9	-1.90000
17	18.8	-1.80000
23	21.5	1.50000
28	26.2	1.80000
33	31.3	1.70000
35	36.9	-1.90000

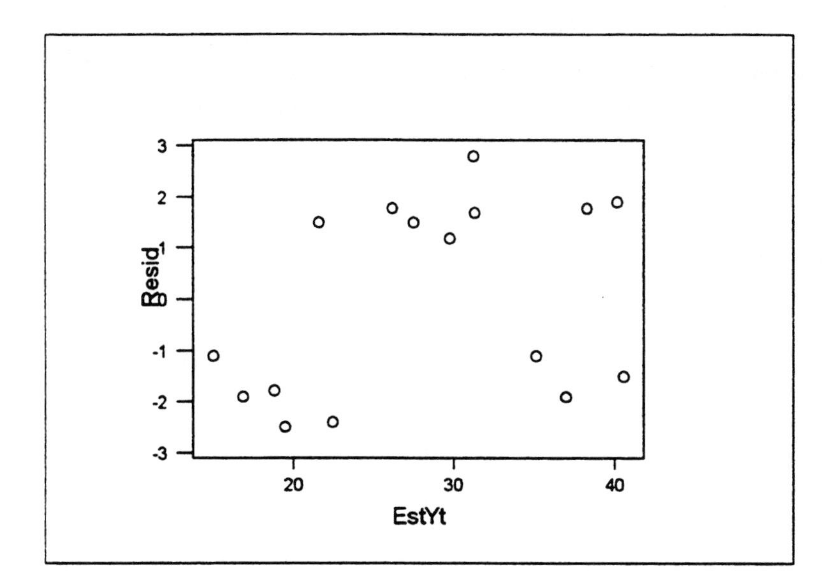

The residuals are positive for middle values of $\hat{y}_t$ and negative for small and large values of $\hat{y}_t$. This information suggests the fitted model may be improved. The residuals are not randomly distributed in a band about zero as the error term assumptions would suggest.

b.
MAPE = 7.171
MAD = 1.775
MSD = 3.369

It is difficult to select a best measure of accuracy without the context of the investigation. The accuracy measures express the forecast errors in different ways and cannot be directly compared.

12.36.
Regression Analysis

```
The regression equation is
RentY = 223 + 0.809 SqFt - 132 No.Bed

Predictor        Coef        Stdev     t-ratio         p
Constant        222.7        100.5        2.22     0.045
SqFt           0.8089       0.1530        5.29     0.000
No.Bed        -132.43        52.92       -2.50     0.026

s = 59.54        R-sq = 72.0%     R-sq(adj) = 67.7%
```

```
Analysis of Variance

SOURCE          DF          SS          MS          F         p
Regression       2      118338       59169      16.69     0.000
Error           13       46078        3544
Total           15      164416

SOURCE          DF      SEQ SS
SqFt             1       96143
No.Bed           1       22195
```

Autocorrelation Function

```
ACF of RESI1

           -1.0 -0.8 -0.6 -0.4 -0.2  0.0  0.2  0.4  0.6  0.8  1.0
            +----+----+----+----+----+----+----+----+----+----+
  1  -0.001                          X
  2  -0.094                        XXX
  3   0.087                          XXX
  4  -0.305                    XXXXXXXXX
```

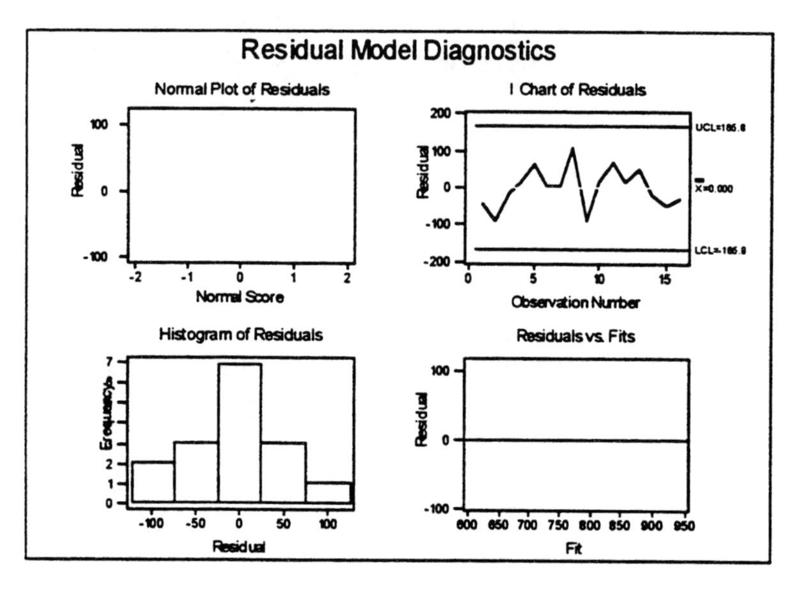

This model appears to provide an adequate explanation of these data. The regression is significant, the individual predictor variables are significant, R^2 is relatively large, and the residual plots (shown here) indicate no serious violations of the regression assumptions.

12.37.

RentY = 222.7 + 0.809 * 1050 - 132.43 * 2 = 807.2

95% intervals:

```
   Fit  Stdev.Fit        95.0% C.I.             95.0% P.I.
 807.2       17.2   (   770.0,    844.4)   (   673.3,    941.2)
```

These intervals are smaller than the ones in Section 12.6. The predictor variable No.Bath contains additional information and the effect is to narrow the intervals.

12.38.
Regression Analysis:

The regression equation is
LifeEx = 50.2 + 0.194 %Urban +0.000377 GDP/Cap - 0.00068 PSqMi + 0.812 LnGDP

Predictor	Coef	Stdev	t-ratio	p	VIF
Constant	50.247	2.313	21.73	0.000	
%Urban	0.19441	0.04530	4.29	0.000	2.7
GDP/Cap	0.0003774	0.0001560	2.42	0.023	2.6
PSqMi	-0.000678	0.001708	-0.40	0.695	1.0
LnGDP	0.8119	0.3728	2.18	0.039	1.3

s = 4.172   R-sq = 84.6%   R-sq(adj) = 82.3%

Analysis of Variance

SOURCE	DF	SS	MS	F	p
Regression	4	2489.51	622.38	35.76	0.000
Error	26	452.49	17.40		
Total	30	2942.00			

SOURCE	DF	SEQ SS
%Urban	1	2273.48
GDP/Cap	1	133.18
PSqMi	1	0.30
LnGDP	1	82.55

Unusual Observations

Obs.	%Urban	LifeEx	Fit	Stdev.Fit	Residual	St.Resid
1	18.0	44.900	54.657	1.741	-9.757	-2.57R
4	17.0	55.100	56.351	3.246	-1.251	-0.48 X
14	63.0	77.700	69.393	0.918	8.307	2.04R
22	10.0	52.500	49.175	3.134	3.325	1.21 X

R denotes an obs. with a large st. resid.
X denotes an obs. whose X value gives it large influence.

Since the P-value for the F-ratio is 0, the regression is significant.
The coefficients have the ‘right’ signs.

Multicollinearity appears to be a problem. The VIF values for %Urban and GDP/Cap are relatively large. Given the t-value, -0.40, and the associated P-values of 0.695, the variable PSqMi can be dropped from the regression function if the other predictor variables remain in the model.

Best Subsets Regression
Response is LifeEx

Vars	R-sq	Adj. R-sq	C-p	s	%Urban	GDP/Cap	PSqMi	LnGDP
1	77.3	76.5	11.4	4.8013	X			
1	67.1	65.9	28.7	5.7800		X		
2	81.8	80.5	5.8	4.3726	X	X		
2	81.1	79.8	6.9	4.4552	X			X
3	84.5	82.8	3.2	4.1061	X	X		X
3	81.8	79.8	7.7	4.4516	X	X	X	
4	84.6	82.3	5.0	4.1717	X	X	X	X

The "best" model, based on R^2, Adj. R^2, and Cp, appears to be the 3 variable model with predictor variables %urban, GDP/Cap and LnGDP.

Eliminating the two influential observations does not change the regression results appreciably. These data points (Bangaladesh and Nepal) should remain in the data set.

12.39.

a.

The all possible regression results are given in 12.38.

Stepwise Regression:

```
 F-to-Enter:        4.00     F-to-Remove:        4.00

 Response is  LifeEx  on  4 predictors, with N =   31

    Step          1         2         3
Constant      51.16     52.45     49.93

%Urban        0.312     0.218     0.197
T-Ratio        9.93      4.76      4.48

GDP/Cap               0.00043   0.00037
T-Ratio                  2.64      2.44

LnGDP                              0.79
T-Ratio                            2.18

S              4.80      4.37      4.11
R-Sq          77.28     81.80     84.53
```

Regression Analysis

The regression equation is
LifeEx = 49.9 + 0.197 %Urban +0.000375 GDP/Cap + 0.794 LnGDP

Predictor	Coef	Stdev	t-ratio	p	VIF
Constant	49.931	2.137	23.37	0.000	
%Urban	0.19723	0.04403	4.48	0.000	2.7
GDP/Cap	0.0003746	0.0001534	2.44	0.021	2.6
LnGDP	0.7938	0.3642	2.18	0.038	1.3

s = 4.106 R-sq = 84.5% R-sq(adj) = 82.8%

Analysis of Variance

SOURCE	DF	SS	MS	F	p
Regression	3	2486.77	828.92	49.16	0.000
Error	27	455.23	16.86		
Total	30	2942.00			

SOURCE	DF	SEQ SS
%Urban	1	2273.48
GDP/Cap	1	133.18
LnGDP	1	80.12

Unusual Observations

Obs.	%Urban	LifeEx	Fit	Stdev.Fit	Residual	St.Resid
1	18.0	44.900	54.428	1.617	-9.528	-2.52R
14	63.0	77.700	69.290	0.866	8.410	2.10R
22	10.0	52.500	49.211	3.083	3.289	1.21 X

R denotes an obs. with a large st. resid.
X denotes an obs. whose X value gives it large influence.

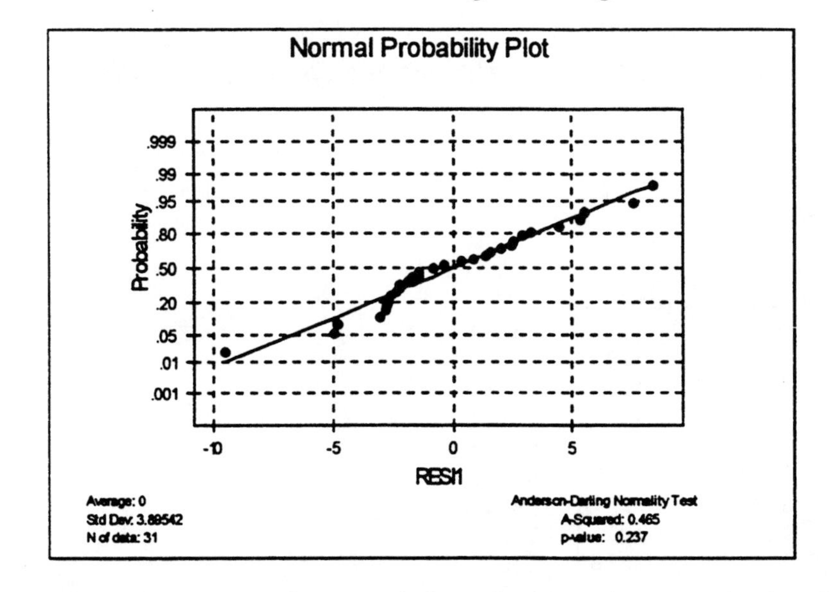

These data are nearly normal since the normal scores plot is a fairly straight line.

b. F = 49.16, and its associated *P*-value =0, show that the regression is significant. The estimated coefficients are all positive as they should be. There may be a problem with multicollinearity since the VIF values for %Urban and GDP/Cap are large. GDP/Cap and LnGDP could be dropped, although the R^2 value would then decrease. There is an influential observation for Nepal. It should, however, remain in the data set.

12.40.
ARIMA Model for Can/US exchange rate:

```
Estimates at each iteration
Iteration        SSE      Parameters
    0        4.11112     0.100     0.090
    1        0.03644     0.074     0.002
    2        0.03378    -0.051    -0.000
    3        0.03376    -0.057    -0.000
    4        0.03376    -0.058    -0.000
    5        0.03376    -0.058    -0.000
Relative change in each estimate less than  0.0010

Final Estimates of Parameters
Type         Estimate      St. Dev.  t-ratio
AR   1      -0.0579       0.0447     -1.30
Constant-0.0001214     0.0003660     -0.33

Differencing: 1 regular difference
No. of obs.:  Original series 504, after differencing 503
Residuals:    SS = 0.0337567  (backforecasts excluded)
              MS = 0.0000674  DF = 501

Modified Box-Pierce (Ljung-Box) chisquare statistic
Lag                   12            24            36            48
Chisquare    10.6(DF=11)   16.8(DF=23)   27.2(DF=35)   43.4(DF=47)
```

The random walk method is reasonable since the AR(1) coefficient is not significantly different from zero and the residual plots (and autocorrelations) are consistent with the model assumptions.

12.41.
ARIMA Model (original series):

```
Estimates at each iteration
Iteration        SSE      Parameters
    0        3602923     0.100     0.100   853.680
    1        2908299     0.250     0.039   758.113
    2        2316554     0.400    -0.023   664.214
    3        1820800     0.550    -0.086   570.943
    4        1418635     0.700    -0.149   477.875
    5        1108793     0.850    -0.212   384.579
    6         890850     1.000    -0.274   290.369
    7         765743     1.150    -0.334   193.667
    8         736009     1.250    -0.367   122.715
    9         735510     1.250    -0.361   113.913
   10         735486     1.249    -0.359   112.155
   11         735485     1.249    -0.358   111.777
   12         735485     1.249    -0.358   111.696
```

```
Relative change in each estimate less than  0.0010

Final Estimates of Parameters
Type         Estimate       St. Dev.  t-ratio
AR   1         1.2489        0.1300      9.61
AR   2        -0.3578        0.1300     -2.75
Constant      111.70          16.13      6.92
Mean          1025.3          148.1

No. of obs.:  55
Residuals:    SS =  729563  (backforecasts excluded)
              MS =   14030  DF = 52

Modified Box-Pierce (Ljung-Box) chisquare statistic
Lag                 12           24            36             48
Chisquare     6.5(DF=10)   14.3(DF=22)   22.1(DF=34)   36.1(DF=46)
```

The fitted model is $\hat{y} = 111.70 + 1.249 y_{t-1} - 0.358 y_{t-2}$.

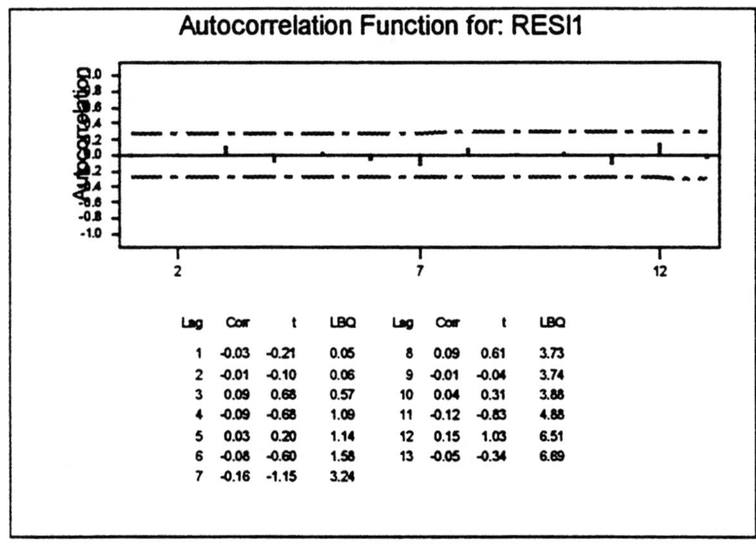

Lag	Corr	t	LBQ	Lag	Corr	t	LBQ
1	-0.03	-0.21	0.05	8	0.09	0.61	3.73
2	-0.01	-0.10	0.06	9	-0.01	-0.04	3.74
3	0.09	0.68	0.57	10	0.04	0.31	3.88
4	-0.09	-0.68	1.09	11	-0.12	-0.83	4.88
5	0.03	0.20	1.14	12	0.15	1.03	6.51
6	-0.08	-0.60	1.58	13	-0.05	-0.34	6.89
7	-0.16	-1.15	3.24				

ARIMA Model (First differences)

```
Estimates at each iteration
Iteration          SSE      Parameters
    0           824939      0.100      0.623
    1           794138      0.250      0.186
    2           792423      0.292     -0.184
    3           792415      0.295     -0.292
    4           792415      0.295     -0.302
    5           792415      0.295     -0.303
    6           792415      0.295     -0.303
Relative change in each estimate less than  0.0010

Final Estimates of Parameters
Type         Estimate       St. Dev.  t-ratio
AR   1         0.2947        0.1328      2.22
Constant        -0.30         16.80     -0.02
```

```
Differencing: 1 regular difference
No. of obs.:  Original series 55, after differencing 54
Residuals:    SS =  792020  (backforecasts excluded)
              MS =   15231  DF = 52

Modified Box-Pierce (Ljung-Box) chisquare statistic
Lag                 12           24           36           48
Chisquare     7.2(DF=11)  13.9(DF=23)  22.2(DF=35)  34.9(DF=47)
```

The fitted model, ignoring the non-significant constant, is $\hat{w}_t = 0.295 w_{t-1}$ or $\hat{y}_t - y_{t-1} = 0.295(y_{t-1} - y_{t-2})$ or $\hat{y}_t = 1.295 y_{t-1} - 0.295 y_{t-2}$.

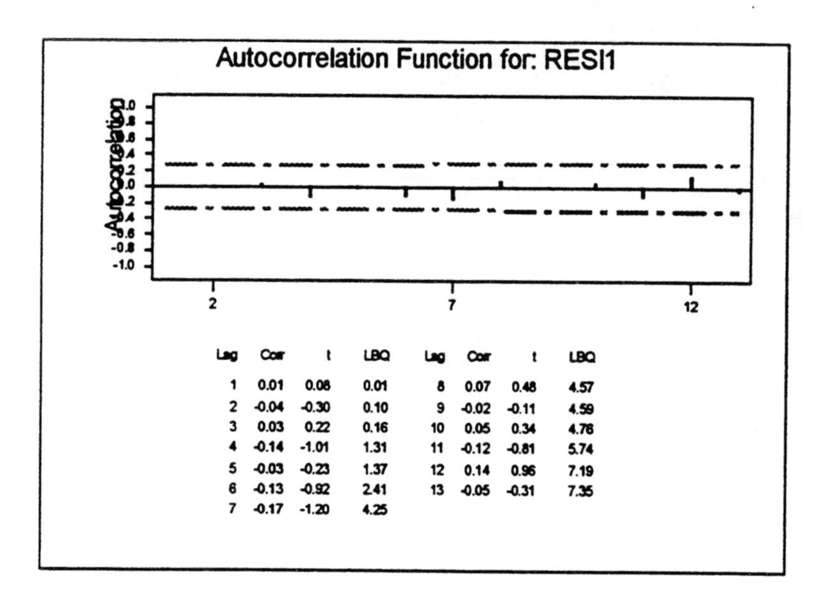

Lag	Corr	t	LBQ	Lag	Corr	t	LBQ
1	0.01	0.08	0.01	8	0.07	0.48	4.57
2	-0.04	-0.30	0.10	9	-0.02	-0.11	4.59
3	0.03	0.22	0.16	10	0.05	0.34	4.78
4	-0.14	-1.01	1.31	11	-0.12	-0.81	5.74
5	-0.03	-0.23	1.37	12	0.14	0.96	7.19
6	-0.13	-0.92	2.41	13	-0.05	-0.31	7.35
7	-0.17	-1.20	4.25				

The two fitted models are very similar. It is difficult to distinguish the models based on the fits. Since one is a stationery model and the other is a non-stationary model, you may have to select one based on the nature of the forecasts.

12.42.

Autocorrelations of the original series and the first difference series suggest the original series is non-stationary. We fit an AR(1) model to the series of differences $w_t = y_t - y_{t-1}$.

ARIMA Model

```
Estimates at each iteration
Iteration        SSE      Parameters
    0          3.25393     0.100     0.089
    1          1.81318    -0.050     0.057
    2          1.07675    -0.200     0.031
    3          0.77932    -0.350     0.008
    4          0.74464    -0.444    -0.001
    5          0.74459    -0.450    -0.001
    6          0.74459    -0.451    -0.001
Relative change in each estimate less than  0.0010
```

```
Final Estimates of Parameters
Type          Estimate       St. Dev.   t-ratio
AR   1         -0.4506         0.0542     -8.31
Constant -0.001301       0.003148     -0.41

Differencing: 1 regular difference
No. of obs.:  Original series 276, after differencing 275
Residuals:    SS = 0.744031  (backforecasts excluded)
              MS = 0.002725  DF = 273

Modified Box-Pierce (Ljung-Box) chisquare statistic
Lag                      12            24             36            48
```

The fitted model, ignoring the non-significant constant, is
$\hat{w}_t = -0.451 w_{t-1}$ or $\hat{y}_t - y_{t-1} = -0.451(y_{t-1} - y_{t-2})$ or $\hat{y}_t = 0.549 y_{t-1} + 0.451 y_{t-2}$.

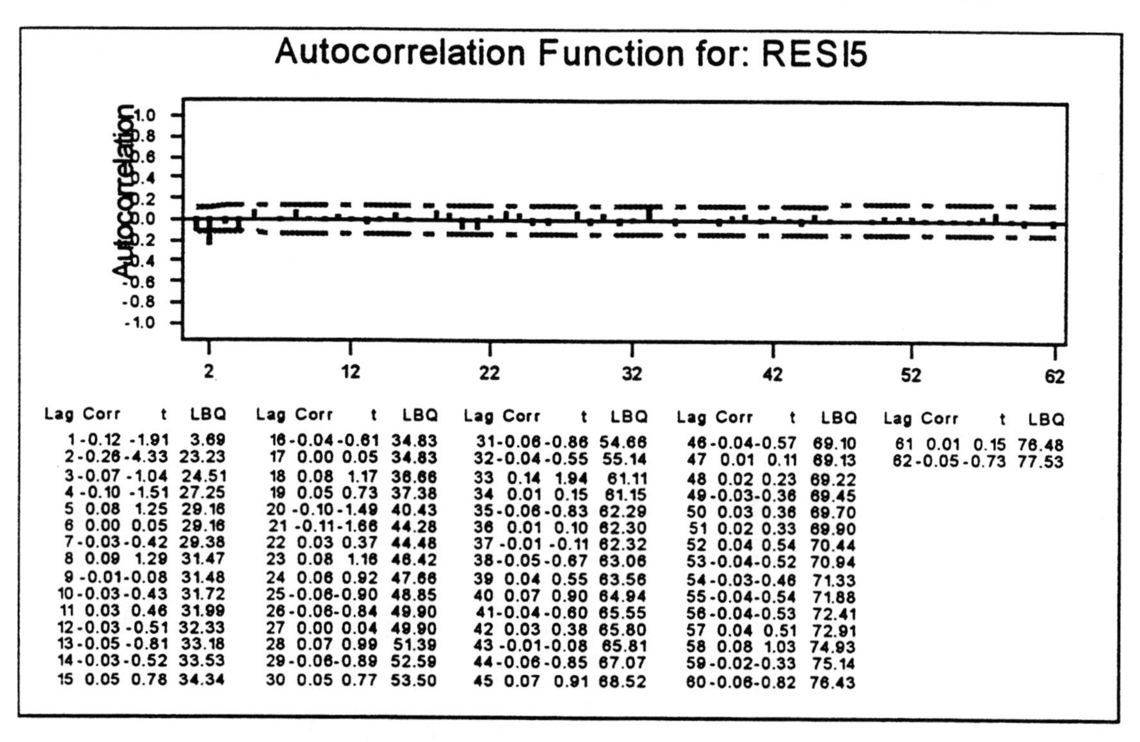

Lag	Corr	t	LBQ
1	-0.12	-1.91	3.69
2	-0.26	-4.33	23.23
3	-0.07	-1.04	24.51
4	-0.10	-1.51	27.25
5	0.08	1.25	29.16
6	0.00	0.05	29.16
7	-0.03	-0.42	29.38
8	0.09	1.29	31.47
9	-0.01	-0.08	31.48
10	-0.03	-0.43	31.72
11	0.03	0.46	31.99
12	-0.03	-0.51	32.33
13	-0.05	-0.81	33.18
14	-0.03	-0.52	33.53
15	0.05	0.78	34.34
16	-0.04	-0.61	34.83
17	0.00	0.05	34.83
18	0.08	1.17	36.66
19	0.05	0.73	37.38
20	-0.10	-1.49	40.43
21	-0.11	-1.66	44.28
22	0.03	0.37	44.48
23	0.08	1.16	46.42
24	0.06	0.92	47.66
25	-0.06	-0.90	48.85
26	-0.06	-0.84	49.90
27	0.00	0.04	49.90
28	0.07	0.99	51.39
29	-0.06	-0.89	52.59
30	0.05	0.77	53.50
31	-0.06	-0.86	54.66
32	-0.04	-0.55	55.14
33	0.14	1.94	61.11
34	0.01	0.15	61.15
35	-0.06	-0.83	62.29
36	0.01	0.10	62.30
37	-0.01	-0.11	62.32
38	-0.05	-0.67	63.06
39	0.04	0.55	63.56
40	0.07	0.90	64.94
41	-0.04	-0.60	65.55
42	0.03	0.38	65.80
43	-0.01	-0.08	65.81
44	-0.06	-0.85	67.07
45	0.07	0.91	68.52
46	-0.04	-0.57	69.10
47	0.01	0.11	69.13
48	0.02	0.23	69.22
49	-0.03	-0.36	69.45
50	0.03	0.36	69.70
51	0.02	0.33	69.90
52	0.04	0.54	70.44
53	-0.04	-0.52	70.94
54	-0.03	-0.46	71.33
55	-0.04	-0.54	71.88
56	-0.04	-0.53	72.41
57	0.04	0.51	72.91
58	0.08	1.03	74.93
59	-0.02	-0.33	75.14
60	-0.06	-0.82	76.43
61	0.01	0.15	76.48
62	-0.05	-0.73	77.53

The fitted model appears to be adequate.

Materials Selected from Applied Management Science

John A. Lawrence, Jr.
Barry A. Pasternak

Chapter 3

Problem Summary

Prob. #	Concepts Covered	Level of Difficulty	Notes
3.1	Graphing constraints, interior points, infeasible points, extreme points, binding, nonbinding, redundant constraints	4	
3.2	Simple maximization problem, binding constraints	3	
3.3	Sensitivity analyses and interpretation -- ranges of optimality and feasibility, shadow prices, reduced costs, 100% rule	5	
3.4	Alternate optimal solutions, redundant constraints	6	
3.5	Minimization problem, evaluating alternatives without using linear programming	3	
3.6	Maximization problem	3	
3.7	Sensitivity analyses, re-solving problem when changes lie outside the ranges of optimality or feasibility, elimination of a binding and nonbinding constraints	5	
3.8	Maximization problem, evaluating slack and surplus	4	
3.9	Sensitivity analyses, interpretation, elimination of a binding constraint	5	
3.10	Maximization problem, redundant constraints, checking the validity of model assumptions	4	
3.11	Sensitivity analyses, elimination of binding constraints, reduced costs	4	
3.12	Different objective functions for same set of constraints	3	
3.13	Minimization problem	4	
3.14	Transforming a 3-variable problem to a 2-variable problem, including a constant in the objective function	4	
3.15	Unbounded feasible region with optimal and unbounded solutions, checking model integrality assumption	4	
3.16	Maximization problem	4	
3.17	Sensitivity analyses, use of shadow prices, 100% rule, checking integrality assumption	5	
3.18	Maximization model, determining the right units for the constraints	5	
3.19	Maximization problem	4	

3.20	Maximization problem, range of optimality, shadow prices, sunk and included costs, 100% rule	5	
3.21	Adding constraints, adding variables, reduced costs	6	
3.22	Minimization problem, interpretation of range of optimality and shadow prices, infeasibility	5	
3.23	Maximization problem, complementary slackness, reduced costs, range of optimality	5	
3.24	Range of feasibility, evaluating the profitability of alternatives using sensitivity analyses	6	
3.25	Maximization problem, calculation of objective function coefficients, evaluating the profitability of alternatives using shadow prices	6	
CD3.1	Maximization problem, interior and infeasible points	4	
CD3.2	Sensitivity analyses, reallocation of slack time and recalculation of optimal solution and shadow prices	6	
CD3.3	Maximization problem, constant in the objective function, checking the integrality assumption	3	
CD3.4	Maximization problem	3	
CD3.5	Sensitivity analyses and interpretation of results	4	
CD3.6	Maximization problem	3	
CD3.7	Range of optimality, use of shadow prices, addition of constraints	6	
CD3.8	Range of optimality, range of feasibility, changes to the left side of a constraint, addition of a constraint	6	
CD3.9	Maximization problem, calculation of objective function coefficients	6	
Case 3.1	Maximization problem, analysis of various alternatives using sensitivity information	7	
Case 3.2	Alternate optimal solutions, infeasibility	6	The text states that in all three cases the solution turns out to be integer; in fact, two of the cases are infeasible.
Case CD3.1	Two 2-variable maximization problems and a 4-variable maximization problem	6	

Problem Solutions

3.1 See file ch3-1.lpp.

X1 = Dozens of baseballs produced daily
X2 = Dozens of softballs produced daily

a.

$$\begin{aligned} X1 \quad\quad\;\; &\leq 500 \quad \text{(Maximum baseballs)} \\ X2 &\leq 500 \quad \text{(Maximum softballs)} \\ 5X1 + 6X2 &\leq 3600 \quad \text{(Cowhide)} \\ X1 + 2X2 &\leq 960 \quad \text{(Minutes)} \\ X1, X2 &\geq 0 \end{aligned}$$

b.

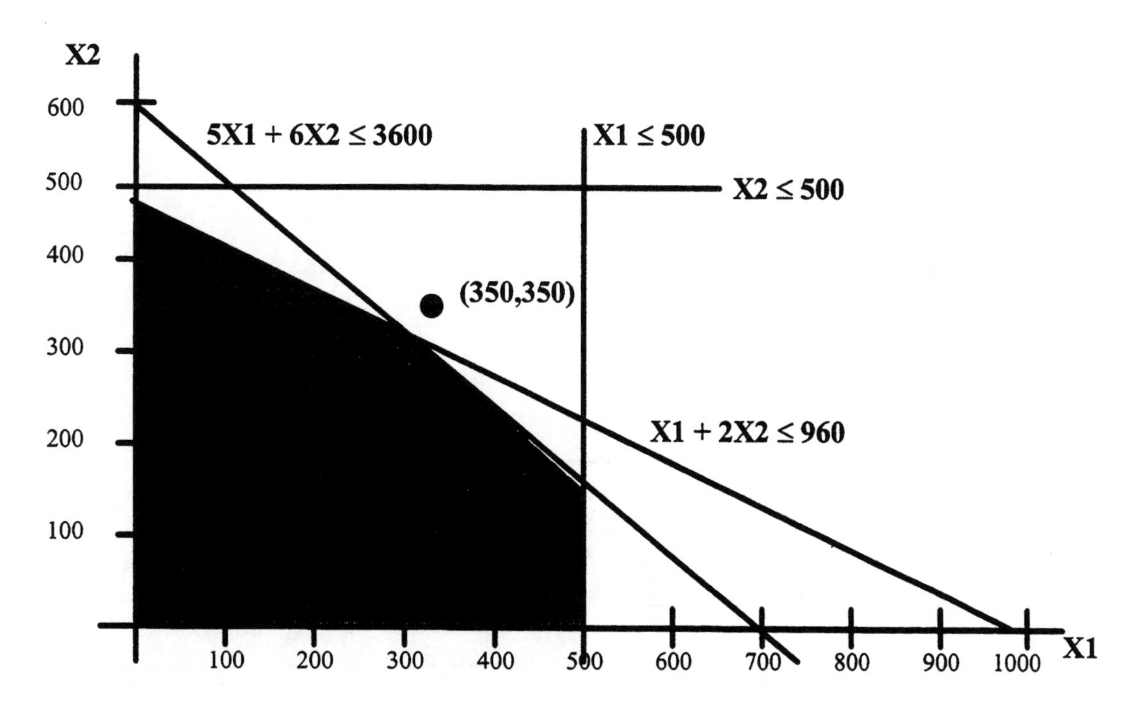

3.1c. (300,300) is an interior point; (350,350) is an infeasible point. Only extreme points and boundary points can be optimal.

d. Add the objective function: MAX 7X1 + 10X2.

	Decision Variable	Solution Value	Unit Cost or Profit c(j)	Total Contribution	Reduced Cost	Basis Status	Allowable Min. c(j)	Allowable Max. c(j)
1	DZ BASE	360	7	2,520.00	0	basic	5	8.3333
2	DZ SOFT	300	10	3,000.00	0	basic	8.4	14
	Objective	Function	(Max.) =	5,520.00				

Produce 360 dozen baseballs and 300 dozen softballs daily. Daily profit = \$5,520

e. The constraints for cowhide and minutes are binding. The other constraints are non-binding. From the graph, note that the nonbinding constraint for the maximum number of baseballs produced daily is also redundant.

3.2 See file ch3-2.lpp.

X1 = Number of GE45 televisions produced per shift

X2 = Number of GE60 televisions produced per shift

MAX 50X1 + 75X2

S.T. $2X1 + 2X2 \leq 300$ (Production hours)

$X1 + 3X2 \leq 240$ (Assembly hours)

$X1, X2 \geq 0$

	Decision Variable	Solution Value	Unit Cost or Profit c(j)	Total Contribution	Reduced Cost	Basis Status	Allowable Min. c(j)	Allowable Max. c(j)
1	GE45	105	50	5,250.00	0	basic	25	75
2	GE60	45	75	3,375.00	0	basic	50	150
	Objective	Function	(Max.) =	8,625.00				
	Constraint	Left Hand Side	Direction	Right Hand Side	Slack or Surplus	Shadow Price	Allowable Min. RHS	Allowable Max. RHS
1	PROD.HRS.	300	<=	300	0	18.75	160	480
2	ASSEM.HRS.	240	<=	240	0	12.5	150	450

a. Produce 105 GE45's and 45 GE60's

b. Profit = \$8625

c. Yes, both have 0 slack.

3.3 Refer to the output for problem 3.2 and files ch3-3de.lpp and ch3-3h.lpp.

a. Range of optimality for GE60 TV's = $50 to $150

b. Shadow price for assembly hours = $12.50 per hour

c. Range of feasibility for assembly hours: 150 to 450 hours

d.

	Decision Variable	Solution Value	Unit Cost or Profit c(j)	Total Contribution	Reduced Cost	Basis Status	Allowable Min. c(j)	Allowable Max. c(j)
1	GE45	0	50	0	-50	at bound	-M	100
2	GE60	80	300	24,000.00	0	basic	150	M
	Objective	Function	(Max.) =	24,000.00				

Produce 80 GE60's only. Profit = $24,000

e. Reduced cost for GE45's = -$50; the profit for GE45's must increase by $50 (to ($100)) before it is profitable to produce them.

f. Optimal solution would change; $90 is outside the range of optimality for GE45's.

g. The optimal solution would not change; $135 is outside the range of optimality for GE60's.

h. C1 = $60 is a $10/$25 = 40% increase from $50.
C2 = $135 is a $60/$75 = 80% increase from $75.
The total % increase is 40% = 80% = 120% so the optimal solution may change.

Re-solving gives the following:

	Decision Variable	Solution Value	Unit Cost or Profit c(j)	Total Contribution	Reduced Cost	Basis Status	Allowable Min. c(j)	Allowable Max. c(j)
1	GE45	105	90	9,450.00	0	basic	45	135
2	GE60	45	135	6,075.00	0	basic	90	270
	Objective	Function	(Max.) =	15,525.00				

Produce 105 GE45's and 45 GE60's. The 100% rule is not violated; it only states that the optimal solution *may* change.

3.4

See files ch3-4a.lpp, ch3-4b.lpp, ch3-4c.lpp, ch3-4d.lpp, ch3-4ei.lpp, ch3-4-eii.lpp, and ch3-4f.lpp.

a. Add the constraint $.5X1 + .75X2 \leq 80$. The problem is now:

$$\begin{aligned} \text{MAX} \quad & 50X1 + 75X2 \\ \text{S.T.} \quad & 2X1 + 2X2 \leq 300 \\ & X1 + 3X2 \leq 240 \\ & .5X1 + .75X2 \leq 80 \\ & X1, X2 \geq 0 \end{aligned}$$

	Decision Variable	Solution Value	Unit Cost or Profit c(j)	Total Contribution	Reduced Cost	Basis Status	Allowable Min. c(j)	Allowable Max. c(j)
1	GE45	80	50	4,000.00	0	basic	25	50
2	GE60	53.3333	75	4,000.00	0	basic	75	150
	Objective	Function	(Max.) =	8,000.00	(Note:	Alternate	Solution	Exists!!)

There are alternate optimal solutions giving an objective function value of $8000.

b. Add the constraint $50X1 + 75X2 = 8000$ to the formulation in part (a) and set the objective to: MAX X1.

	Decision Variable	Solution Value	Unit Cost or Profit c(j)	Total Contribution	Reduced Cost	Basis Status	Allowable Min. c(j)	Allowable Max. c(j)
1	GE45	130	1	130	0	basic	0	M
2	GE60	20	0	0	0	basic	-M	1.5
	Objective	Function	(Max.) =	130				

Produce 130 GE45's and 20 GE60's

c. Add the constraint $50X1 + 75X2 = 8000$ to the formulation in part (a) and set the objective to: MAX X2.

	Decision Variable	Solution Value	Unit Cost or Profit c(j)	Total Contribution	Reduced Cost	Basis Status	Allowable Min. c(j)	Allowable Max. c(j)
1	GE45	80	0	0	0	basic	-M	0.6667
2	GE60	53.3333	1	53.3333	0	basic	0	M

Produce 80 GE45's and 53.33 GE60's (.33 would be work in progress at the end of a shift.)

3.4d. Add the constraint -3X1 + X2 = 0 to the formulation in part (a).

	Decision Variable	Solution Value	Unit Cost or Profit c(j)	Total Contribution	Reduced Cost	Basis Status	Allowable Min. c(j)	Allowable Max. c(j)
1	GE45	106.6667	50	5,333.33	0	basic	-25	M
2	GE60	35.5556	75	2,666.67	0	basic	-150	M

Produce 106.67 GE45's and 35.56 GE60's (the .67 GE45's and .56 GE60's would be work in progress at the end of a shift.)

e. (i)

	Decision Variable	Solution Value	Unit Cost or Profit c(j)	Total Contribution	Reduced Cost	Basis Status	Allowable Min. c(j)	Allowable Max. c(j)
1	GE45	80	49	3,920.00	0	basic	25	50
2	GE60	53.3333	75	4,000.00	0	basic	73.5	147
	Objective	Function	(Max.) =	7,920.00				

Produce 80 GE45's and 53.33 GE60's; Profit = $7920

(ii)

	Decision Variable	Solution Value	Unit Cost or Profit c(j)	Total Contribution	Reduced Cost	Basis Status	Allowable Min. c(j)	Allowable Max. c(j)
1	GE45	130	51	6,630.00	0	basic	50	75
2	GE60	20	75	1,500.00	0	basic	51	76.5
	Objective	Function	(Max.) =	8,130.00				

Produce 130 GE45's and 20 GE60's   Profit = $8130

f. $.5X1 + .75X2 \leq 120$ is a non-binding constraint. It is, in fact, redundant, as can be seen if one generates the graph for the problem from the Solve and Analyze menu of WINQSB.

3.5 See file ch3-5.lpp.

a.

	Per ounce Cow Chow	Required Per Cow	Minimum Ounces Required Per Cow
Calcium	1	100	100
Calories	400	2000	50
Protein	20	1500	75

Thus Dan currently needs 100 ounces of Cow Chow per cow @ $0.015 per ounce = $1.50 per cow. For 100 cows, this is $150.

b.

	Per ounce Moo Town	Required Per Cow	Minimum Ounces Required Per Cow
Calcium	2	100	50
Calories	250	2000	80
Protein	20	1500	75

Thus Dan would need 80 ounces of Moo Town Buffet per cow @ $0.02 per ounce = $1.60 per cow. For 100 cows, this would be $160; so Dan would keep using Cow Chow.

c. Let X1 = number of ounces of Cow Chow in the mixture
X2 = number of ounces of Moo Town Buffet in the mixture

MIN .015X1 + .02X2

S.T.
$$X1 + 2X2 \geq 100 \text{(Calcium)}$$
$$400X1 + 250X2 \geq 2000 \quad \text{(Calories)}$$
$$20X1 + 20X2 \geq 1500 \quad \text{(Protein)}$$
$$X1, X2 \geq 0$$

	Decision Variable	Solution Value	Unit Cost or Profit c(j)	Total Contribution	Reduced Cost	Basis Status	Allowable Min. c(j)	Allowable Max. c(j)
1	COW CHOW	50	0.015	0.75	0	basic	0.01	0.02
2	MOO TOWN	25	0.02	0.5	0	basic	0.015	0.03
	Objective	Function	(Min.) =	1.25				

The mixture consists of 50 ounces of Cow Chow, 25 ounces of Moo Town Buffet. The cost is $1.25 per cow; for 100 cows; this would be $125.

d. Save $150 - $125 = $25 per day

3.6 See file ch3-6.lpp.

X1 = number of skilled workers used
X2 = number of apprentices used

$$\begin{array}{llll} \text{MAX } 480X1 + 320X2 & & & \\ \text{S.T.} \quad X1 & \geq & 25 & \text{(Minimum skilled)} \\ X1 - X2 & \geq & 15 & \text{(Skilled exceeds apprentices by at least 15)} \\ 3840X1 + 2400X2 & \leq & 200000 & \text{(Budget)} \\ X1, X2 \geq 0 & & & \end{array}$$

	Decision Variable	Solution Value	Unit Cost or Profit c(j)	Total Contribution	Reduced Cost	Basis Status	Allowable Min. c(j)	Allowable Max. c(j)
1	SKILLED	37.8205	480	18,153.85	0	basic	-320	512
2	APPRENTICE	22.8205	320	7,302.56	0	basic	300	M
	Objective	Function	(Max.) =	25,456.41				
	Constraint	Left Hand Side	Direction	Right Hand Side	Slack or Surplus	Shadow Price	Allowable Min. RHS	Allowable Max. RHS
1	MIN SKILL	37.8205	>=	25	12.8205	0	-M	37.8205
2	SKILL-APP	15	>=	15	0	-12.3077	-18.3333	52.0833
3	BUDGET	200,000.00	<=	200,000.00	0	0.1282	120,000.00	M

Use 37.8205 skilled, 22.8205 apprentices. Total production = 25,456.41 trusses.

3.7 See the output for problem 3.6 and files ch3-7e.lpp and ch3-7f.lpp.

a. The shadow price for more skilled workers = 0; for the minimum skilled workers can exceed apprentices = -12.3077; for the budget = .1282. (i) The shadow price for the first restriction is 0 because it is not a binding constraint (there is surplus); (ii) the shadow price for the second constraint is negative because increasing the right side of a binding "≥" constraint will be more restrictive and cause a reduction in the objective function value, while increasing the right side of a "≤" constraint is less restrictive and will allow the objective function value to increase.

b. Eliminating this constraint will have no effect since it is non-binding.

c. (i) Decrease in truss production = 2(12.3077) = 24.61 trusses
New number of trusses = 25456.41 - 24.61 = 25,431.80
(ii) Increase in truss production = 2(12.3077) = 24.61 trusses
New number of trusses = 25456.41 + 24.61 = 25,481.02

3.7d. (i) Decrease in truss production = (.1282)*(50,000) = 6410 to 19,046.41 trusses.

(ii) Cannot tell -- $100,000 is outside the range of feasibility for this constraint; but since the lower bound of the range of feasibility is $120,000 ($80,000 below the current value of $200,000), it will decrease by *at least* (.1282)*(80,000) = 10,256 to 15,200.41 trusses.

e. The second constraint is binding, thus the problem must be re-solved.

	Decision Variable	Solution Value	Unit Cost or Profit c(j)	Total Contribution	Reduced Cost	Basis Status	Allowable Min. c(j)	Allowable Max. c(j)
1	SKILLED	25	480	12,000.00	0	basic	-M	512
2	APPRENTICE	43.3333	320	13,866.67	0	basic	300	M
	Objective	Function	(Max.) =	25,866.67				

Use 25 skilled workers, 43.33 apprentices producing 25,866.67 trusses.

f. The new problem is:

$$
\begin{array}{llcrcr}
\text{MIN} & 3840X1 & + & 24000X2 & & \\
\text{S.T.} & X1 & & & \geq & 25 \\
 & X1 & - & X2 & \geq & 15 \\
 & 480X1 & + & 320X2 & \geq & 30000 \\
 & X1, X2 \geq 0 & & & &
\end{array}
$$

Solving gives:

	Decision Variable	Solution Value	Unit Cost or Profit c(j)	Total Contribution	Reduced Cost	Basis Status	Allowable Min. c(j)	Allowable Max. c(j)
1	SKILLED	43.5	3,840.00	167,040.00	0	basic	3,600.00	M
2	APPRENTICE	28.5	2,400.00	68,400.00	0	basic	-3,840.00	2,560.00
	Objective	Function	(Min.) =	235,440.00				

Use 43.5 skilled workers, 28.5 apprentices; the budget is $235,440, which is $35,440 over the current $200,000 budget.

3.8 See file ch3-8.lpp.

X1 = number of KCU's produced during the production cycle
X2 = number of KCP's produced during the production cycle

MAX 100X1 + 250X2

S.T.

$2X1 +$	$X2 \leq 1800$	(Floppy drives)
	$X2 \leq 700$	(Hard drives)
$X1 +$	$X2 \leq 1000$	(Tower cases)
$.4X1 +$	$.6X2 \geq 480$	(Production hours)
$X1$	≥ 300	(Contract)

$X1, X2 \geq 0$

	Decision Variable	Solution Value	Unit Cost or Profit c(j)	Total Contribution	Reduced Cost	Basis Status	Allowable Min. c(j)	Allowable Max. c(j)
1	KCU	300	100.00	30,000.00	0	basic	-M	166.6667
2	KCP	600	250.00	150,000.00	0	basic	150.00	M
	Objective	Function	(Max.) =	180,000.00				

	Constraint	Left Hand Side	Direction	Right Hand Side	Slack or Surplus	Shadow Price	Allowable Min. RHS	Allowable Max. RHS
1	FLOPPY	1,200.00	<=	1,800.00	600	0	1,200.00	M
2	HARD	600.00	<=	700.00	100	0	600.00	M
3	CASES	900.00	<=	1,000.00	100	0	900.00	M
4	PROD. TIME	480	<=	480	0	416.6667	120	540
5	TEXAS ST.	300	>=	300	0	-66.6667	150	600.0001

Produce 300 KCU's, 600 KCP's, Profit = \$180,000. Klone will use 1200 floppy drives, 600 hard drives, and 900 cases. This leaves 600 floppy drives, 100 hard drives, and 100 cases. All 480 production hours are used.

3.9 See output for problem 3.8 and file ch3-9f.lpp.

a. Range of optimality for profit of KCU's: $-\infty$ to \$166.67; for KCP's \$150 to ∞. Within this range the current optimal solution will not change.

b. If the profit for KCU's is \$150 -- this is within its range of optimality -- the optimal solution will not change; however since the profit on KCU's has increased by \$50 per unit, the optimal profit will increase by \$50(3000) = \$15,000. If the profit for KCU's is \$200 -- this is outside its range of optimality and the optimal solution and the profit will change.

3.9c. Since KCU's use two floppy drives, its unit profit will increase by 2($15) = $30; since KCP's use only one floppy drive, it unit profit will increase by $15. Since both profit coefficients change, we use the 100% rule. The total % change is (30/66.67) + (15/∞) = 45% + 0% = 45%. Since this is less than 100% the optimal solution will stay the same.

d.

Resource	Shadow Price	Range of Feasibility
Floppy Drives	$0	1200 to ∞
Hard Drives	$0	600 to ∞
Cases	$0	900 to ∞
Production Hours	$416.67	120 to 540

It would not pay to purchase extra floppy drives, hard drives or cases. Each extra production hour (within its range of feasibility) will increase the profit by $416.67.

e. Shadow price for minimum production of KCU's = -66.67; its range of feasibility is from 150 - 600 units.

f. This was a binding constraint so the problem must be re-solved.

	Decision Variable	Solution Value	Unit Cost or Profit c(j)	Total Contribution	Reduced Cost	Basis Status	Allowable Min. c(j)	Allowable Max. c(j)
1	KCU	150	100	15,000.00	0	Basic	0	166.6667
2	KCP	700	250	175,000.00	0	Basic	150	M
	Objective	Function	(Max.) =	190,000.00				

Produce 150 KCU's and 700 KCP's giving an optimal profit of $19,000; this is an increase of $10,000. The minimum value in the range of feasibility is 150 (a decrease of 150 from 300) for the Texas State contract. At this value, the increase in profit is: (-$66.67)(-150) = $10,000. Since this is the increase attained by eliminating the constraint altogether, for values lower than 150, this constraint is not binding.

3.10 See file ch3-10.lpp.

X1 = the number of weekday ads placed
X2 = the number of Sunday ads placed

a.

$$\begin{array}{llll} \text{MAX} & 30000X1 + 80000X2 & & \\ \text{S.T.} & X1 & \leq 26 & \text{(Max. daily ads)} \\ & X2 & \leq 5 & \text{(Max. Sunday ads)} \\ & X1 & \geq 8 & \text{(Min. daily ads)} \\ & X2 & \geq 2 & \text{(Min. Sunday ads)} \\ & 2000X1 + 8000X2 & \leq 40000 & \text{(Budget)} \\ & X1, X2 \geq 0 & & \end{array}$$

b. Given the budget a maximum of 20 daily ads or 5 Sunday ads are already guaranteed.

c. Many of the same people probably read the newspaper each day and much of the Sunday readership will include its weekday readership. Since they "interact", PMA may not truly get the given anticipated increase in exposure.

d.

	Decision Variable	Solution Value	Unit Cost or Profit c(j)	Total Contribution	Reduced Cost	Basis Status	Allowable Min. c(j)	Allowable Max. c(j)
1	DAILY	12	30,000.00	360,000.00	0	basic	20,000.00	M
2	SUNDAY	2	80,000.00	160,000.00	0	basic	-M	120,000.00
	Objective	Function	(Max.) =	520,000.00				
	Constraint	Left Hand Side	Direction	Right Hand Side	Slack or Surplus	Shadow Price	Allowable Min. RHS	Allowable Max. RHS
1	MAX DAILY	12	<=	26.00	14	0	12	M
2	MAX SUNDAY	2	<=	5	3	0	2	M
3	MIN DAILY	12	>=	8.00	4	0	-M	12
4	MIN SUNDAY	2	>=	2	0	-40,000.00	0	3
5	BUDGET	40,000.00	<=	40,000.00	0	15	32,000.00	68,000.00

Place 12 daily ads and 2 Sunday ads. The total exposure = 520,000.

3.11 See the output for problem 3.10 and file ch3-11c.lpp.

a. The range of optimality for daily circulation is 20,000 to ∞ and for Sunday -∞ to 120,000.

b. The shadow prices for maximum daily ads, maximum Sunday ads, and minimum daily ads are 0 since there is slack or surplus for each of these constraints. Requiring an extra Sunday ad will decrease exposure by 40,00. Each extra advertising dollar spent (up to $68,000 (or an additional $28,000)) will increase the exposure by 15.

c. If the constraint, X2 ≥ 2 were eliminated, the problem must be re-solved.

	Decision Variable	Solution Value	Unit Cost or Profit c(j)	Total Contribution	Reduced Cost	Basis Status	Allowable Min. c(j)	Allowable Max. c(j)
1	DAILY	20	30,000.00	600,000.00	0	basic	20,000.00	M
2	SUNDAY	0	80,000.00	0.00	-40,000.00	at bound	-M	120,000.00
	Objective	Function	(Max.) =	600,000.00				

Place 20 daily ads only. The reduced cost for Sunday ads is now -40,000 meaning exposure for Sunday newspapers would have to increase by 40,000 to 120,00 before it would be profitable to place Sunday ads.

3.12 See files ch3-12a.lpp, ch3-12b.lpp, and ch3-12c.lpp.

X1 = the number of tricycles displayed
X2 = the number of bicycles displayed

The constraints are:

$$\begin{aligned} X1 \quad\quad\quad &\geq 10 \quad \text{(Min. tricycles displayed)} \\ X2 &\geq 8 \quad \text{(Min. bicycles displayed)} \\ 3X1 + 5X2 &\leq 100 \quad \text{(Linear feet)} \\ X1, X2 &\geq 0 \end{aligned}$$

a. MAX .10(40)X1 + .12(80)X2, or MAX 4X1 + 9.6X2

	Decision Variable	Solution Value	Unit Cost or Profit c(j)	Total Contribution	Reduced Cost	Basis Status	Allowable Min. c(j)	Allowable Max. c(j)
1	TRICYCLES	10	4.00	40.00	0	basic	-M	5.76
2	BICYCLES	14	9.60	134.40	0.00	basic	6.6667	M
	Objective	Function	(Max.) =	174.40				

Display 10 tricycles, 14 bicycles, $174 expected profit.

3.12b. MAX .10X1 + .12 X2

	Decision Variable	Solution Value	Unit Cost or Profit c(j)	Total Contribution	Reduced Cost	Basis Status	Allowable Min. c(j)	Allowable Max. c(j)
1	TRICYCLES	20	0.1	2	0	basic	0.072	M
2	BICYCLES	8	0.12	0.96	0	basic	-M	0.1667
	Objective	Function	(Max.) =	2.96				

Display 20 tricycles, 8 bicycles, 2.96 expected sales.

c. MIN X1 + X2

	Decision Variable	Solution Value	Unit Cost or Profit c(j)	Total Contribution	Reduced Cost	Basis Status	Allowable Min. c(j)	Allowable Max. c(j)
1	TRICYCLES	10	1	10	0	basic	0	M
2	BICYCLES	8	1	8	0	basic	0	M
	Objective	Function	(Min.) =	18				

Display 10 tricycles, 8 bicycles; 18 units are displayed.

3.13 See file ch3-13.

X1 = amount invested in a certificate of deposit
X2 = amount invested in the venture capital project

MIN $X1 + X2$
S.T. $.06X1 + .10X2 \geq 4000$ (Potential return)
$.06X1 + .03X2 \geq 2000$ (Guaranteed return)
$X1 + X2 \leq 60000$ (Maximum investment)
$X1, X2 \geq 0$

	Decision Variable	Solution Value	Unit Cost or Profit c(j)	Total Contribution	Reduced Cost	Basis Status	Allowable Min. c(j)	Allowable Max. c(j)
1	CD	19,047.62	1	19,047.62	0	basic	0.6	2
2	VENT CAP	28,571.43	1	28,571.43	0	basic	0.5	1.67
	Objective	Function	(Min.) =	47,619.05				

Invest $19,047.62 in the C/D, and $28,571.43 in the venture capital project. He can keep $12,380.95 for personal use.

3.14 See file ch3-14.lpp.

X3 = amount invested in the oil exploration company

a.

$$
\begin{array}{llllll}
\text{MAX} & .06X1 + & .10X2 + & X3 + & 60,000 & \\
\text{S.T.} & X1 + & X2 + & X3 & = 60,000 & \text{(Total investment)} \\
 & & & X3 & \leq 30,000 & \text{(Maximum -- oil exploration)} \\
 & X1 & & & \geq 20,000 & \text{(Minimum -- c/d)} \\
 & 1.06X1 + 1.03X2 & & & \geq 40,000 & \text{(Minimum value next year)} \\
 & & X1, X2, X3 \geq 0 & & &
\end{array}
$$

b. Substituting X3 = 60,000 - X1 + X2 in the above gives

$$
\begin{array}{llll}
\text{MAX} & -.94X1 - & .90X2 + 120,000 & \\
\text{S.T.} & X1 + & X2 & \geq 30000 \\
 & X1 & & \geq 20000 \\
 & 1.06X1 + 1.03X2 & & \geq 40000 \\
 & X1 + & X2 & \leq 60000 \\
 & & X1, X2 \geq 0 &
\end{array}
$$

c.

	Decision Variable	Solution Value	Unit Cost or Profit c(j)	Total Contribution	Reduced Cost	Basis Status	Allowable Min. c(j)	Allowable Max. c(j)
1	CD	20,000.00	-0.94	-18,800.00	0	basic	-M	-0.93
2	VENT CAP	18,252.43	-0.9	-16,427.19	0	basic	-0.91	0
	Objective	Function	(Max.) =	-35,227.19				

Invest \$20,000 in the C/D, \$18,252.43 in the venture capital project and \$11,747.57 in the oil exploration company; the maximum Potential Return = \$60,000 - \$35,277.19 = \$24,772.81.

Maximum Potential Value Next Year = \$24,772.81 + \$60,000 = \$84,772.81
Minimum Potential Value Next Year = \$40,000 (0 slack on third constraint).

3.15 See files ch3-15c.lpp and ch3-15d.lpp.

X1 = the number of machines leased
X2 = the number of machines purchased

a.

MIN	3000X1 + 1000X2		
S.T.	X2	≤ 5	(Maximum number purchased)
	2000X1 + 800X2	≥ 10,000	(Minimum frames)
	X1, X2 ≥ 0		

b. There is no upper limit on X1; the feasible region is unbounded.

c.

	Decision Variable	Solution Value	Unit Cost or Profit c(j)	Total Contribution	Reduced Cost	Basis Status	Allowable Min. c(j)	Allowable Max. c(j)
1	LEASE	3.00	3,000.00	9,000.00	0	basic	2,500.00	M
2	BUY	5.00	1,000.00	5,000.00	0	basic	-M	1,200.00
	Objective	Function	(Min.) =	14,000.00				

Lease 3 machines and purchase 5. The monthly payment = $14,000.

d. The new objective function is now MAX 2000X1 + 8000X2. Re-solving, the problem is now unbounded.

e. The number of machines leased or purchased must be integer-values; thus the problem is really an integer programming problem.

3.16 See file ch3-16.lpp.

X1 = the number of 8 square foot crates shipped
X2 = the number of 4 square foot insulated crates shipped

MAX 80X1 + 60X2
S.T.

X1	$\leq$ 140	(Total number of 8 sq. ft. crates)
X2	$\leq$ 100	(Total number of 4 sq. ft. crates)
.2X1 + .4X2	$\leq$ 48	(Loading time)
8X1 + 4X2	$\leq$ 1200	(Cargo space)
X1, X2 $\geq$ 0		

	Decision Variable	Solution Value	Unit Cost or Profit c(j)	Total Contribution	Reduced Cost	Basis Status	Allowable Min. c(j)	Allowable Max. c(j)
1	8-SQFT	120.00	80.00	9,600.00	0	basic	30.00	120
2	4-SQFT	60.00	60.00	3,600.00	0	basic	40	160.00
	Objective	Function	(Max.) =	13,200.00				
	Constraint	Left Hand Side	Direction	Right Hand Side	Slack or Surplus	Shadow Price	Allowable Min. RHS	Allowable Max. RHS
1	8FT CRATES	120.00	<=	140.00	20.00	0	120.00	M
2	4FT CRATES	60.00	<=	100.00	40	0	60.00	M
3	LOADING	48.00	<=	48.00	0.00	66.67	36.00	60.00
4	CARGO SPACE	1,200.00	<=	1,200.00	0.00	8.33	720.00	1,320.00

Ship 120 8-square foot crates, 60 4-square foot crates; the profit is $13,200. All loading hours and space available are used (no slack). 20 8-square foot crates and 40 4-square foot crates will not be shipped.

3.17 See the output for problem 3.16 and files ch3-17d.lpp and ch3-17f.lpp.

a. Ranges of optimality:

8-square foot crates $30 to $120
4-square foot crates $40 to $160

b. Ranges of feasibility

Square footage	720 to 1320
Loading hours	36 to 60
8-square foot crates	120 to ∞
4-square foot crates	60 to ∞

3.17c. Shadow prices

Square footage	\$ 8.33/sq. ft.
Loading hours	\$66.67/ hour
8-square foot crates	\$0
4-square foot crates	\$0

Square footage is an included cost, loading time is a sunk cost. Extra available square feet are worth \$8.33 more than the \$10 charge; thus as long as extra space could be leased for less than \$18.33, it should be leased. Extra loading time hours will add \$66.67 to profit. No additional profit can be made for having more than 140 8-square foot crates or more than 100 4-square foot crates ready for shipment.

d. Crates should be integer-valued; for example if one additional square foot were available, the new optimal solution would be to ship 120.1667 8-square foot crates and 59.9167 4-square foot crates (see file ch3-17d.lpp).

e. Percent changes: for 8-square foot crates = ((80-64)/(80-30)) = 32%
for 4-square foot crates = ((52-40)/(60-40)) = 60%
Total = 92%

Thus the optimal solution will not change.

f. It reduces it to \$6.33 so TTC would still pay \$6.33 + \$12 = \$18.33 per square foot for additional cargo space; this makes sense. (You can check this using file ch3-17f.lpp.)

3.18 See files ch3-18a.lpp and ch3-18b.lpp.

X1 = the number of volunteer hours allocated to phones
X2 = the number of volunteer hours allocated to door to door canvassing

MAX 30X1 + 18X2

S.T.

X1 + X2	≤ 6250	(=25hrs./volunteer*250)
X1	≤ 1680	(=14 hrs./day*6days*20)
X2	≥ 2083.33	(=6250hrs./3)
X1	≥ 1260	(=14 hrs./day*6days*15)

X1, X2 ≥ 0

a.

	Decision Variable	Solution Value	Unit Cost or Profit c(j)	Total Contribution	Reduced Cost	Basis Status	Allowable Min. c(j)	Allowable Max. c(j)
1	PHONES	1,680.00	30.00	50,400.00	0	basic	18.00	M
2	DOOR TO DOOR	4,570.00	18.00	82,260.00	0	basic	0	30.00
	Objective	Function	(Max.) =	132,660.00				

Use 1680 phone hours, 4570 door to door hours; value = 132,660

3.18b. Change the objective function to: MAX 30X1 + 36X2.

	Decision Variable	Solution Value	Unit Cost or Profit c(j)	Total Contribution	Reduced Cost	Basis Status	Allowable Min. c(j)	Allowable Max. c(j)
1	PHONES	1,260.00	30.00	37,800.00	0	basic	-M	36
2	DOOR TO DOOR	4,990.00	36.00	179,640.00	0	basic	30	M
	Objective	Function	(Max.) =	217,440.00				

Use 1260 phone hours, 4990 door to door canvassing hours; value = 217,440

3.19 See file ch3-19.lpp.

X1 = amount invested in Tater, Inc.
X2 = amount invested in Lakeside Resorts

$$
\begin{array}{llll}
\text{MAX} & .10X1 + .12X2 & & \\
\text{S.T.} & X1 + X2 & = 100000 & \text{(Investment)} \\
& .000002X1 + .000004X2 & \geq .25 & \text{(Minimum potential)} \\
& .000003X1 + .000008X2 & \leq .50 & \text{(Maximum risk)} \\
& X1, X2 \geq 0 & &
\end{array}
$$

	Decision Variable	Solution Value	Unit Cost or Profit c(j)	Total Contribution	Reduced Cost	Basis Status	Allowable Min. c(j)	Allowable Max. c(j)
1	TATER	60,000.00	0.10	6,000.00	0	basic	-M	0.12
2	LAKESIDE	40,000.00	0.12	4,800.00	0	basic	0.1	M
	Objective	Function	(Max.) =	10,800.00				

Invest $60,000 in Tater, Inc. and $40,000 in Lakeside Resorts;
Expected annual return = $10,800.

3.20 See file ch3-20.lpp.

X1 = the number of dozen custard pies baked
X2 = the number of dozen fruit pies baked

a.

$$\begin{aligned} \text{MAX} \quad & 15X1 + 25X2 \\ \text{S.T.} \quad & 12X1 + 10X2 \le 150 && \text{(Flour)} \\ & 50X1 + 40X2 \le 500 && \text{(Eggs)} \\ & 5X1 + 10X2 \le 90 && \text{(Sugar)} \\ & 15X2 \le 120 && \text{(Fruit mixture)} \\ & X1, X2 \ge 0 \end{aligned}$$

b.

	Decision Variable	Solution Value	Unit Cost or Profit c(j)	Total Contribution	Reduced Cost	Basis Status	Allowable Min. c(j)	Allowable Max. c(j)
1	DZ CUST	4.67	15.00	70.00	0	basic	12.50	31.25
2	DZ FRUIT	6.67	25.00	166.67	0	basic	12	30.00
	Objective	Function	(Max.) =	236.67				
	Constraint	Left Hand Side	Direction	Right Hand Side	Slack or Surplus	Shadow Price	Allowable Min. RHS	Allowable Max. RHS
1	FLOUR	122.67	<=	150.00	27.33	0	122.67	M
2	EGGS	500.00	<=	500.00	0	0.08	420.00	617.14
3	SUGAR	90.00	<=	90.00	0.00	2.17	50.00	98.00
4	FRUIT MIX	100.00	<=	120.00	20.00	0	100.00	M

Bake 56 (4 2/3 dozen) custard pies and 80 (6 2/3 dozen) fruit pies; Profit = $236.67

c. $30 is within the range of optimality -- no change.

d. Percent changes:

Custard pies	3/16.25	= 18.46%
Fruit pies	3/5	= 60.00%
	Total	= 78.46%

78.46% < 100%; there is no change to optimal solution.

e. This means .10(120) = 12 pounds of fruit mixture is not available; but there is slack of 20 pounds and thus the optimal solution will not be affected.

f. The shadow price for sugar is $2.17. Since sugar is an included cost, extra sugar is worth $2.17 + $0.50 = $2.67. Since this is greater than $2.25, Mary Custard's should purchase it.

3.21 See the output for problem 3.20.

a. The current solution is X1 = 4 2/3, X2 = 6 2/3.
Add the constraint: $12X1 + 12X2 \leq 200$. $12(4\ 2/3) + 12(6\ 2/3) = 134 < 200$.
Since this constraint is satisfied the optimal solution will not change.

b. Add the constraint $12X1 + 12X2 \leq 100$.
Since 134 > 100, the constraint is now violated and the optimal solution will change.

c. The marginal profit per dozen chocolate pies is:

$$27 - (15(0) + 30(.0833) + 12(2.1667) + 0(0)) = -\$1.50$$

Since the marginal profit is negative, no chocolate pies should be baked.

d. The price per dozen chocolate pies must increase to \$27 + \$1.50 = \$28.50.

3.22 See files ch3-22a.lpp, ch3-22c.lpp, and ch3-22e.lpp.

X1 = 100's of pounds of La Paz ore purchased
X2 = 100's of pounds of Sucre ore purchased

a.

$$\begin{array}{lll} \text{MIN} & 100X1 + 140X2 & \\ \text{S.T.} & 20X1 + 40X2 \geq 800 & \text{(Copper)} \\ & 20X1 + 25X2 \geq 600 & \text{(Zinc)} \\ & 20X1 + 10X2 \geq 500 & \text{(Iron)} \\ & X1, X2 \geq 0 & \end{array}$$

	Decision Variable	Solution Value	Unit Cost or Profit c(j)	Total Contribution	Reduced Cost	Basis Status	Allowable Min. c(j)	Allowable Max. c(j)
1	LA PAZ	20.00	100.00	2,000.00	0	basic	70.00	280
2	SUCRE	10.00	140.00	1,400.00	0	basic	50	200.00
	Objective	Function	(Min.) =	3,400.00				
	Constraint	Left Hand Side	Direction	Right Hand Side	Slack or Surplus	Shadow Price	Allowable Min. RHS	Allowable Max. RHS
1	COPPER	800.00	>=	800.00	0.00	3	700.00	2,000.00
2	ZINC	650.00	>=	600.00	50	0	-M	650.00
3	IRON	500.00	>=	500.00	0.00	2.00	400.00	800.00

Purchase 2000 pounds of La Paz ore and 1000 pounds of Sucre ore; Cost = \$3,400.

3.22b. Range of Optimality for cost of La Paz ore: \$70 to \$280; Sucre ore: \$50 to \$200. Within these limits the optimal solution will not change from that in part (a).

c. \$250 is outside the range of optimality. The reduced cost is \$250 - \$200 = \$50 as verified by file ch3-22c.lpp.

d.

	Shadow Price	Range of Feasibility
Copper	\$3.00	700 to 2000
Zinc	\$ 0	-∞ to 650
Iron	\$2.00	400 to 600

On an individual basis, as long as the needs for these resources fall within their range of feasibility, the shadow prices will not change.

e. Add the constraint: $40X1 + 25X2 \leq 1000$; the problem is now infeasible (as shown in file ch3-22e.lpp).

3.23 See file ch3-23.lpp.

X1 = the number of in-house workers assigned to the design division
X2 = the number of secretaries hired from Techhelp for the design division

a.

$$\begin{aligned} &\text{MAX } 40X1 + 30X2 \\ &\text{S.T.} \quad 2X1 + X2 \leq 8 \quad \text{(In house)} \\ &\quad X1 + X2 \leq 10 \quad \text{(Techhelp)} \\ &\quad X1, X2 \geq 0 \end{aligned}$$

	Decision Variable	Solution Value	Unit Cost or Profit c(j)	Total Contribution	Reduced Cost	Basis Status	Allowable Min. c(j)	Allowable Max. c(j)
1	IN-HOUSE	0	40	0	-20	at bound	-M	60
2	TECHHELP	8	30	240	0	basic	20	M
	Objective	Function	(Max.) =	240				
	Constraint	Left Hand Side	Direction	Right Hand Side	Slack or Surplus	Shadow Price	Allowable Min. RHS	Allowable Max. RHS
1	ASSISTANCE	8	<=	8	0	30	0	10
2	WORKSTATIONS	8	<=	10	2	0	8	M

Hire 8 secretaries from Techhelp only giving 240 pages per day.

3.23b. Complementary slackness indicates that the input for in-house workers must increase before it is economically feasible to transfer them to the design division. The reduced cost is -$20. Since the solution is to hire 8 Techhelp secretaries, no additional incentive is needed (the reduced cost for Techhelp pages is 0).

c. Range of optimality for in-house workers is $-\infty$ to 60 pages and for Techhelp secretaries is 20 pages to ∞. 40 pages per day was insufficient to use in-house employees. A reduction in pages of any amount would make this option even less attractive.

3.24 See the out put for problem 3.23 and file ch3-24c.lpp.

a. The range of feasibility for the number of workstations is 8 to ∞. Since not all workstations were used, having additional workstations will not improve the output.

The range of feasibility for supervisor assistance is 0 to 10 hrs. Within this range, each extra hour of assistance corresponds to an increase in output of 30 pages per day.

b. Two additional supervisor hours will add 30(2) = 60 pages worth $60. Two hours at $25 per hour is only $50. Thus the profit would increase by $60 - $50 = $10 per day with two additional hours.

c. Change the first constraint to: $2X1 + X2 \leq 16$. Re-solving gives the output below.

	Decision Variable	Solution Value	Unit Cost or Profit c(j)	Total Contribution	Reduced Cost	Basis Status	Allowable Min. c(j)	Allowable Max. c(j)
1	IN-HOUSE	6	40	240	0	basic	30	60
2	TECHHELP	4	30	120	0	basic	20	40
	Objective	Function	(Max.) =	360				
	Constraint	Left Hand Side	Direction	Right Hand Side	Slack or Surplus	Shadow Price	Allowable Min. RHS	Allowable Max. RHS
1	ASSISTANCE	16	<=	16	0	10	10	20
2	WORKSTATIONS	10	<=	10	0	20	8	16

The new optimal daily output of 360 pages, an increase of 360 - 240 = 120 pages or $120. But 8 additional hours of supervisor help costs $25(8) = $200. This is greater than the gross increase in profit due to the increased page output.

3.25 See file ch3-25.lpp.

X1 = the number of C15 units produced each week
X2 = the number of UN8 units produced each week

a. Unit profit for C15's: $11.10 - .40 - 8(.40) - 2(.75) = $6.00
Unit profit for UN8's: $12.40 - .90 - 5(.40) - 2(.75) = $8.00

b. MAX 6X1 + 8X2

S.T. 8X1 + 5X2 $\leq$ 15,000 (Particle board)

2X1 + 2X2 $\leq$ 4,500 (Assemblies)

4X1 + 6X2 $\leq$ 12,000 (Minutes)

X1, X2 $\geq$ 0

	Decision Variable	Solution Value	Unit Cost or Profit c(j)	Total Contribution	Reduced Cost	Basis Status	Allowable Min. c(j)	Allowable Max. c(j)
1	C15	750	6	4,500.00	0	basic	5.33	8
2	UN8	1,500.00	8	12,000.00	0	basic	6	9
	Objective	Function	(Max.) =	16,500.00				
	Constraint	Left Hand Side	Direction	Right Hand Side	Slack or Surplus	Shadow Price	Allowable Min. RHS	Allowable Max. RHS
1	PARTICLE BD	13,500.00	<=	15,000.00	1,500.00	0	13,500.00	M
2	ASSEMBLIES	4,500.00	<=	4,500.00	0	1	4,000.00	4,714.29
3	MINUTES	12,000.00	<=	12,000.00	0	1	11,000.00	13,500.00

Produce 750 C15 units and 1,500 UN8 units weekly; Weekly profit = $16,500

c. An extra 200 slide assemblies gives a total of 4700. This is within its range of feasibility. Thus Compaids should be willing to pay up to the shadow price of $1 above the current retail price of $0.75 or up to $1.75 per assembly or 200($1.75) = $350.

d. One additional worker adds 2400 minutes per week giving a total of 14,400 minutes per week. This is above the upper limit of its range of feasibility of 13,500. Thus the shadow prices will change.

e. The upper limit of the range of optimality for the profit of C15's is $8; this is $2 more than its current value. Thus the price could increase by $2 to $13.10.

CD3.1 See file cd3-1d.lpp.

$X1$ = the number of standard models produced daily

$X2$ = the number of trimline models produced daily

a.

$$7X1 + 2X2 \leq 1400 \quad \text{(Manufacturing)}$$
$$3X1 + 7X2 \leq 2100 \quad \text{(Assembly)}$$
$$X1 + 4X2 \leq 1000 \quad \text{(Quality Control)}$$
$$X1, X2 \geq 0$$

b. $X1 = 100$, $X2 = 300$ is an infeasible point -- it violates the assembly and the quality control constraints; thus it cannot be the optimal solution.

c. Substituting $X1 = 100$, $X2 = 200$ into the constraints gives 1100 manufacturing minutes, 1700 assembly minutes, and 900 quality control minutes. Since it satisfies none of the functional or nonnegativity constraints with equality, $X1 = 100$, $X2 = 200$ is an interior point and thus cannot be the optimal solution.

d. Add the objective function MAX $2X1 + 4X2$. See file cd3-1d.lpp.

	Decision Variable	Solution Value	Unit Cost or Profit c(j)	Total Contribution	Reduced Cost	Basis Status	Allowable Min. c(j)	Allowable Max. c(j)
1	STANDARD	138.46	2	276.92	0	basic	1	14
2	TRIMLINE	215.38	4	861.54	0	basic	0.57	8
	Objective	Function	(Max.) =	1,138.46				
	Constraint	Left Hand Side	Direction	Right Hand Side	Slack or Surplus	Shadow Price	Allowable Min. RHS	Allowable Max. RHS
1	MANUFACTURE	1,400.00	<=	1,400.00	0	0.15	500	2,320.00
2	ASSEMBLY	1,923.08	<=	2,100.00	176.92	0	1,923.08	M
3	QC	1,000.00	<=	1,000.00	0	0.92	200	1,106.98

Produce 138.46 standard phones and 215.38 trimline phones daily.
Optimal profit = $1,138.46.

CD3.2 Refer to the output for CD3.1 and files cd3-2ci, cd3-2cii, and cd3-2d.

a.

	Reduced Cost	Range of Optimality
Standard	$0	$1.00 to $14.00
Trimline	$0	$0.57 to $ 8.00

b.

	Shadow Prices	Range of Feasibility
Manufacturing Minutes	$0.15	500.00 to 2320.00
Assembly Minutes	$ 0	1923.08 to $+\infty$
Quality Control Minutes	$0.92	200.00 to 1106.98

c. There is a slack of 176.92 assembly minutes.

(i) Allocating all 176.92 minutes to manufacturing changes the minutes for manufacturing, assembly and quality control to 1576.92, 1923.08, and 1000 respectively.

	Decision Variable	Solution Value	Unit Cost or Profit c(j)	Total Contribution	Reduced Cost	Basis Status	Allowable Min. c(j)	Allowable Max. c(j)
1	STANDARD	167.26	2	334.52	0	Basic	1.71	14
2	TRIMLINE	203.04	4	812.17	0	Basic	0.57	4.67
	Objective	Function	(Max.) =	1,146.69				
	Constraint	Left Hand Side	Direction	Right Hand Side	Slack or Surplus	Shadow Price	Allowable Min. RHS	Allowable Max. RHS
1	MANUFACTURE	1,576.92	<=	1,576.92	0	0.05	1,400.02	4,487.19
2	ASSEMBLY	1,923.08	<=	1,923.08	0	0.56	675.82	1,957.10
3	QC	979.43	<=	1,000.00	20.57	0	979.43	M

<u>Daily Production Schedule</u>

Standard	167.26
Trimline	203.04
Daily Profit	$1,146.69

<u>Shadow Prices Per Minute</u>

Manufacturing	$0.05
Assembly	$0.56
Quality Control	$ 0

CD3.2c(ii) Allocating all 176.92 minutes to quality control changes the minutes for manufacturing, assembly and quality control to 1400, 1923.08, and 1176.92 respectively.

	Decision Variable	Solution Value	Unit Cost or Profit c(j)	Total Contribution	Reduced Cost	Basis Status	Allowable Min. c(j)	Allowable Max. c(j)
1	STANDARD	138.46	2	276.92	0	basic	1.71	14
2	TRIMLINE	215.39	4	861.54	0	basic	0.57	4.67
	Objective	Function	(Max.) =	1,138.46				
	Constraint	Left Hand Side	Direction	Right Hand Side	Slack or Surplus	Shadow Price	Allowable Min. RHS	Allowable Max. RHS
1	MANUFACTURE	1,400.00	<=	1,400.00	0	0.05	549.45	4,487.19
2	ASSEMBLY	1,923.08	<=	1,923.08	0	0.56	600	2,215.68
3	QC	1,000.00	<=	1,176.92	176.92	0	1,000.00	M

Daily Production Schedule	
Standard	138.46
Trimline	215.39
Daily Profit	$1,138.46

Shadow Prices Per Minute	
Manufacturing	$0.05
Assembly	$0.56
Quality Control	$ 0

d. Allocating 88.46 minutes to manufacturing and 88.46 minutes to quality control changes the minutes for manufacturing, assembly and quality control to 1488.46, 1923.08, and 1088.46 respectively.

	Decision Variable	Solution Value	Unit Cost or Profit c(j)	Total Contribution	Reduced Cost	Basis Status	Allowable Min. c(j)	Allowable Max. c(j)
1	STANDARD	152.86	2	305.72	0	basic	1.71	14
2	TRIMLINE	209.21	4	836.85	0	basic	0.57	4.67
	Objective	Function	(Max.) =	1,142.58				
	Constraint	Left Hand Side	Direction	Right Hand Side	Slack or Surplus	Shadow Price	Allowable Min. RHS	Allowable Max. RHS
1	MANUFACTURE	1,488.46	<=	1,488.46	0	0.05	639.26	4,487.19
2	ASSEMBLY	1,923.08	<=	1,923.08	0	0.56	637.91	2,086.39
3	QC	989.72	<=	1,088.46	98.74	0	989.72	M

Daily Production Schedule	
Standard	152.86
Trimline	209.21
Daily Profit	$1,142.58

Shadow Prices Per Minute	
Manufacturing	$0.05
Assembly	$0.56
Quality Control	$ 0

CD3.3 See files cd3-3c.lpp and cd3-3d.lpp.

For the model in part c)

MAXIMIZE 9000 - 450X1 - 672X2

S.T. X1 ≥ 4

X2 ≥ 2

X1, X2 ≥ 0

The 9000 is a constant, so maximize -450X1 -672X2

	Decision Variable	Solution Value	Unit Cost or Profit c(j)	Total Contribution	Reduced Cost	Basis Status	Allowable Min. c(j)	Allowable Max. c(j)
1	DAY	4	-450	-1,800	0	basic	-M	0
2	NIGHT	2	-672	-1,344	0	basic	-M	0
	Objective	Function	(Max.) =	-3,144				

Use 4 day guards, 2 night guards; profit = \$9,000 - 3,144 = \$5,856.

d. MAXIMIZE 10000 -450X1 - 672X2 - 1000, or

MAXIMIZE 9000 -450X1 - 672X2

S.T. X1 ≥ 4.4

X2 ≥ 2

X1, X2 ≥ 0 and integer

Without including the integer restrictions X1 turns out to be a fractional value (4.4). Including the integer restrictions: use 5 day guards, 2 night guards; profit = \$9,000 - 3,594 = \$5,406.

CD3.4 See file cd3-4.lpp.

X1 = the number of 19-inch lawn mowers produced weekly
X2 = the number of 21-inch lawn mowers produced weekly

MAX	$50X1 + 60X2$	
S.T.	$40X1 + 60X2 \le 9600$	(Minutes)
	$X1 + X2 \le 200$	(Engines)
	$X2 \le 100$	(Assemblies)
	$X1, X2 \ge 0$	

	Decision Variable	Solution Value	Unit Cost or Profit c(j)	Total Contribution	Reduced Cost	Basis Status	Allowable Min. c(j)	Allowable Max. c(j)
1	19INCH	120.00	50.00	6000.00	0.00	basic	40.00	60.00
2	21INCH	80.00	60.00	4800.00	0.00	basic	50.00	75.00
	Objective	Function	(Max.) =	10,800.00				
	Constraint	Left Hand Side	Direction	Right Hand Side	Slack or Surplus	Shadow Price	Allowable Min. RHS	Allowable Max. RHS
1	TIME	9,600.00	<=	9,600.00	0.00	0.50	8,000.00	10,000.00
2	ENGINES	200.00	<=	200.00	0.00	30.00	190.00	240.00
3	VAR. SPD.	80.00	<=	100.00	20.00	0.00	80.00	M

Manufacture 120 19-inch and 80 21-inch lawn mowers; weekly profit = $10,800.

CD3.5 See output from problem CD3-4 and see file cd3-5d.lpp.

a.

Objective Function Coefficient For	Range of Optimality
19-inch mowers	$40 to $60
21-inch mowers	$50 to $75

A single change to a profit coefficient that gives a value within its range of optimality results in the same optimal solution.

CD3.5b. Engines -- \$30/engine
Assemblies -- \$0/assembly
Production time -- \$0.50/min. = \$30/hr.

Assuming all other coefficients remain the same, each extra available engine will add \$30 to profit within its range of feasibility and each extra production hour will add \$30 within its range of feasibility; additional assemblies will not affect the production quantities or profit within its range of feasibility.

c.

	Range of feasibility
Engines	190 to 240
Assemblies	80 to ∞
Production time	8,000 to 10,000 min. or 133 1/3 to 166 2/3 hr.

Assuming all other coefficients remain the same, any single change in the availability of engines, assemblies, or production time within their ranges of feasibility, will not affect the shadow prices.

d. Change the amount of production minutes to 8000; produce 200 19-inch models only.

	Decision Variable	Solution Value	Unit Cost or Profit c(j)	Total Contribution	Reduced Cost	Basis Status	Allowable Min. c(j)	Allowable Max. c(j)
1	19INCH	200.00	50.00	10000.00	0.00	basic	40.00	60.00
2	21INCH	0.00	60.00	0.00	0.00	basic	50.00	75.00
	Objective	Function	(Max.) =	10,000.00				

CD3.6 See file cd3-6.lpp.

X1 = the number of one-hour sections taught by part-time instructors
X2 = the number of one-hour sections taught by graduate assistants

$$
\begin{array}{lll}
\text{MAX} & 30X1 + 25X2 & \\
\text{S.T.} & 18X1 + 12X2 \leq 540 & \text{(Budget per week)} \\
 & X1 \leq 20 & \text{(Total part-time hours per week)} \\
 & X2 \leq 30 & \text{(Total graduate assistant hours per week)} \\
 & X1, X2 \geq 0 &
\end{array}
$$

	Decision Variable	Solution Value	Unit Cost or Profit c(j)	Total Contribution	Reduced Cost	Basis Status	Allowable Min. c(j)	Allowable Max. c(j)
1	INS. SEC.	10.00	30.00	300.00	0.00	basic	0.00	37.50
2	GA SEC.	30.00	25.00	750.00	0.00	basic	20.00	M
	Objective	Function	(Max.) =	1,050.00				
	Constraint	Left Hand Side	Direction	Right Hand Side	Slack or Surplus	Shadow Price	Allowable Min. RHS	Allowable Max. RHS
1	BUDGET	540.00	<=	540.00	0.00	1.67	360.00	720.00
2	HRS. INS.	10.00	<=	20.00	10.00	0.00	10.00	M
3	HRS. GA'S	30.00	<=	30.00	0.00	5.00	15.00	45

Staff 10 sections with part-time instructors and 30 sections with graduate assistants; this will accommodate 1050 students.

CD3.7 See output from problem CD3.6 and file cd3.7e.lpp.

a.

Objective Function Coefficient For	Range of Optimality
Instructor Section Size	0 to 37.5
Grad. Ass't. Section Size	20 to ∞

b. Shadow price = 1.67 students per dollar; if the budget were $5,000 ($500 per week), this is within the range of feasibility and hence this $40 per week decrease in budget would result in 40(1.67) = 67 fewer students being accommodated.

c. The shadow price for the number of sections taught by part-time instructors = 0; this equals 0 per instructor. The shadow price for the number of sections taught by graduate assistants is 5; this is 5(6) = 30 students per graduate assistant.

CD3.7d. Add the constraint: X1 + X2 ≤ 44. The current solution already meets this condition; there will be no change to the optimal solution.

e. Add the constraint: X1 + X2 ≤ 36. The current solution does not meet this condition.

	Decision Variable	Solution Value	Unit Cost or Profit c(j)	Total Contribution	Reduced Cost	Basis Status	Allowable Min. c(j)	Allowable Max. c(j)
1	INS. SEC.	18.00	30.00	540.00	0.00	basic	25.00	37.50
2	GA SEC.	18.00	25.00	450.00	0.00	basic	20.00	30.00
	Objective	Function	(Max.) =	990.00				

California Union should now staff 18 sections with part-time instructors and 18 sections with graduate assistants. The total number of students accommodated is now 990.

CD3.8 See files cd3-8.lpp (the original problem from section 3.11), cd3-8a.lpp, cd3-8b.lpp, and cd3-8d.lpp.

The output from Section 3.11 is:

	Decision Variable	Solution Value	Unit Cost or Profit c(j)	Total Contribution	Reduced Cost	Basis Status	Allowable Min. c(j)	Allowable Max. c(j)
1	TEXFOODS	1.5000	0.6000	0.9000	0.0000	basic	0.5000	2.5000
2	CALRATION	2.5000	0.5000	1.2500	0.0000	basic	0.1200	0.6000
	Objective	Function	(Min.) =	2.1500				
	Constraint	Left Hand Side	Direction	Right Hand Side	Slack or Surplus	Shadow Price	Allowable Min. RHS	Allowable Max. RHS
1	VITAMIN A	155.0000	>=	100.0000	55.0000	0.0000	-M	155.0000
2	VITAMIN D	100.0000	>=	100.0000	0.0000	0.0190	76.0870	250.0000
3	IRON	100.0000	>=	100.0000	0.0000	0.0025	40.0000	173.3333

a. $0.75 is within the range of optimality for the cost of Texfoods, so the optimal solution will not change, but the minimum cost is now $2.375 and the shadow price for Vitamin D is now $0.0175 and for iron is now $0.0063.

b. 120% is within the range of feasibility for iron, so the shadow prices will not change but the optimal solution will change to 2 2-oz. portions of Texfoods and 2 2-oz. of Calration and the minimum cost will increase to $2.20.

	Decision Variable	Solution Value	Unit Cost or Profit c(j)	Total Contribution	Reduced Cost	Basis Status	Allowable Min. c(j)	Allowable Max. c(j)
1	TEXFOODS	2.0000	0.6000	1.2000	0.0000	basic	0.5000	2.5000
2	CALRATION	2.0000	0.5000	1.0000	0.0000	basic	0.1200	0.6000
	Objective	Function	(Min.) =	2.2000				
	Constraint	Left Hand Side	Direction	Right Hand Side	Slack or Surplus	Shadow Price	Allowable Min. RHS	Allowable Max. RHS
1	VITAMIN A	140.0000	>=	100.0000	40.0000	0.0000	-M	140.0000
2	VITAMIN D	100.0000	>=	100.0000	0.0000	0.0190	82.6087	300.0000
3	IRON	120.0000	>=	120.0000	0.0000	0.0025	40.0000	173.3333

CD3.8c. No change since there is already slack for Vitamin A.

d. Add the constraint: $18X1 + 32X2 \leq 100$. For the current solution of X1 = 1.5, X2 = 2.5, 18(1.5) + 32(2.5) = 107. The level of 100% is exceeded so the optimal solution will change. Re-solving, the optimal solution is: 2 2-oz. portions of Texfoods and 2 2-oz. of Calration and the minimum cost will increase to $2.20.

	Decision Variable	Solution Value	Unit Cost or Profit c(j)	Total Contribution	Reduced Cost	Basis Status	Allowable Min. c(j)	Allowable Max. c(j)
1	TEXFOODS	2.0000	0.6000	1.2000	0.0000	basic	0.5000	M
2	CALRATION	2.0000	0.5000	1.0000	0.0000	basic	-M	0.6000
	Objective	Function	(Min.) =	2.2000				
	Constraint	Left Hand Side	Direction	Right Hand Side	Slack or Surplus	Shadow Price	Allowable Min. RHS	Allowable Max. RHS
1	VITAMIN A	140.0000	>=	100.0000	40.0000	0.0000	-M	140.0000
2	VITAMIN D	100.0000	>=	100.0000	0.0000	0.0291	95.0704	138.8889
3	IRON	120.0000	>=	100.0000	20.0000	0.0000	-M	120.0000
4	FAT	100.0000	<=	100.0000	0.0000	-0.0071	81.3333	107.0000

CD3.9 See file cd3-9.lpp.

X1 = the number of gallons of Bugoff in the mixture
X2 = the number of gallons of Weedaway in the mixture

Profit per gallon of Bugoff = 20 - 4(.20) - .25(12) - .50(8) - .25(9) = \$9.95
Profit per gallon of Weedaway = 16 - 4(.30) - .60(12) - .10(8) - .30(9) = \$4.10

$$
\begin{array}{llll}
\text{MAX} & 9.95X1 + 4.10X2 & & \\
\text{S.T.} & X1 & \geq 250 & \text{(1000 quarts = 250 gallons Bugoff)} \\
& X2 & \geq 250 & \text{(1000 quarts = 250 gallons Weedaway)} \\
& .25X1 + .60X2 & \leq 1000 & \text{(Xylothon)} \\
& .50X1 + .10X2 & \leq 1000 & \text{(Diazon)} \\
& .25X1 + .30X2 & \leq 1000 & \text{(Sulferious)} \\
& X1, X2 \geq 0 & &
\end{array}
$$

	Decision Variable	Solution Value	Unit Cost or Profit c(j)	Total Contribution	Reduced Cost	Basis Status	Allowable Min. c(j)	Allowable Max. c(j)
1	BUGOFF	1818.18	9.95	18090.91	0.00	basic	1.71	20.50
2	WEEDAWAY	909.09	4.10	3727.27	0.00	basic	1.99	23.88
	Objective	Function	(Max.) =	21818.18				
	Constraint	Left Hand Side	Direction	Right Hand Side	Slack or Surplus	Shadow Price	Allowable Min. RHS	Allowable Max. RHS
1	MIN BUGOFF	1818.18	>=	250.00	1568.18	0.00	-M	1818.18
2	MIN WEEDAWAY	909.09	>=	250.00	659.09	0.00	-M	909.09
3	MAX XYLOTHON	1000.00	<=	1000.00	0.00	3.84	637.50	1600.00
4	MAX DIAZON	1000.00	<=	1000.00	0.00	17.98	281.25	1725.00
5	MAX SULFERIOUS	727.27	<=	1000.00	272.73	0.00	727.27	M

a. 1818.18 gallons of Bugoff, 909.09 gallons of Weedaway; to net profit = \$21,818.18.

b. It will use all 1000 gallons of xylothon and diazon, and will use 727.27 gallons of sulferious, leaving only 273.73 gallons of sulferious.

Case 3.1 See file ca3-1.lpp.

			Pioneer		Heritage
Revenue:	Sales Price		\$12.75		\$18.00
Costs:					
	Straw	1 lb. =	\$ 1.50	1.5 lb. =	\$ 2.25
	Handle		\$ 0.75		\$ 0.75
		Net Profit:	\$10.50		\$15.00

X1 = the number of Pioneer brooms produced per day
X2 = the number of Heritage brooms produced per day

$$
\begin{aligned}
\text{MAX} \quad & 10.50X1 + 15.00X2 \\
\text{S.T.} \quad & X1 + 1.5\,X2 \leq 350 \quad \text{(Straw)} \\
& X1 + X2 \leq 300 \quad \text{(Handles)} \\
& .25X1 + .40X2 \leq 80 \quad \text{(Production time)} \\
& X1, X2 \geq 0
\end{aligned}
$$

	Decision Variable	Solution Value	Unit Cost or Profit c(j)	Total Contribution	Reduced Cost	Basis Status	Allowable Min. c(j)	Allowable Max. c(j)
1	PIONEER	266.67	10.50	2800.00	0.00	basic	9.38	15.00
2	HERITAGE	33.33	15.00	500.00	0.00	basic	10.50	16.80
	Objective	Function	(Max.) =	3300.00				
	Constraint	Left Hand Side	Direction	Right Hand Side	Slack or Surplus	Shadow Price	Allowable Min. RHS	Allowable Max. RHS
1	STRAW	316.67	<=	350.00	33.33	0.00	316.67	M
2	HANDLES	300.00	<=	300.00	0.00	3.00	200.00	320.00
3	PROD. TIME	80.00	<=	80.00	0.00	30.00	75.00	90.00

Produce 266 2/3 Pioneer brooms and 33 1/3 Heritage brooms per day. Fixed costs are \$2800 per day; thus the total profit = \$3300 - \$2800 = \$500 per day.

As long as the profit for Pioneer brooms is between \$9.38 and \$15.00, or the profit for Heritage brooms is between \$10.50 and \$16.80, there will be no change in the recommendation.

Analysis of Options

1. Straw

Seeking additional straw is of no value; the amount of straw already available is not used up. (Shadow price = \$0.)

2. Delivery of an extra box of handles

Each extra available handle will add \$3 (shadow price = \$3) to profit; thus a box of 10 handles will add \$30 to profit. (Note that 10 additional handles brings the total number of handles to 310 and this is within the range of feasibility for handles; thus the shadow price is valid for all 10 handles.) This is an "included" cost so they are worth up to \$30.00 + \$7.50 = \$37.50 per box. Thus paying \$25 will add \$37.50 - \$25 = \$12.50 per day to overall profit. This option should be considered.

3. Adding a half-time worker

Since in this model production costs are "sunk" costs, additional production hours will increase profit by \$30 each (the shadow price for production hours). Adding a worker for 4 hours per day brings the total number of production hours available to 84, which is within the range of feasibility for production hours. Thus adding a half-time worker for \$50 per day will net 4(\$30) - \$50 = \$70 per day in additional profit. This option should also be considered.

How about implementing both options 2 and 3?

Using the 100% rule, the % change for handles = 10/20 = 50% and for workers = 4/10 = 40%. This sum is 90%. Since this is less than 100% the shadow prices will still not change from that used in the above analyses, i.e. both analyses are valid. Thus implementing both will add \$82.50 per day to profit.

Case 3.2 See files ca3-2.lpp, ca3-2p.lpp (no vans), ca3-2v.lpp (no pickups), ca3-2s.lpp (same number of vans as pickups), ca3-2m.lpp (minimum number of vehicles).

X1 = the number of 1-ton pickup trucks purchased
X2 = the number of 2 1/2-ton vans purchased

MIN $24{,}000\,X1 + 60{,}000X2$

S.T.

$X1 + 2.5X2 \geq 36$ (Minimum trucking capacity)

$X1 + X2 \leq 40$ (Maximum facilities)

$X1 + 4X2 \leq 48$ (Workers)

$X1, X2 \geq 0$

	Decision Variable	Solution Value	Unit Cost or Profit c(j)	Total Contribution	Reduced Cost	Basis Status	Allowable Min. c(j)	Allowable Max. c(j)
1	PICKUPS	16	24,000.00	384,000.00	0	basic	24,000.00	M
2	VANS	8	60,000.00	480,000.00	0	basic	-M	60,000.00
	Objective	Function	(Min.) =	864,000.00	(Note:	Alternate	Solution	Exists!!)
		Left Hand		Right Hand	Slack	Shadow	Allowable	Allowable
	Constraint	Side	Direction	Side	or Surplus	Price	Min. RHS	Max. RHS
1	CAPACITY	36	>=	36	0	24,000.00	30	44
2	FACILITIES	24	<=	40	16	0	24	M
3	WORKERS	48	<=	48	0	0	36	57.6

There are alternate optimal solutions giving a total cost of $864,000.

1. Purchasing pickups only -- Add the constraint X2 = 0; purchase 36 pickups.

   Purchasing vans only -- Add the constraint X1 = 0. The problem is infeasible.

2. Purchasing the same number of pickups and vans -- Add the constraint X1 - X2 = 0. The problem is infeasible.

3. Purchasing the minimum number of trucks. Change the objective function to: MIN X1 + X2 and add the constraint 24000X1 + 60000X2 = 864,000.

   The result is that given when the problem was first solved; purchase 16 pickups and 8 vans.

Case CD3.1 See files cac3-1e.lpp (Eastside), cac3-1w.lpp (Westside), and cac3-1c.lpp (Combined).

Eastside

X1 = the number of tables produced per week
X2 = the number of chairs produced per week

$$\begin{array}{llll} \text{MAX} & 70X1 + 30X2 & & \\ \text{S.T.} & 20X1 + 12.5X2 & \leq 6000 & \text{(Pressed wood)} \\ & X1 + X2 & \leq 400 & \text{(Aluminum fittings)} \\ & 1.2X1 + .3X2 & \leq 300 & \text{(Labor)} \\ & X1, X2 \geq 0 & & \end{array}$$

	Decision Variable	Solution Value	Unit Cost or Profit c(j)	Total Contribution	Reduced Cost	Basis Status	Allowable Min. c(j)	Allowable Max. c(j)
1	TABLES	216.67	70.00	15,166.67	0	basic	48.00	120
2	CHAIRS	133.33	30.00	4,000.00	0	basic	17.5	43.75
	Objective	Function	(Max.) =	19,166.67				

Produce 216.67 tables and 133.33 chairs weekly. Weekly profit = $19,166.67

Westside

X3 = the number of desks produced per week
X4 = the number of workstations produced per week

$$\begin{array}{llll} \text{MAX} & 100X3 + 125X4 & & \\ \text{S.T.} & 20X3 + 30X4 & \leq 6000 & \text{(Pressed wood)} \\ & X3 + X4 & \leq 400 & \text{(Aluminum fittings)} \\ & 1.5X3 + 2X4 & \leq 300 & \text{(Labor)} \\ & X3, X4 \geq 0 & & \end{array}$$

	Decision Variable	Solution Value	Unit Cost or Profit c(j)	Total Contribution	Reduced Cost	Basis Status	Allowable Min. c(j)	Allowable Max. c(j)
1	DESKS	200	100.00	20,000.00	0	basic	93.75	M
2	WORKST.	0	125.00	0.00	-8.33	at bound	-M	133.33
	Objective	Function	(Max.) =	20,000.00				

Produce 200 desks and 0 workstations weekly. Weekly profit = $20,000.00

Thus current combined total weekly profit = $39,166.67

Combined Operation

$$\begin{aligned}
\text{MAX}\quad & 70X1 + 30X2 + 100X3 + 125X4 \\
\text{S.T.}\quad & 20X1 + 12.5X2 + 20X3 + 30X4 \leq 12{,}000 \text{ (Pressed wood)} \\
& X1 + X2 + X3 + X4 \leq 800 \text{ (Aluminum fittings)} \\
& 1.2X1 + .3X2 + 1.5X3 + 2X4 \leq 600 \text{ (Labor)} \\
& X1, X2, X3, X4 \geq 0
\end{aligned}$$

	Decision Variable	Solution Value	Unit Cost or Profit c(j)	Total Contribution	Reduced Cost	Basis Status	Allowable Min. c(j)	Allowable Max. c(j)
1	TABLES	0	70.00	0.00	-18.6957	at bound	-M	88.6957
2	CHAIRS	417.391	30.00	12,521.74	0	basic	21.4286	62.50
3	DESKS	339.13	100	33,913.04	0	basic	86.0938	140
4	WORKST	0	125	0.00	-19.3478	at bound	-M	144.3478
	Objective	Function	(Max.) =	46,434.78				

Produce only 417.39 chairs and 339.13 desks per week.
Total weekly profit = $46,434.78.

This is a $7,268.11 increase, which exceeds the $5,000 renovation cost. Thus Franklin should combine the operations.

Chapter 4

Problem Summary

Prob. #	Concepts Covered	Level of Difficulty	Notes
4.1	Maximization production problem, percentage constraints	4	
4.2	Interpretation of output, shadow prices and range of optimality	4	
4.3	Maximization financial mix problem, percentage constraint, determining current rate of return and future rate of return, shadow prices	6	
4.4	Maximization retailing problem, addition of lower bound constraints	4	
4.5	Evaluating changes to objective function coefficients, the profitability of added resources, and the addition of a constraint	5	
4.6	Minimization diet problem, alternate optimal solutions, interpretation of shadow prices, the effects of deleting a constraint	5	
4.7	Maximization production problem, allocating resources optimally	5	Variable definition requires thought.
4.8	Maximization production problem, allocating resources optimally	6	Variable definition requires thought.
4.9	Maximization problem, calculation of net profit	5	
4.10	Sensitivity analyses, evaluation of purchasing additional resources	6	Answer to part (e) in the back of the book is wrong ; it should be "Buy cheese".
4.11	Maximization production problem	4	Make sure units of variables are reflected consistently in constraints.
4.12	Addition of constraints, infeasibility, reduced costs	4	
4.13	Sensitivity analyses, addition of constraints	4	Make sure units are consistent.
4.14	Maximization finance problem	4	
4.15	Blending problem, alternate optimal solutions, shadow prices, determining the acceptability of the solution	6	Check variable definitions.
4.16	Maximization agriculture problem	7	The objective function coefficients involve calculation, and units in constraints must be consistent.
4.17	Sensitivity analyses, evaluation of an alternative based on shadow prices	6	The problem is dependent on formulating problem 4.16 correctly.
4.18	Maximization model, alternate optimal solutions	5	Problem is straightforward if variables are defined correctly.
4.19	Maximization financial problem	6	The constraint for the maximum invested in business loans is $X4 \leq .49$(all investments *except* the investment in securities).
4.20	Interpretation of a computer printout, sensitivity analyses, 100% rule	5	

4.21	Maximization advertising problem	8	10 functional constraints, some not straightforward. Could be used as a mini-case.
4.22	Sensitivity analyses, adding constraints, changing the RHS values	7	Depends on a correct formulation of problem 4.21.
4.23	Maximization production problem	7	Formulation is made easier by using a definitional variable for the total number of 2-oz. bars produced daily.
4.24	Minimization transportation problem	5	This is a straightforward transportation problem that will be discussed in Chapter 6; the definition of the variables is what's tricky if the student is seeing this type of problem for the first time.
4.25	Maximization problem involving calculation of some of the coefficients and the use of accounting data, discussion of sensitivity	10	The problem only has four constraints but much information must be drawn from the data presented.
CD4.1	Maximization problem	4	Definition of the variables is important here.
CD4.2	Maximization production problem, analysis of why a high profit item is not produced	4	
CD4.3	Sensitivity analyses, evaluation of purchasing additional resources	6	
CD4.4	Minimization problem	6	Straightforward but the problem has 8 variables and 14 functional constraints.
CD4.5	Basic minimization transportation problem, evaluation of infeasibility	5	
CD4.6	Range of optimality, positive and negative shadow prices in a minimization transportation problem	4	Part b refers to the original distance of 49.
CD4.7	Maximization educational scheduling problem, "rounding" solution to "academic time"	6	
CD4.8	Classic trim loss problem, alternate optimal solutions	6	Definition of variables is the key to trim loss problems.
CD4.9	Maximization advertising problem, checking the linear programming assumptions, elimination of a nonbinding constraint	6	
CD4.10	Minimization marketing problem	7	Variables and constraints require some thought.
Case 4.1	Maximization production problem with input in Excel, many percentage constraints, analysis of output	8	This is a very good "project" case.
Case 4.2	Maximization financial mix problem, percentage constraints, analysis of output	8	You may wish to give student a hint on how to formulate the risk and liquidity constraints. This is a "smaller" case than Case 4.1.
Case CD4.1	This is a classic "on the job training" problem with many "linking" constraints	9	Defining the variables can be challenging
Case CD4.2	Maximization production/assignment problem	9	You may wish to show the student how to define one of the 32 variables and show the calculation of its profit coefficient. Then, although there are 15 constraints, the problem is now more straightforward.

Problem Solutions

4.1 See file ch4-1.lpp

X1 = the number of standard Z345's produced weekly
X2 = the number of industrial Z345's produced weekly
X3 = the number of standard W250's produced weekly
X4 = the number of industrial W250's produced weekly
X5 = the total number of products produced weekly

$$
\begin{array}{lrrrrrrl}
\text{MAX} & 400X1 + & 560X2 + & 500X3 + & 700X4 & & & \\
\text{S.T.} & 25X1 + & 46X2 + & 16X3 + & 34X4 & & \leq 2500 & \text{(zinc)} \\
 & 50X1 + & 30X2 + & 28X3 + & 12X4 & & \leq 2800 & \text{(iron)} \\
 & X1 + & X2 & & & & \geq 20 & \text{(Min Z345's)} \\
 & X1 + & X2 + & X3 + & X4 - & X5 & = 0 & \text{(X5 definition)} \\
 & & X2 + & & X4 - & .50X5 & \geq 0 & \text{(Industrial min.)} \\
 & X1 + & X2 & & - & .75X5 & \leq 0 & \text{(Max Z345's)} \\
 & & & X3 + & X4 - & .75X5 & \leq 0 & \text{(Max W250's)} \\
 & & & & & & & X1, X2, X3, X4, X5 \geq 0
\end{array}
$$

	Decision Variable	Solution Value	Unit Cost or Profit c(j)	Total Contribution	Reduced Cost	Basis Status	Allowable Min. c(j)	Allowable Max. c(j)
1	STD Z345	22.9358	400	9,174.31	0	basic	299.4643	716
2	IND Z345	0	560	0	-103.3027	At bound	-M	663.3027
3	STD W250	22.9358	500	11,467.89	0	basic	345.6693	606.2264
4	IND W250	45.8716	700	32,110.09	0	basic	602.087	968.4932
5	TOTAL	91.7431	0	0	0	basic	-272.2222	M
	Objective	Function	(Max.) =	52,752.29				
	Constraint	Left Hand Side	Direction	Right Hand Side	Slack or Surplus	Shadow Price	Allowable Min. RHS	Allowable Max. RHS
1	ZINC	2,500.00	<=	2,500.00	0	21.1009	2,180.00	2,992.16
2	IRON	2,339.45	<=	2,800.00	460.5504	0	2,339.45	M
3	MIN Z345	22.9358	>=	20	2.9358	0	-M	22.9358
4	DEF-X5	0	=	0	0	-127.5229	-3.8095	17.3103
5	MIN IND.	0	>=	0	0	-179.8165	-14.0223	17.7778
6	MAX Z345	-45.8716	<=	0	45.8716	0	-45.8716	M
7	MAX W250	0	<=	0	0	289.9083	-21.1864	3.2

Produce 22.9358 standard Z345's, 22.9358 standard W250's, 45.8716 industrial W250's (fractional production quantities are work in progress carried over from one week to the next). Weekly profit = $52,752.29

4.2 See out put for problem 4.1 and files ch4-2iii.lpp and ch4-2iv.lpp

a. % W250's = (22.9358 + 45.8716)/91.7431 = 75%. If this restriction is loosened or eliminated, the weekly profit will increase.

b. The shadow price of zinc is $21.1009, which is valid for an additional 492.15 pounds.

i) 100 pounds is worth $2110.09 > $1500; yes purchase 100 additional pounds.

ii) $2110.09 < $2600, no, 100 additional pounds should not be purchased.

iii) Cannot tell without resolving since 800 additional pounds is outside the range of feasibility for the shadow price. Re-solving with 3300 pounds of zinc gives:

	Decision Variable	Solution Value	Unit Cost or Profit c(j)	Total Contribution	Reduced Cost	Basis Status	Allowable Min. c(j)	Allowable Max. c(j)
1	STD Z345	28.8268	400	11,530.73	0	basic	333.5208	828.9474
2	IND Z345	0	560	0	-66.0335	at bound	-M	626.0335
3	STD W250	20.0559	500	10,027.93	0	basic	345.6693	554.3948
4	IND W250	66.4246	700	46,497.21	0	basic	644.6888	968.4932
5	TOTAL	115.3073	0	0	0	basic	-272.2222	2,462.50
	Objective	Function	(Max.) =	68,055.87				

The profit increases by $15,303.87 to $68,055.87. Since $15,303.87 > $10,000, yes, 800 additional pounds should be purchased.

iv-v) Cannot tell without resolving since 900 additional pounds is outside the range of feasibility for the shadow price. Re-solving with 3400 pounds of zinc gives:

	Decision Variable	Solution Value	Unit Cost or Profit c(j)	Total Contribution	Reduced Cost	Basis Status	Allowable Min. c(j)	Allowable Max. c(j)
1	STD Z345	29.2737	400	11,709.50	0	basic	333.5208	828.9474
2	IND Z345	0	560	0	-66.0335	at bound	-M	626.0335
3	STD W250	17.6536	500	8,826.82	0	basic	345.6693	554.3948
4	IND W250	70.1676	700	49,117.32	0	basic	644.6888	968.4932
5	TOTAL	117.095	0	0	0	basic	-272.2222	2,462.50
	Objective	Function	(Max.) =	69,653.63				

The profit increases by $16,901.34 to $69,653.63. Thus if 900 pounds can be purchased for (iv) $12,000, they should be purchased; for (v) $25,000, they should not be purchased.

c. The range of optimality extends fairly far on either side of the objective function coefficients; thus the statement is correct.

4.3 See file ch4-3.lpp

X1 = amount invested in EAL stock
X2 = amount invested in BRU stock
X3 = amount invested in TAT stock
X4 = amount invested in long term bonds
X5 = amount invested in short term bonds
X6 = amount invested in the tax deferred annuity
X7 = the total amount invested in stocks only

MAX $.15X1 + .12\,X2 + .09X3 + .11X4 + .085X5 + .06X6$
S.T.

$$
\begin{array}{lllllllcll}
X1\,+ & X2\,+ & X3\,+ & X4\,+ & X5\,+ & X6\,+ & & = & 50{,}000 & \text{(Total)} \\
 & & & & & X6 & & \geq & 10{,}000 & \text{(TDA)} \\
 & & X3\,- & & & & .25X7 & \geq & 0 & \text{(Min EAL)} \\
-X1\,- & X2\,- & X3\,+ & X4\,+ & X5 & & & \geq & 0 & \text{(Bond/stock)} \\
 & & X3\,+ & & X5\,+ & X6 & & \leq & 12{,}500 & \text{(Low \%)} \\
X1\,+ & X2\,+ & X3 & & & & -X7 & = & 0 & \text{(Stocks)}
\end{array}
$$

$X1, X2, X3, X4, X5, X6, X7 \geq 0$

	Decision Variable	Solution Value	Unit Cost or Profit c(j)	Total Contribution	Reduced Cost	Basis Status	Allowable Min. c(j)	Allowable Max. c(j)
1	EAL	7,500.00	0.15	1,125.00	0	basic	0.12	M
2	BRU	0	0.12	0	-0.03	at bound	-M	0.15
3	TAT	2,500.00	0.09	225	0	basic	-0.01	M
4	LONG	30,000.00	0.11	3,300.00	0	basic	-M	0.135
5	SHORT	0	0.08	0	-0.13	at bound	-M	0.21
6	TDA	10,000.00	0.06	600	0	basic	-M	0.21
7	STOCKS	10,000.00	0	0	0	basic	-0.025	M
	Objective	Function	(Max.) =	5,250.00				
	Constraint	Left Hand Side	Direction	Right Hand Side	Slack or Surplus	Shadow Price	Allowable Min. RHS	Allowable Max. RHS
1	TOT INV	50,000.00	=	50,000.00	0	0.11	30,000.00	M
2	MIN TDA	10,000.00	>=	10,000.00	0	-0.15	7,142.86	12,500.00
3	MIN TAT	0	>=	0	0	-0.16	-2,500.00	1,875.00
4	MIN BONDS	20,000.00	>=	0	20,000.00	0	-M	20,000.00
5	MAX LOW YIELD	12,500.00	<=	12,500.00	0	0.1	10,000.00	15,000.00
6	STOCKS	0	=	0	0	0.04	-7,500.00	10,000.00

Invest in:

EAL	$ 7,500
TAT	$ 2,500
Long Term Bonds	$30,000
Tax Deferred Annuity	$10,000

Total return: $5,250

b. Rate of return for this investment = ($5,250/$50,000) = 10.5%
The rate of return for additional funds = shadow price for the total investment constraint (the first constraint above) = 11% which is valid to $+\infty$.

c. The return on EAL cannot fall below 12% from 15%; the return on BRU cannot increase above 15% from 12%; the rate on long term bonds cannot increase above 13.5% from 11%.

d. Shadow price for:
Total investment = .11 -- each additional dollar invested will earn 11%

Minimum invested in taxed deferred annuity -- $0.15 lost for each extra dollar required to be invested in tax deferred annuities.

Minimum invested in low risk stock -- a $0.16 decrease in return for each extra dollar required to be in the low risk stock above 25%.

Minimum investment in bonds -- no change in return for requiring bond investment to exceed stock investment by at least $1.

Maximum invested in low yield investments -- $0.10 additional return for each extra dollar allowed to be invested in investments with returns less than 10%.

4.4 See files ch4-4a.lpp and ch4-4b.lpp

a. X1 = number of polyester suits ordered for the season
X2 = the number of wool suits ordered for the season
X3 = the number of cotton suits ordered for the season
X4 = the number of imported suits ordered for the season

$$\begin{aligned}
\text{MAX } & 35X1 + 47X2 + 30X3 + 90X4 \\
\text{S.T. } & .4X1 + .5X2 + .3X3 + X4 \le 1{,}800 \text{ (Salesperson hours)} \\
& 2X1 + 4X2 + 3X3 + 9X4 \le 15{,}000 \text{ (Advertising budget)} \\
& X1 + 1.5X2 + 1.25X3 + 3X4 \le 18{,}000 \text{ (Square footage)} \\
& X1, X2, X3, X4 \ge 0
\end{aligned}$$

	Decision Variable	Solution Value	Unit Cost or Profit c(j)	Total Contribution	Reduced Cost	Basis Status	Allowable Min. c(j)	Allowable Max. c(j)
1	POLYESTER	1,500.00	35	52,499.99	0	basic	34	40
2	WOOL	0	47	0	-0.5	at bound	-M	47.5
3	COTTON	4,000.00	30	120,000.00	0	basic	29.5	52.5
4	IMPORTED	0	90	0	-7.5	at bound	-M	97.5
	Objective	Function	(Max.) =	172,500.00				
	Constraint	Left Hand Side	Direction	Right Hand Side	Slack or Surplus	Shadow Price	Allowable Min. RHS	Allowable Max. RHS
1	SALES HRS.	1,800.00	<=	1,800.00	0	75	1,500.00	3,000.00
2	AD$	15,000.00	<=	15,000.00	0	2.5	9,000.00	18,000.00
3	DISPLAY	6,500.00	<=	18,000.00	11,500.00	0	6,500.00	M

Order 1500 polyester suits, 4000 cotton suits --Profit = $172,500

b. Add lower bound constraints: $X1 \geq 200$, $X2 \geq 200$, $X3 \geq 200$, $X4 \geq 200$.

	Decision Variable	Solution Value	Unit Cost or Profit c(j)	Total Contribution	Reduced Cost	Basis Status	Allowable Min. c(j)	Allowable Max. c(j)
1	POLYESTER	1,300.00	35	45,499.99	0	basic	34	40
2	WOOL	200	47	9,400.00	-0.5	at bound	-M	47.5
3	COTTON	3,266.67	30	98,000.00	0	basic	29.5	52.5
4	IMPORTED	200	90	18,000.00	-7.5	at bound	-M	97.5
	Objective	Function	(Max.) =	170,900.00				

Order 1300 polyester, 200 wool, 3266.67(rounded to 3266) cotton, and 200 imported suits; the profit is now $170,900. It decreases because the problem is now more constrained.

4.5 See output from problem 4.4 and files ch4-5a3.lpp and ch4-5c.lpp

a.

	New Objective Coefficient	Within the Range of Optimality?	Change to Solution	Profit
(i)	$34	Yes	None (alternate optima)	$171,000
(ii)	$36	Yes	None	$174,000
(iii)	$33	No	3000 wool, 1000 cotton*	$171,000
(iv)	$37	Yes	None	$175,500

*OUTPUT for a (iii):

	Decision Variable	Solution Value	Unit Cost or Profit c(j)	Total Contribution	Reduced Cost	Basis Status	Allowable Min. c(j)	Allowable Max. c(j)
1	POLYESTER	0.00	33	0.00	-1	at bound	-M	34
2	WOOL	3,000.00	47	141,000.00	0	basic	46.5	50
3	COTTON	1,000.00	30	30,000.01	0	basic	28.2	30.5
4	IMPORTED	0	90	0.00	-7	at bound	-M	97
	Objective	Function	(Max.) =	171,000.00				

b. (i) No, there is a slack on the square footage constraint.
(ii) Yes, $400 will net 400(2.5) = $1000 additional profit.
(iii) Yes, 260 salesperson hours will add 260($75) = $19,500 for a cost of $3,600.

c. Add the constraint $X1 + X2 + X3 + X4 \leq 5000$.

	Decision Variable	Solution Value	Unit Cost or Profit c(j)	Total Contribution	Reduced Cost	Basis Status	Allowable Min. c(j)	Allowable Max. c(j)
1	POLYESTER	999.9999	35	35,000.00	0	basic	34	38.2
2	WOOL	999.9999	47	47,000.00	0	basic	45.7692	47.5
3	COTTON	3,000.00	30	90,000.01	0	basic	29.5	33.2
4	IMPORTED	0	90	0	-5.3333	at bound	-M	95.3333
	Objective	Function	(Max.) =	172,000.00				

Order 1000 polyester, 1000 wool and 3000 cotton suits giving a $172,000 profit.

4.6 See files ch4-6.lpp and ch4-6d.lpp

X1 = the number of ounces of Multigrain Cheerios in the mixture
X2 = the number of ounces of Grape Nuts in the mixture
X3 = the number of ounces of Product 19 in the mixture
X4 = the number of ounces of Frosted Bran in the mixture
X5 = the total number of ounces in the mixture

$$
\begin{array}{llr}
\text{MIN} & 12X1 + 9X2 + 9X3 + 15X4 & \\
\text{S.T.} & 30X1 + 30X2 + 20X3 + 20X4 \geq 50 & \text{(Vitamin A)} \\
 & 25X1 + 2X2 + 100X3 + 25X4 \geq 50 & \text{(Vitamin C)} \\
 & 25X1 + 25X2 + 25X3 + 25X4 \geq 50 & \text{(Vitamin D)} \\
 & 25X1 + 25X2 + 100X3 + 25X4 \geq 50 & \text{(Vitamin B6)} \\
 & 45X1 + 45X2 + 100X3 + 25X4 \geq 50 & \text{(Iron)} \\
 & X1 + X2 + X3 + X4 - X5 = 0 & \text{(Total)} \\
 & X1 - .1X5 \geq 0 & (\geq 10\% \text{ M/G Cheerios}) \\
 & X2 - .1X5 \geq 0 & (\geq 10\% \text{ Grape Nuts}) \\
 & X3 - .1X5 \geq 0 & (\geq 10\% \text{ Product 19}) \\
 & X4 - .1X5 \geq 0 & (\geq 10\% \text{ Frosted Bran}) \\
 & X1, X2, X3, X4, X5 \geq 0 &
\end{array}
$$

From the WINQSB outputs below, these are alternate optimal solutions giving 19.8 grams of sugar; any weighted average of the following is also optimal:

.2 oz. Multigrain Cheerios	.2 oz. Multigrain Cheerios
1.2245 oz. Grape Nuts	.8 oz. Grape Nuts
.3755 oz. Product 19	.8 oz. Product 19
.2 oz. Frosted Bran	.2 oz. Frosted Bran
Total 2.0 oz.	Total 2.0 oz.

OUTPUT 1:

	Decision Variable	Solution Value	Unit Cost or Profit c(j)	Total Contribution	Reduced Cost	Basis Status	Allowable Min. c(j)	Allowable Max. c(j)
1	CHEERIOS	0.2	12	2.40	0	basic	9	M
2	GRAPENUTS	1.2245	9	11.02	0	basic	-2.4141	9
3	PRODUCT19	0.38	9	3.38	0	basic	9	21.7826
4	FROSTBRAN	0.2	15	3	0	basic	9	M
5	TOTAL	2	0	0	0	basic	-9.9	M
	Objective	Function	(Min.) =	19.8	(Note:	Alternate	Solution	Exists!!)

	Constraint	Left Hand Side	Direction	Right Hand Side	Slack or Surplus	Shadow Price	Allowable Min. RHS	Allowable Max. RHS
1	VIT A	54.24	>=	50.00	4.2449	0	-M	54.24
2	VIT C	50.00	>=	50.00	0.00	0	32.80	91.6
3	VIT D	50.00	>=	50.00	0	0.396	46.42	76.22
4	VIT B6	78.1633	>=	50	28.1633	0	-M	78.1633
5	IRON	106.6531	>=	50	56.6531	0	-M	106.6531
6	TOTAL	0	=	0	0	-0.9	-1.1944	2
7	MIN CH	0	>=	0	0	3	-0.2	0.7478
8	MIN GN	1.0245	>=	0	1.0245	0	-M	1.0245
9	MIN P19	0.1755	>=	0	0.1755	0	-M	0.1755
10	MIN FB	0	>=	0	0	6	-0.2	0.5547

OUTPUT 2:

	Decision Variable	Solution Value	Unit Cost or Profit c(j)	Total Contribution	Reduced Cost	Basis Status	Allowable Min. c(j)	Allowable Max. c(j)
1	CHEERIOS	0.2	12	2.4	0	basic	9	M
2	GRAPENUTS	0.8	9	7.20	0	basic	9	12
3	PRODUCT19	0.8	9	7.2	0	basic	5.5862	9
4	FROSTBRAN	0.2	15	3	0	basic	9	M
5	TOTAL	2	0	0	0	basic	-9.9	M
	Objective	Function	(Min.) =	19.80	(Note:	Alternate	Solution	Exists!!)
	Constraint	Left Hand Side	Direction	Right Hand Side	Slack or Surplus	Shadow Price	Allowable Min. RHS	Allowable Max. RHS
1	VIT A	50	>=	50	0	0	44	54.2449
2	VIT C	91.6	>=	50	41.6	0	-M	91.6
3	VIT D	50	>=	50	0	0.396	46.4237	56.8182
4	VIT B6	110	>=	50	60	0	-M	110
5	IRON	130	>=	50	80	0	-M	130
6	TOTAL	0	=	0	0	-0.9	-3	2
7	MIN CH	0	>=	0	0	3	-0.2	0.6
8	MIN GN	0.6	>=	0	0.6	0	-M	0.6
9	MIN P19	0.6	>=	0	0.6	0	-M	0.6
10	MIN FB	0	>=	0	0	6	-0.2	0.5547

b. Total = 2 oz. of cereal and 2 (1/2) = 1 cup of skim milk

c. Extra % above 50% required of vitamin D adds .396 grams of sugar
Extra ounce above 10% required of Multigrain Cheerios adds 3 grams of sugar
Extra ounce above 10% required of Frosted Bran adds 6 grams of sugar

4.6d. Product 19 has less sugar and gives percentages that are at least as large as those for Frosted Bran for every vitamin and iron requirement. Re-solving gives the following alternate optimal solutions with 18 grams of sugar.

OUTPUT 1:

	Decision Variable	Solution Value	Unit Cost or Profit c(j)	Total Contribution	Reduced Cost	Basis Status	Allowable Min. c(j)	Allowable Max. c(j)
1	CHEERIOS	0	12	0	3	at bound	9	M
2	GRAPENUTS	1.5306	9	13.78	0	basic	0.18	9
3	PRODUCT19	0.4694	9	4.2245	0	basic	9	21.7826
4	FROSTBRAN	0	15	0	6	at bound	9	M
5	TOTAL	2	0	0	0	basic	-9	M
	Objective	Function	(Min.) =	18.00	(Note:	Alternate	Solution	Exists!!)

OUTPUT 2:

	Decision Variable	Solution Value	Unit Cost or Profit c(j)	Total Contribution	Reduced Cost	Basis Status	Allowable Min. c(j)	Allowable Max. c(j)
1	CHEERIOS	0	12	0	3	at bound	9	M
2	GRAPENUTS	1	9	9.00	0	basic	9	12
3	PRODUCT19	1	9	9	0	basic	6	9
4	FROSTBRAN	0	15	0	6	at bound	9	M
5	TOTAL	2	0	0	0	basic	-9	M
	Objective	Function	(Min.) =	18.00	(Note:	Alternate	Solution	Exists!!)

Any weighted average of the following is optimal:

	1.53 oz. Grape Nuts		1.0 oz. Grape Nuts
	.47 oz. Product 19		1.0 oz. Product 19
Total	2.00 oz.	Total	2.0 oz.

4.7 See files ch4-7a.lpp and ch4-7b.lpp

a. X1 = the number of standard doors manufactured weekly
X2 = the number of high glazed doors manufactured weekly
X3 = the number of engraved doors manufactured weekly

$$\begin{aligned}
\text{MAX} \quad & 45X1 + 90X2 + 120X3 \\
\text{S.T.} \quad & .5X1 + .5X2 + X3 \leq 250 \text{ (Manufacturing hours)} \\
& .25X1 + .5X2 + .5X3 \leq 150 \text{ (Finishing hours)} \\
& X1, X2, X3 \geq 0
\end{aligned}$$

	Decision Variable	Solution Value	Unit Cost or Profit c(j)	Total Contribution	Reduced Cost	Basis Status	Allowable Min. c(j)	Allowable Max. c(j)
1	STANDARD	0	45	0	-15	at bound	-M	60
2	HIGHGLAZE	100	90	9,000.00	0	basic	60	120
3	ENGRAVED	200	120	24,000.00	0	basic	90	180
	Objective	Function	(Max.) =	33,000.00				

Manufacture 100 high glazed and 200 engraved doors weekly; profit = $33,000.

b. Define X4 = the number of hours assigned to manufacturing weekly
X5 = the number of hours assigned to finishing weekly

$$\begin{aligned}
\text{MAX} \quad & 45X1 + 90X2 + 120X3 \\
\text{S.T.} \quad & .5X1 + .5X2 + X3 - X4 \leq 0 \text{ (Manufacturing hours)} \\
& .25X1 + .5X2 + .5X3 - X5 \leq 0 \text{ (Finishing hours)} \\
& X4 + X5 = 400 \text{ (Total hours)} \\
& X1, X2, X3, X4, X5 \geq 0
\end{aligned}$$

	Decision Variable	Solution Value	Unit Cost or Profit c(j)	Total Contribution	Reduced Cost	Basis Status	Allowable Min. c(j)	Allowable Max. c(j)
1	STANDARD	0	45	0	-22.5	at bound	-M	67.5
2	HIGHGLAZE	400	90	36,000.00	0	basic	80	M
3	ENGRAVED	0	120	0.00	-15	at bound	-M	135
4	MAN HRS.	200	0	0	0	basic	-180	60
5	FIN. HRS.	200	0	0.00	0	basic	-60	180
	Objective	Function	(Max.) =	36,000.00				

Manufacture only 400 high glazed doors weekly;
Allocate 200 hours weekly to manufacturing and 200 hours weekly to finishing;
Weekly profit = $36,000

4.8 See files ch4-8a.lpp and ch4-8b.lpp

a. X1 = the number of standard doors manufactured this week
X2 = the number of high glazed doors manufactured this week
X3 = the number of engraved doors manufactured this week
X4 = the number of premanufactured doors sold as standard doors
X5 = the number of premanufactured doors sold as high glazed doors

$$\begin{array}{llr}
\text{MAX} & 45X1 + 90X2 + 120X3 + 15X4 + 50X5 & \\
\text{S.T.} & .5X1 + .5X2 + X3 & \leq 250 \text{ (Manufacturing hours)} \\
& .25X1 + .5X2 + .5X3 + .1X4 + .5X5 & \leq 150 \text{ (Finishing hours)} \\
& X1 + X4 & \geq 280 \text{ (Standard doors)} \\
& X2 + X5 & \geq 120 \text{ (High glazed doors)} \\
& X3 & \geq 100 \text{ (Engraved doors)} \\
& X1, X2, X3, X4, X5 \geq 0 &
\end{array}$$

	Decision Variable	Solution Value	Unit Cost or Profit c(j)	Total Contribution	Reduced Cost	Basis Status	Allowable Min. c(j)	Allowable Max. c(j)
1	STANDARD	0	45	0	-6	at bound	-M	51
2	HIGHGLAZE	120	90	10,800.00	0	basic	50	120
3	ENGRAVED	124	120	14,880.00	0	basic	100	M
4	PRE-STD	280	15	4,200.00	0	basic	9	24
5	PRE-H/G	0	50	0.00	-40	at bound	-M	90
	Objective	Function	(Max.) =	29,880.00				

Manufacture 120 high glazed doors in house, 124 engraved doors in house, and produce 280 standard models from premanufactured doors; total weekly profit = $29,880.

b. Define X6 = the number of hours assigned to manufacturing for the week
X7 = the number of hours assigned to finishing the week

$$\begin{array}{llr}
\text{MAX} & 45X1 + 90X2 + 120X3 + 15X4 + 50X5 & \\
\text{S.T.} & .5X1 + .5X2 + X3 - X6 & \leq 0 \text{ (Manufacturing hours)} \\
& .25X1 + .5X2 + .5X3 + .1X4 + .5X5 - X7 & \leq 0 \text{ (Finishing hours)} \\
& X6 + X7 & = 400 \text{ (Total hours)} \\
& X1 + X4 & \geq 280 \text{ (Standard doors)} \\
& X2 + X5 & \geq 120 \text{ (High glazed doors)} \\
& X3 & \geq 100 \text{ (Engraved doors)} \\
& X1, X2, X3, X4, X5, X6, X7 \geq 0 &
\end{array}$$

	Decision Variable	Solution Value	Unit Cost or Profit c(j)	Total Contribution	Reduced Cost	Basis Status	Allowable Min. c(j)	Allowable Max. c(j)
1	STANDARD	0	45	0	-67.5	at bound	-M	112.5
2	HIGHGLAZE	0	90	0.00	-35	at bound	-M	125
3	ENGRAVED	100	120	12,000.00	0	basic	-M	225
4	PRE-STD	1,900.00	15	28,500.00	0	basic	10	M
5	PRE-H/G	120	50	6,000.00	0	basic	15	75
6	MAN HRS.	100	0	0	0	basic	-M	70
7	FIN. HRS.	300.00	0	0.00	0	basic	-70.00	M
	Objective	Function	(Max.) =	46,500.00				

Manufacture 100 engrave doors in house, produce 1900 standard and 120 engraved doors from premanufactured doors. Allocate 100 hours to manufacturing and 300 hours to finishing. The weekly profit is $46,500.

4.9 See file ch4-9.lpp

X1 = the number of Turkey De-Lite sandwiches made daily
X2 = the number of Beef Boy sandwiches made daily
X3 = the number of Hungry Ham sandwiches made daily
X4 = the number of Club sandwiches made daily
X5 = the number of All Meat sandwiches made daily

$$\begin{aligned}
\text{MAX}\quad & 2.75X1 + 3.5X2 + 3.25X3 + 4X4 + 4.25X5 \\
\text{S.T.}\quad & 4X1 + 2X4 + 3X5 \le 384 \text{ (Turkey)} \\
& 4X2 + 2X4 + 3X5 \le 576 \text{ (Beef)} \\
& 4X3 + 2X4 + 3X5 \le 480 \text{ (Ham)} \\
& X1 + X2 + 2X3 + 2X4 \le 384 \text{ (Cheese)} \\
& X1 + X2 + X3 + X4 + X5 \le 300 \text{ (Rolls)} \\
& X1, X2, X3, X4, X5 \ge 0
\end{aligned}$$

	Decision Variable	Solution Value	Unit Cost or Profit c(j)	Total Contribution	Reduced Cost	Basis Status	Allowable Min. c(j)	Allowable Max. c(j)
1	TURK DE-LIT	52	2.75	143	0	basic	1.25	4.375
2	BEEF BOY	100	3.5	350.00	0	basic	2	4.6
3	HUNG HAM	76	3.25	247.00	0	basic	1.75	4.35
4	CLUB	40.00	4	160.00	0	basic	3.6	4.75
5	ALL MEAT	32	4.25	136.00	0	basic	2.625	5.25
	Objective	Function	(Max.) =	1,036.00				

	Constraint	Left Hand Side	Direction	Right Hand Side	Slack Or Surplus	Shadow Price	Allowable Min. RHS	Allowable Max. RHS
1	TURKEY	384	<=	384	0	0.2292	246.8571	748.8
2	BEEF	576	<=	576	0	0.4167	438.8571	825.6
3	HAM	480	<=	480	0	0.2708	288	792
4	CHEESE	384	<=	384	0	0.3333	336	480
5	ROLLS	300	<=	300	0	1.5	248	320

Make 52 Turkey De-Lites, 100 Beef Boys, 76 Hungry Hams, 40 Clubs, 32 All Meat

Total Daily Revenue = $1,036
- Daily Supplies = $ 700
Net Daily Profit = $ 336 Net Annual Profit = 200($336) = $67,200

4.10 See output for problem 4.9 and files ch4-10d.lpp, and ch4-10e.lpp. Note the answer in the back of the text for part (e) is wrong.

a. Ranges of Optimality:

Turkey De-Lite	$1.25 - $4.375
Beef Boy	$2.00 - $4.60
Hungry Ham	$1.75 - $4.35
Club	$3.60 - $4.75
All Meat	$2.625 - $5.25

b.

Ingredient	Shadow Price	Range of Feasibility
Turkey	$0.2292	246.8571 - 748.8
Beef	$0.4167	438.8571 - 825.6
Ham	$0.2708	288.0000 - 792.0
Cheese	$0.3333	336.0000 - 480.0
Rolls	$1.50	248.0000 - 320.0

As long as a resource (individually) is within its range of optimality the shadow prices do not change, i.e. extra ounces of turkey will $0.2292 to profit, extra ounces of beef will add $0.4167 to profit, extra ounces of ham will add $0.2708 to profit, extra ounces of cheese will add $0.3333 to profit and extra rolls (or space available on the truck) will add $1.50 to profit.

c. Using the 100% rule, the total percent increase is (.10/1.625) + (.10/1.10) + (.10/1.10) + (.10/.75) + (.10/1) = 6.15% + 9.09% + 9.09% + 13.33% + 10% = 47.66% < 100%; thus the optimal solution will not change, but the daily revenue will increase by 300(.10) = $30. (Check results from ch4-10c.lpp.) You should assume demand does not change and Jami can still sell all the sandwiches made.

4.10d. Using the 100% rule, the total percent increase is (.25/1.625) + (.25/1.10) + (.25/1.10) + (.25/.75) + (.25/1) = 15.38% + 22.72% + 22.72% + 33.33% + 25% = 119.15% >100%; thus the optimal solution may change.

	Decision Variable	Solution Value	Unit Cost or Profit c(j)	Total Contribution	Reduced Cost	Basis Status	Allowable Min. c(j)	Allowable Max. c(j)
1	TURK DE-LIT	52	3	156	0	basic	1.25	4.625
2	BEEF BOY	100	3.75	375.00	0	basic	2	4.85
3	HUNG HAM	76	3.5	266.00	0	basic	1.75	4.6
4	CLUB	40.00	4.25	170.00	0	basic	3.85	5.125
5	ALL MEAT	32	4.5	144.00	0	basic	2.875	5.5
	Objective	Function	(Max.) =	1,111.00				

The optimal solution is the same; the optimal profit is now $1,111.

e. Turkey: Additional 8 lbs. = 128 oz. is within its range of feasibility, so this will add: 128(.2292) = $29.33 to revenue - $20 cost = $9.33 net additional profit
Beef: Additional 12 lbs. = 192 oz. is within its range of feasibility, so this will add: 192(.4167) = $80 to revenue - $42 cost = $38 net additional profit
Ham: Additional 10 lbs. = 160 oz. is within its range of feasibility, so this will add: 160(.2708) = $43.33 to revenue - $30 cost = $13.33 net additional profit
Cheese: Additional 8 lbs. = 128 oz. is NOT within its range of feasibility. The problem must be re-solved.

Output for 8 additional pounds of cheese:

	Decision Variable	Solution Value	Unit Cost or Profit c(j)	Total Contribution	Reduced Cost	Basis Status	Allowable Min. c(j)	Allowable Max. c(j)
1	TURK DE-LIT	36	2.85	102.6	0	basic	1.25	4.6
2	BEEF BOY	84	3.6	302.40	0	basic	2	4.85
3	HUNG HAM	60	3.35	201.00	0	basic	1.75	4.6
4	CLUB	120.00	4.1	492.00	0	basic	3.7	4.9
5	ALL MEAT	0	4.35	0.00	-1	at bound	-M	5.35
	Objective	Function	(Max.) =	1,098.00				

The new optimal revenue = $1098. This is an increase of $1098- $1036 = $62 in additional revenue - $18 cost = $44 net additional profit.

Buy the cheese.

4.11 See file ch4-11.lpp

X1 = 100's of men's jackets produced in the week
X2 = 100's of women's jackets produced in the week
X3 = 100's of men's pants produced in the week
X4 = 100's of women's pants produced in the week

$$
\begin{aligned}
\text{MAX} \quad & 2000X1 + 2800X2 + 1200X3 + 1500X4 \\
\text{S.T.} \quad & 150X1 + 125X2 + 200X3 + 150X4 \leq 2500 \text{ (Denim)} \\
& 3X1 + 4X2 + 2X3 + 2X4 \leq 36 \text{ (Cutting)} \\
& 4X1 + 3X2 + 2X3 + 2.5X4 \leq 36 \text{ (Stitching)} \\
& .75X1 + .75X2 + .50X3 + .50X4 \leq 8 \text{ (Boxing)} \\
& X1, X2, X3, X4, \geq 0
\end{aligned}
$$

	Decision Variable	Solution Value	Unit Cost or Profit c(j)	Total Contribution	Reduced Cost	Basis Status	Allowable Min. c(j)	Allowable Max. c(j)
1	M-JACKETS	0	2,000.00	0	-275	at bound	-M	2,275.00
2	W-JACKETS	4.5	2,800.00	12,600.00	0	basic	1,800.00	3,000.00
3	M-PANTS	0	1,200.00	0.00	-250	at bound	-M	1,450.00
4	W-PANTS	9.00	1,500.00	13,500.00	0	basic	1,400.00	2,333.33
	Objective	Function	(Max.) =	26,100.00				
	Constraint	Left Hand Side	Direction	Right Hand Side	Slack or Surplus	Shadow Price	Allowable Min. RHS	Allowable Max. RHS
1	DENIM	1,912.50	<=	2,500.00	587.5	0	1,912.50	M
2	CUTTING	36	<=	36	0	625	28.8	37.33
3	STITCHING	36	<=	36	0	100	27	37
4	BOXING	7.88	<=	8	0.13	0	7.88	M

Produce 450 women's jackets, 900 women's pants; weekly profit = $26,100

4.12 See file ch4-12a.lpp and ch4-12b.lpp

a. Add to the formulation $X1 \geq 5$, $X2 \geq 5$, $X3 \geq 5$, $X4 \geq 5$; the problem is now infeasible. 500 of each requires a minimum of 55 cutting hours and 57.5 stitching hours which exceeds the limit of 36 hours each.

4.12b. Add to the formulation X1 ≥ 3, X2 ≥ 3, X3 ≥ 3, X4 ≥ 3.

	Decision Variable	Solution Value	Unit Cost or Profit c(j)	Total Contribution	Reduced Cost	Basis Status	Allowable Min. c(j)	Allowable Max. c(j)
1	M-JACKETS	3	2,000.00	6,000.00	0	basic	-M	3,733.33
2	W-JACKETS	3.5	2,800.00	9,800.00	0	basic	1,800.00	M
3	M-PANTS	3	1,200.00	3,600.00	0	basic	-M	1,866.67
4	W-PANTS	3.00	1,500.00	4,500.00	0	basic	-M	2,333.33
	Objective	Function	(Max.) =	23,900.00				

Produce 300 men's jackets, 350 women's jackets, 300 men's pants, and 300 women's pants; weekly profit = $23,900

c. The reduced cost for all objective function coefficients are 0 since there is production of each item (complementary slackness).

Note: If the variables were defined to be the amount above 300 produced of each item and the right hand sides of the constraints had been adjusted to take into account this minimum production, only the objective function coefficient for women's jackets would have had a 0 reduced cost since this is the only product that exceeds 300.

4.13 See output for problem 4.11 and file ch4-13c2.lpp

a. The following are the reduced costs and ranges of optimality per 100 items -- to get a per item value, divide by 100.

Item	Reduced Cost	Range of Optimality
Men's Jackets	-$275	-∞ to $2,275
Women's Jackets	$0	$1,800 to $3,000
Men's Pants	-$250	-∞ to $1,450
Women's Pants	$0	$1,400 to $2,333.33

b.

Resource	Shadow Price	Range of Feasibility
Denim	$0	1912.5 to ∞
Cutting hours	$625	28.8 to 37.33
Stitching hours	$100	27 to 37
Boxing hours	$0	7.875 to ∞

4.13c. Currently all items produced are women's items. Thus adding a constraint requiring that at least 50% of the items produced be women's items would be a nonbinding constraint and the solution would not change.

Adding a constraint that at least 50% of the items produced be men's items ($X1 + X3 \geq .5(X1 + X2 + X3 + X4)$ or $.5X1 - .5X2 + .5X3 - .5X4 \geq 0$) is binding, giving:

	Decision Variable	Solution Value	Unit Cost or Profit c(j)	Total Contribution	Reduced Cost	Basis Status	Allowable Min. c(j)	Allowable Max. c(j)
1	M-JACKETS	5.1429	2,000.00	10,285.71	0	basic	1,900.00	2,275.00
2	W-JACKETS	5.1429	2,800.00	14,400.00	0	basic	2,600.00	2,900.00
3	M-PANTS	0	1,200.00	0.00	-66.6667	at bound	-M	1,266.67
4	W-PANTS	0.00	1,500.00	0.00	0	basic	1,428.57	1,600.00
	Objective	Function	(Max.) =	24,685.71				

Produce 514 men's jackets and 514 women's jackets (rounded) for a profit of $24,672.

4.14 See file ch4-14.lpp.

X1 = amount invested in Bonanza Gold
X2 = amount invested in Cascade Telephone
X3 = amount invested in the money market account
X4 = amount invested in two-year treasury bonds

$$
\begin{array}{llr}
\text{MAX} & .15X1 + .09X2 + .07X3 + .08X4 & \\
\text{S.T.} & X1 + X2 \leq 50{,}000 & \text{(Max stocks)} \\
& X1 + X2 + X3 \geq 60{,}000 & \text{(Min potential 9\%)} \\
& X4 \leq 30{,}000 & \text{(Max treasury bonds)} \\
& -.5X1 + .03X2 + .06X3 + .08X4 \geq 4{,}000 & \text{(Min return)} \\
& X1 + X2 + X3 + X4 = 100{,}000 & \text{(Total)} \\
& X1, X2, X3, X4 \geq 0 &
\end{array}
$$

	Decision Variable	Solution Value	Unit Cost or Profit c(j)	Total Contribution	Reduced Cost	Basis Status	Allowable Min. c(j)	Allowable Max. c(j)
1	BONANZA	2,075.47	0.15	311.3207	0	basic	0.09	0.4433
2	CASCADE	47,924.53	0.09	4,313.21	0	basic	0.0743	0.15
3	MONEY MKT	20,000.00	0.07	1,400.00	0	basic	-M	0.0823
4	TREAS. BND	30,000.00	0.08	2,400.00	0	basic	0.0677	M
	Objective	Function	(Max.) =	8,424.53				

Invest $2,075.14 in Bonanza Gold, $47,924.53 is Cascade Telephone, $20,000 in the money market account, $30,000 in two year treasury bonds. The expected return is $8,424.53 (8.42453%).

4.15 See files ch4-15.lpp and ch4-15c.lpp

Xij = number of gallons of crude i blended into grade j
i = p (Pacific), g (Gulf), m (Middle East) j = r (Regular), p (Premium)
Xj = total amount of grade j produced

Example of profit coefficient:
Selling price of regular = \$0.52, purchase cost of Pacific crude = (\$14.28)/42 = \$0.34; thus profit on a gallon of Pacific crude = \$0.52 - \$0.34 = \$0.18

$$
\begin{array}{lllllllll}
\text{MAX} & \multicolumn{8}{l}{.18Xpr + .16Xgr + .05Xmr + .26Xpp + .24Xgp + .13Xmp} \\
\text{S.T.} & Xpr & & & + Xpp & & & \leq 126{,}000 & \text{(Pacific)} \\
 & & Xgr & & & + Xgp & & \leq 84{,}000 & \text{(Gulf)} \\
 & & & Xmr & & & + Xmp & \leq 336{,}000 & \text{(M. East)} \\
 & Xpr + & Xgr + & Xmr & & & & \geq 200{,}000 & \text{(Min Reg)} \\
 & & & & Xpp + & Xgp + & Xmp & \geq 100{,}000 & \text{(Min Pre)} \\
 & Xpr + & Xgr + & Xmr - Xr & & & & = 0 & \text{(Regular)} \\
 & & & & Xpp + & Xgp + & Xmp - Xp & = 0 & \text{(Premium)} \\
 & 85Xpr + & 87Xgr + & 95Xmr - 87Xr & & & & \geq 0 & \text{(Reg Oct.)} \\
 & & & & 85Xpp + & 87Xgp + & 95Xmp - 91Xp & \geq 0 & \text{(Pre Oct)} \\
 & \multicolumn{8}{l}{Xpr, Xgr, Xmr, Xr, Xpp, Xgp, Xmp, Xp \geq 0}
\end{array}
$$

From file ch4-15, there are alternate optimal solutions giving a profit of \$61,620. Here are three:

Pacific - Regular	126,000	96,000	126,000
Gulf - Regular	42,500	0	0
Mid East - Regular	31,500	104,000	74,000
Pacific - Premium	0	30,000	0
Gulf - Premium	41,500	84,000	84,000
Mid East - Premium	158,500	86,000	116,000

b. Shadow price = \$0.13 per gallon; Thus 50,000 gallons is worth 50,000(\$0.13) = \$6,500 > \$5,000 -- Yes Caloco should secure the extra refining capacity.

c. All solutions have 190,000 gallons of Mid East Oil @ \$0.47/gallon = \$89,300.
8000 barrels @ \$16.80 per barrel = \$134,400 -- the Middle East distributors receive more.

To see if it would be profitable for Caloco, the problem must be re-solved. \$16.80 per barrel = \$0.40 per gallon. Thus the new profit coefficients for Xmr and Xmp would be \$0.12 and \$0.20 respectively. Change the third constraint to Xmr + Xmp = 336,000.

	Decision Variable	Solution Value	Unit Cost or Profit c(j)	Total Contribution	Reduced Cost	Basis Status	Allowable Min. c(j)	Allowable Max. c(j)
1	PAC-REG	64,000.00	0.18	11,520.00	0	basic	0.18	M
2	GULF-REG	0.00	0.16	0.00	-0.02	at bound	-M	0.18
3	MIDE-REG	136,000.00	0.12	16,320.00	0	basic	-M	0.12
4	PAC-PREM	0.00	0.26	0.00	0	at bound	-M	0.26
5	GULF-PREM	0	0.24	0	-0.02	at bound	-M	0.26
6	MIDE-PREM	200,000.00	0.2	40,000.00	0	basic	0.2	M
7	TOTAL-REG	200,000.00	0	0.00	0	basic	-M	0.08
8	TOTAL-PREM	200,000.00	0	0.00	0	basic	-0.08	M
	Objective	Function	(Max.) =	67,840.00				

This gives Caloco a profit of $67,840 -- thus it would be profitable for Caloco to accept this offer. There is now no Gulf crude purchased and only slightly more than 1500 barrels of Pacific crude purchased. These distributors would have to cut their prices to stay competitive.

4.16 See file ch4-16.lpp.

X1 = the number of acres of wheat planted
X2 = the number of acres of corn planted
X3 = the number of acres of oats planted
X4 = the number of acres of soybeans planted

Profit coefficients are 210($3.20) - $50 = $622, 300($2.55) - $75 = $690, 180($1.45) - $30 = $231, and 240($3.10) - $60 = $684 respectively.

$$
\begin{array}{llr l}
\text{MAX} & 622X1 + 690X2 + 231X3 + 684X4 & & \\
\text{S.T.} & 4X1 + 5X2 + 3X3 + 10X4 \le & 1,800 & \text{(Labor hours)} \\
& 50X1 + 75X2 + 30X3 + 60X4 \le & 25,000 & \text{(Expenses)} \\
& 2X1 + 6X2 + X3 + 4X4 \le & 1,200 & \text{(Water)} \\
& 210X1 \ge & 30,000 & \text{(Min. Wheat)} \\
& 300X2 \ge & 25,000 & \text{(Min. Corn)} \\
& 180X3 \le & 25,000 & \text{(Max Oats)} \\
& X1 + X2 + X3 + X4 \le & 300 & \text{(Total acres)} \\
& X1, X2, X3, X4 \ge 0 & &
\end{array}
$$

	Decision Variable	Solution Value	Unit Cost or Profit c(j)	Total Contribution	Reduced Cost	Basis Status	Allowable Min. c(j)	Allowable Max. c(j)
1	WHEAT	142.86	622	88,857.15	0	basic	-M	678
2	CORN	142.86	690	98,571.43	0	basic	684	746
3	OATS	0.00	231	0.00	-444	at bound	-M	675
4	SOYBEANS	14.29	684	9,771.43	0	basic	656	690
	Objective	Function	(Max.) =	197,200.00				
	Constraint	Left Hand Side	Direction	Right Hand Side	Slack or Surplus	Shadow Price	Allowable Min. RHS	Allowable Max. RHS
1	LABOR	1,428.57	<=	1,800.00	371.43	0	1,428.57	M
2	EXPENSES	18,714.29	<=	25,000.00	6,285.71	0	18,714.29	M
3	WATER	1,200.00	<=	1,200.00	0	3	1,114.29	1,228.57
4	MIN WHEAT	30,000.00	>=	30,000.00	0	-0.27	22,909.09	31,500.00
5	MIN CORN	42,857.14	>=	30,000.00	12,857.14	0.00	-M	42,857.14
6	MAX OATS	0	<=	25,000.00	25,000.00	0	0	M
7	ACRES	300.00	<=	300.00	0.00	672	295.24	318.57

Plant 142.86 acres of wheat, 142.86 acres of corn and 14.29 acres of soybeans; profit = $197,200.

4.17 See the output for problem 4.16 and files ch4-17b.lpp and ch4-17c.lpp.

a. The net profit must rise to $675; adding the $30/acre in expenses = $705.
$705 per acre/$1.45 per bushel = a yield of 486.21 bushels per acre.

b. Yes, corn would still be planted; there is currently slack on the corn constraint. If corn were not grown, the problem must be re-solved.

	Decision Variable	Solution Value	Unit Cost or Profit c(j)	Total Contribution	Reduced Cost	Basis Status	Allowable Min. c(j)	Allowable Max. c(j)
1	WHEAT	200.00	622	124,400.00	0	basic	295.71	684
2	OATS	0.00	231	0.00	-380.67	at bound	-M	611.67
3	SOYBEANS	100.00	684	68,400.00	0	basic	622	1,555.00
	Objective	Function	(Max.) =	192,800.00				

Plant 200 acres of wheat and 100 acres of soybeans for a profit of $92,800 -- a decrease of $4,400.

c. The range of feasibility for acres is only valid up to 318.57 acres (18.57 additional acres.) Thus the problem must be re-solved.

	Decision Variable	Solution Value	Unit Cost or Profit c(j)	Total Contribution	Reduced Cost	Basis Status	Allowable Min. c(j)	Allowable Max. c(j)
1	WHEAT	181.82	622	113,090.90	0	basic	341.17	678
2	CORN	101.82	690	70,254.55	0	basic	632.33	746
3	OATS	0.00	231	0.00	-370.18	at bound	-M	601.18
4	SOYBEANS	56.36	684	38,552.73	0	basic	656	1,030.00
	Objective	Function	(Max.) =	221,898.20				

Plant 181.82 acres of wheat, 101.82 acres of corn, and 56.36 acres of soybeans for a profit of $221,898.20 -- an increase of $24,698.20. Yes Bill should lease this property for $2,000.

4.18 See file ch4-18.lpp.

$X1$ = the weight given to teaching
$X2$ = the weight given to research
$X3$ = the weight given to professional activities
$X4$ = the weight given to service

The average teaching score is (90+75+90)/3 = 85, the average research score is (60+60+75)/3 = 65, the average professional score is (90+95+85)/3 = 90, and the average service score is (80+95+95) = 90.

$$
\begin{array}{llllll}
\text{MAX} & \multicolumn{5}{l}{85X1 + 65X2 + 90X3 + 90X4} \\
\text{S.T.} & -X1 + X2 & & & \leq 0 & \text{(Teaching} \geq \text{research)} \\
 & -X1 & + X3 & & \leq 0 & \text{(Teaching} \geq \text{professional)} \\
 & -X1 & & + X4 & \leq 0 & \text{(Teaching} \geq \text{service)} \\
 & X2 & & & \geq .25 & \text{(Min. research)} \\
 & X1 + X2 & & & \geq .75 & \text{(Teaching + research} \geq .75) \\
 & X1 + X2 & & & \leq .90 & \text{(Teaching + research} \leq .90) \\
 & & X3 - & X4 & \leq 0 & \text{(Service} \geq \text{professional)} \\
 & & X3 & & \geq .05 & \text{(Professional} \geq .05) \\
 & X1 + X2 + & X3 + & X4 & = 1 & \text{(Total weights)} \\
 & \multicolumn{5}{l}{X1, X2, X3, X4 \geq 0}
\end{array}
$$

From file ch4-18.lpp, there are alternate optimal solutions. Two are:

Teaching	.50	.50
Research	.25	.25
Professional	.125	.05
Service	.125	.20

The optimal value of either solution is 81.25 < 85; Professor Anna Sung should not get tenure.

4.19 See file ch4-19.lpp.

X1 = Amount invested in first trust deeds
X2 = Amount invested in second trust deeds
X3 = Amount invested in automobile loans
X4 = Amount invested in business loans
X5 = Amount invested in securities

MAX $.09X1 + .105X2 + .1225X3 + .1175X4 + .0675X5$

S.T.

$$
\begin{array}{llllll}
 & & & & X5 \le 3{,}333{,}333.33 & \text{(Max sec.)} \\
X1 + & X2 & & & - X5 \le 0 & \text{(Trust} \le \text{sec)} \\
-.49X1 - & .49X2 - & .49X3 + & .51X4 & \le 0 & \text{(Max bus. loan)} \\
-.50X1 - & .50X2 + & X3 & & \le 0 & \text{(Max auto loan)} \\
X1 + & X2 + & X3 + & X4 + & X5 = 10{,}000{,}000 & \text{(Total)} \\
\end{array}
$$

$$X1, X2, X3, X4, X5 \ge 0$$

	Decision Variable	Solution Value	Unit Cost or Profit c(j)	Total Contribution	Reduced Cost	Basis Status	Allowable Min. c(j)	Allowable Max. c(j)
1	1ST TRUST	0.00	0.09	0.00	-0.015	at bound	-M	0.105
2	2ND TRUST	2,537,314.00	0.105	266,417.90	0	basic	0.09	0.165
3	AUTO LOAN	1,268,657.00	0.1225	155,410.50	0	basic	0.0761	0.24
4	BUS LOAN	3,656,717.00	0.1175	429,664.20	0	basic	0.0935	M
5	SECURITIES	2,537,314.00	0.0675	171,268.70	0	basic	-0.3356	0.1141
	Objective	Function	(Max.) =	1,022,761.00				

Invest $2,537,313.50 in second trust deeds, $1,268,656.75 in auto loans, $3,656,716.50 in business loans, $2,537,313.50 in securities. Total return = $1,022,761.25.

4.20 a. Production costs are $37.15 and profit per medium grade boots is now $7.85. This is greater than the upper limit of its range of optimality of $7.67, so they should change their recommendation and recommend producing medium grade boots.

b. The new profit on walking shoes is now $5.25; this is within its range of optimality. Thus the optimal solution will not change but the optimal profit will increase by ($0.25)(1000) = $250.

c. The new profit coefficient for medium grade boots would now be $6.50; this is within its range of optimality. The optimal solution and profit will not change (since no medium grade boots are produced).

d. % increase for walking shoes = (.25/.33) = 75.8%
% decrease for medium grade boots = (.50/1.67) = 26.8%
Total % change = 102.6% > 100%

The optimal solution may change.

4.21 See file ch4-21.lpp.

$X1$ = advertising on ketchup only
$X2$ = advertising on spaghetti sauce only
$X3$ = advertising on taco sauce only
$X4$ = joint advertising

$$\begin{array}{lrrrrl}
\text{MAX} & 1.20X1 + & 1.12X2 + & 1.10X3 + & 1.05X4 & \\
\text{S.T.} & X1 + & X2 + & X3 + & X4 & \leq 2{,}000{,}000 \text{ (Total advertising)} \\
 & & & & X4 & \leq 400{,}000 \text{ (Max joint adv.)} \\
 & & & & X4 & \geq 100{,}000 \text{ (Min joint adv.)} \\
 & & & X3 + & X4 & \geq 1{,}000{,}000 \text{ (Min total taco sauce)} \\
 & X1 & & & & \geq 250{,}000 \text{ (Min ketchup only)} \\
 & & X2 & & & \geq 250{,}000 \text{ (Min spag sauce only)} \\
 & & & X3 & & \geq 750{,}000 \text{ (Min taco sauce only)} \\
 & X1 & & + & X4 & \geq 700{,}000 \text{ (Min total ketchup)} \\
 & & X2 & + & X4 & \geq 700{,}000 \text{ (Min total spag sauce)} \\
 & 4X1 + & 3.2X2 + & 11X3 + & 4.2X4 & \geq 7{,}500{,}000 \text{ (Min bottles sold)} \\
 & & & X1, X2, X3, X4 \geq 0 & &
\end{array}$$

	Decision Variable	Solution Value	Unit Cost or Profit c(j)	Total Contribution	Reduced Cost	Basis Status	Allowable Min. c(j)	Allowable Max. c(j)
1	KETCHUP	550,000	1.2	660,000	0	basic	1.17	M
2	SPAG. SAUCE	450,000	1.12	504,000	0	basic	1.05	1.15
3	TACO SAUCE	750,000	1.1	825,000	0	basic	-M	1.13
4	JOINT	250,000	1.05	262,500	0	basic	1.02	1.12
	Objective	Function	(Max.) =	2,251,500				
	Constraint	Left Hand Side	Direction	Right Hand Side	Slack or Surplus	Shadow Price	Allowable Min. RHS	Allowable Max. RHS
1	TOTAL ADV	2,000,000	<=	2,000,000	0	1.2	1,900,000	M
2	MAX JOINT	250,000	<=	400,000	150,000	0	250,000	M
3	MIN JOINT	250,000	>=	100,000	150,000	0	-M	250,000
4	MIN TACO	1,000,000	>=	1,000,000	0	-0.07	900,000	1,150,000
5	MIN KET ONLY	550,000	>=	250,000	300,000	0.00	-M	550,000
6	MIN SPAG ONLY	450,000	>=	250,000	200,000	0	-M	450,000
7	MIN TACO ONLY	750,000	>=	750,000	0	-0.03	600,000	800,000
8	KET YR TO YR	800,000	>=	700,000	100,000	0	-M	800,000
9	SPAG YR TO YR	700,000	>=	700,000	0	-0.08	500,000	800,000
10	TOTAL BOT	12,940,000	>=	7,500,000	5,440,000	0	-M	12,940,000

Advertising budget: \$550,000 ketchup only, \$450,000 spaghetti sauce only, \$750,000 taco sauce only, \$250,000 joint advertising

Total return = \$2,251,500 or (2,251,500/2,000,000) = 112.575%

4.22 See output for problem 4.21.

a. Shadow price = $1.20

b. Range of optimality for taco sauce = -∞ to $1.13 (current value = $1.10) and for joint advertising $1.02 to $1.12 (current value = $1.05). Since the current coefficients are close to a boundary of its range of optimality, the estimates of their values should be studied closely.

c. No effect since these are not binding constraints.

d. Shadow price = -.03. Since 700,000 is within its range of feasibility, the profit would increase by (-50,000)(-$0.03) = $1500.

4.23 See file ch4-23.lpp.

X1 = the number of 2-oz. Go bars produced daily
X2 = the number of 2-oz. Power bars produced daily
X3 = the number of 2-oz. Energy bars produced daily
X4 = the number of 8-oz. Energy bars produced daily
X5 = the total number of 2-oz. bars produced daily

Costs: Protein concentrate $0.40 per 2oz.
Sugar substitute $0.175 per 2oz.
Carob $0.325 per 2 oz.

Unit Profits:

2 oz. Go bars:	.68 - .03 - .2(.40) - .6(.175) - .2(.325)	= $0.40
2 oz. Power bars:	.84 - .03 - .5(.40) - .3(.175) - .2(.325)	= $0.4925
2 oz. Energy bars:	.76 - .03 - .3(.40) - .4(.175) - .3(.325)	= $0.4425
8 oz. Energy bars:	3.00 - .05 - .3(1.60) - .4(.70) - .3(1.30)	= $1.80

$$
\begin{array}{llllll}
\text{MAX} & .40X1 + .4925X2 + .4425X3 + 1.80X4 & & & & \\
\text{S.T.} & .4X1 + & X2 + & .6X3 + 2.4X4 & & \le 9{,}600 \text{ (Oz. protein)} \\
 & 1.2X1 + & .6X2 + & .8X3 + 3.2X4 & & \le 16{,}000 \text{ (Oz. sugar sub.)} \\
 & .4X1 + & .4X2 + & .6X3 + 2.4X4 & & \le 12{,}800 \text{ (Oz. carob)} \\
 & X1 + & X2 + & X3 + X4 & - X5 & = 0 \text{ (Total definition)} \\
 & & & & X5 & \le 25{,}000 \text{ (Max 2-oz.)} \\
 & & & X4 & & \le 2{,}000 \text{ (Max 8-oz.)} \\
 & X1 & & & & \ge 2{,}500 \text{ (Min 2-oz. Go)} \\
 & & X2 & & & \ge 2{,}500 \text{ (Min 2-oz. Power)} \\
 & & & X3 & & \ge 2{,}500 \text{ (Min 2-oz. Energy)} \\
 & X1 & & & -.5X5 & \le 0 \text{ (Max 2-oz. Go)} \\
 & & X2 & & -.5X5 & \le 0 \text{ (Max 2-oz. Power)} \\
 & & & X3 & -.5X5 & \le 0 \text{ (Max 2-oz. Energy)} \\
 & & & 2X3 + 4X4 & - X5 & \le 0 \text{ (Max Energy*)} \\
 & & X1, X2, X3, X4, X5 \ge 0 & & &
\end{array}
$$

*The Maximum Energy Bar constraint is formulated as follows:

Total Weight Energy Bars: $2X3 + 8X4$ Total Weight All Bars: $8X4 + 2X5$

Thus, $2X3 + 8X4 \le .5(8X4 + 2X5)$ or $2X3 + 4X4 - X5 \le 0$.

	Decision Variable	Solution Value	Unit Cost or Profit c(j)	Total Contribution	Reduced Cost	Basis Status	Allowable Min. c(j)	Allowable Max. c(j)
1	GO2	7,550	0.4000	3,020.0000	0	basic	0.3075	0.5348
2	POWER2	2,500	0.4925	1,231.2500	0	basic	-M	0.6407
3	ENERGY2	5,050	0.4425	2,234.6250	0	basic	0.35	0.45
4	ENERGY8	437.5	1.8000	787.5000	0	basic	1.77	2.02
5	TOTAL2	15,100.00	0	0.0000	0	basic	-0.0463	0.0075
	Objective	Function	(Max.) =	7,273.3750				

Daily production, giving daily profit of $7,273.375:

2-oz. Go:	7,550
2-oz. Power:	2,500
2-oz. Energy:	5,050
8-oz. Energy	437.5

4.24 See file ch4-24.

Xij = the number of Matey 20 catamarans made in plant I and sold to dealership j
i = 1 (San Diego), 2 (Santa Ana), 3 (San Jose)
j = 1 (Newport Beach), 2 (Long Beach), 3 (Ventura), 4 (San Luis Obispo) 5 (San Francisco)

MIN 1265X11 + 1285X12 + 1345X13 + 1390X14 + 1565X15 + 1130X21 + 1130X22 + 1285X23 + 1355X24 + 1405X25 + 1365X31 + 1340X32 + 1275X33 + 1225X34 + 1075X35

S.T.
X11 + X12 + X13 + X14 + X15 ≤ 38 (San Diego)
X21 + X22 + X23 + X24 + X25 ≤ 45 (Santa Ana)
X31 + X32 + X33 + X34 + X35 ≤ 58 (San Jose)
X11 + X21 + X31 = 42 (Newport Beach)
X12 + X22 + X32 = 33 (Long Beach)
X13 + X23 + X33 = 14 (Ventura)
X14 + X24 + X34 = 10 (San Luis Obispo)
X15 + X25 + X35 = 22 (San Francisco)
All Xij's ≥ 0

	Decision Variable	Solution Value	Unit Cost or Profit c(j)	Total Contribution	Reduced Cost	Basis Status	Allowable Min. c(j)	Allowable Max. c(j)
1	SD-NB	30	1,265.00	37,950.00	0	basic	1,130.00	1,285.00
2	SD-LB	0	1,285.00	0	20	at bound	1,265.00	M
3	SD-V	0	1,345.00	0	70	at bound	1,275.00	M
4	SD-SLO	0	1,390.00	0	165	at bound	1,225.00	M
5	SD-SF	0	1,565.00	0	490	at bound	1,075.00	M
6	SA-NB	12	1,130.00	13,560.00	0	basic	1,110.00	1,265.00
7	SA-LB	33	1,130.00	37,290.00	0	basic	-M	1,150.00
8	SA-V	0	1,285.00	0	145	at bound	1,140.00	M
9	SA-SLO	0	1,355.00	0	265	at bound	1,090.00	M
10	SA-SF	0	1,405.00	0	465	at bound	940	M
11	SJ-NB	0	1,365.00	0	100	at bound	1,265.00	M
12	SJ-LB	0	1,340.00	0	75	at bound	1,265.00	M
13	SJ-V	14	1,275.00	17,850.00	0	basic	-M	1,345.00
14	SJ-SLO	10	1,225.00	12,250.00	0	basic	-M	1,390.00
15	SJ-SF	22	1,075.00	23,650.00	0	basic	-M	1,540.00
	Objective	Function	(Min.) =	142,550.00				

Make 30 in San Diego - ship all to Newport Beach
Make 45 in Santa Ana - ship 12 to Newport Beach, 33 to Long Beach
Make 46 in San Jose - ship 14 to Ventura, 10 to San Luis Obispo, 22 to San Francisco
Total Cost = $142,550

4.25 See file ch4-25.lpp.

X1 = the number of toms of Alpha fuel produced next month
X2 = the number of tons of Beta fuel produced next month
X3 = the number of tons of Gamma fuel produced next month

Raw Propellant Cost = \$10/pound = \$20,000/ton
Expenses Affecting Cash Position = \$50,000 + \$20,000 + \$30,000 = \$100,000
Available Cash = \$1,600,000 - Expenses = \$1,600,000 - \$100,000 = \$1,500,000

	Material + Labor Costs (Per Ton)	Profit (Before Tax) Per Ton
Alpha:	3(\$20,000) + \$60,000 = \$120,000	\$200,000 - \$120,000 = \$80,000
Beta:	1(\$20,000) + \$40,000 = \$ 60,000	\$120,000 - \$ 60,000 = \$60,000
Gamma:	4(\$20,000) + \$50,000 = \$130,000	\$195,000 - \$130,000 = \$65,000

$$
\begin{array}{llll}
\text{MAX} & 80000X1 + 60000X2 + 65000X3 & & \\
\text{S.T.} & 3X1 + X2 + 4X3 \le & 30 & \text{(Propellant)} \\
 & 30X1 + 10X2 + 20X3 \le & 250 & \text{(Processing time)} \\
 & 40X1 + 30X2 + 30X3 \le & 170 & \text{(QC/Test time)} \\
 & 120000X1 + 60000X2 + 130000X3 \le & 1{,}500{,}000 & \text{(Cash)} \\
 & X1, X2, X3 \ge 0 & &
\end{array}
$$

	Decision Variable	Solution Value	Unit Cost or Profit c(j)	Total Contribution	Reduced Cost	Basis Status	Allowable Min. c(j)	Allowable Max. c(j)
1	ALPHA	0	80,000.00	0.00	-6,666.67	at bound	-M	86,666.66
2	BETA	0	60,000.00	0	-5,000.00	at bound	-M	65,000.00
3	GAMMA	5.67	65,000.00	368,333.30	0	basic	60,000.00	M
	Objective	Function	(Max.) =	368,333.30				
	Constraint	Left Hand Side	Direction	Right Hand Side	Slack or Surplus	Shadow Price	Allowable Min. RHS	Allowable Max. RHS
1	PROPELLANT	22.67	<=	30	7.33	0	22.67	M
2	PROCESSING	113.33	<=	250	136.67	0	113.33	M
3	QC/TEST	170	<=	170	0	2,166.67	0.00	225
4	CASH	736,666.60	<=	1,500,000.00	763,333.30	0	736,666.70	M

Produce 5.6667 tons of Gamma fuel only.

Before tax profit		\$368,333.30
Taxes @ 32%\$	-	\$117,866.66
After Tax Profit		\$250,466.64

Sensitivity Discussion:

1. Solution will remain optimal unless profit for Gamma fuel falls below $60,00 per ton or profit for Alpha fuel increases to over $86,666.66 per ton or profit for Beta fuel increases to over $65,000 per ton.

2. If Alpha or Beta fuel is produced, Clyadetics will lose $6,666.67 and $5,000 in pre-tax profits respectively for each ton of Alpha or Beta fuel produced.

3. Only 22.67 tons of raw propellant and 113.33 processing hours are used and the cash expended amounts to only $736,666.66. There are 7.33 tons of raw propellant and 136.67 hours of processing time left over and there is an additional $763,333.33 that could have been used during the moth.

4. All 179 quality control/test hours are used. Additional hours will gross the company an additional $2,166,67 in pre-tax profits for 55 additional hours during which time these resources would still be only dedicated to the production of Gamma fuel.

CD4.1 See file cd4-1.lpp

X1 = the number of production runs
X2 = 100,000's of grapes processed and sold

$$
\begin{array}{lrrrl}
\text{MAX} & 70000X1 + 25000X2 & & & \\
\text{S.T.} & 600X1 + 200X2 & \le & 3300 & \text{(Hours)} \\
 & X1 & \le & 5 & \text{(Max production runs)} \\
 & X2 & \le & 5 & \text{(Max grapes processed)} \\
 & 2.4X1 + X2 & \le & 14 & \text{(Max grapes available)} \\
 & X1, X2 \ge 0 & & &
\end{array}
$$

	Decision Variable	Solution Value	Unit Cost or Profit c(j)	Total Contribution	Reduced Cost	Basis Status	Allowable Min. c(j)	Allowable Max. c(j)
1	PROD. RUNS	4.17	70,000.00	291,666.70	0	basic	60,000.00	75,000.00
2	GRAPES	4	25,000.00	99,999.95	0	basic	23,333.33	29,166.67
	Objective	Function	(Max.) =	391,666.70				

4 1/6 production runs = 41,666 2/3 cases of wine; 400,000 pounds of grapes processed
Profit = $391,666.70.

CD4.2 See file cd4-2.lpp

X1 = the number of PAD assemblies produced daily
X2 = the number of PAW assemblies produced daily
X3 = the number of PAT assemblies produced daily

$$
\begin{array}{lrrrl}
\text{MAX} & 800X1 + 900X2 + 600X3 & & & \\
\text{S.T.} & X1 + X2 + X3 & \le & 7 & \text{(X70686 chips)} \\
 & 2X1 + X2 + X3 & \le & 8 & \text{(Production hours)} \\
 & 80X1 + 160X2 + 80X3 & \le & 480 & \text{(Quality minutes)} \\
 & X1, X2, X3 \ge 0 & & &
\end{array}
$$

	Decision Variable	Solution Value	Unit Cost or Profit c(j)	Total Contribution	Reduced Cost	Basis Status	Allowable Min. c(j)	Allowable Max. c(j)
1	PAD	2	800.00	1,600.00	0	basic	600.00	900.00
2	PAW	0	900.00	0.00	-100	at bound	-M	1,000.00
3	PAT	4	600	2,400.00	0	basic	566.6667	800
	Objective	Function	(Max.) =	4,000.00				

Produce 2 PAD's and 4 PAT's; daily profit = $4,000. No PAW's are produced because the use too much quality control time compared to its slightly higher unit profit.

CD4.3 See the output for problem CD4.2.

a. Reduced cost = -100; thus minimum profit for production = \$1,000. Since costs are \$900, this implies a minimum price of \$1,900.

b. 6; there is a slack of 1.

c. (i) There is slack on X70686 chips, thus additional profit is \$0.

(ii) Production hours are sunk costs. Since the shadow price for production hours is \$200 and 3 additional hours is within the range of feasibility, these 3 hours gross 3(\$200) = \$600 additional profit. Thus the net additional profit = \$600 - \$525 = \$75.

(iii) Quality control hours are also sunk costs. Since 1 additional hour = 60 additional minutes is also within its range of feasibility, the \$5 shadow price is valid for all 60 minutes grossing 60(\$5) = \$300 additional profit. The net additional profit is \$300 - \$200 = \$100.

OPTION (iii) is of the most value.

CD4.4 See file cd4-4.lpp.

X1 = the number in group I contacted by telephone
X2 = the number in group II contacted by telephone
X3 = the number in group III contacted by telephone
X4 = the number in group IV contacted by telephone
X5 = the number in group I contacted in person
X6 = the number in group II contacted in person
X7 = the number in group III contacted in person
X8 = the number in group IV contacted in person

$$
\begin{array}{llr}
\text{MIN} & 15X1 + 12X2 + 20X3 + 18X4 + 35X5 + 30X6 + 50X7 + 40X8 & \\
\text{S.T.} & X1 + X2 + X3 + X4 + X5 + X6 + X7 + X8 = 2000 & \text{(Total)} \\
& X1 + X2 + X5 + X6 \geq 1000 & \text{(W\&R)} \\
& X5 + X6 + X7 + X8 \geq 500 & \text{(In pers)} \\
& -.5X1 + .5X5 \geq 0 & \text{(W\&R,ip)} \\
& X2 + X4 + X6 + X8 \leq 800 & \text{(Small)} \\
& -.25X2 - .25X4 + .75X6 + .75X8 \leq 0 & \text{(Small,ip)} \\
& X1 + X5 \geq 200 & \text{(Min I)} \\
& X1 + X5 \leq 1000 & \text{(Max I)} \\
& X2 + X6 \geq 200 & \text{(Min II)} \\
& X2 + X6 \leq 1000 & \text{(Max II)} \\
& X3 + X7 \geq 200 & \text{(Min III)} \\
& X3 + X7 \leq 1000 & \text{(Max III)} \\
& X4 + X8 \geq 200 & \text{(Min IV)} \\
& X4 + X8 \leq 1000 & \text{(Max IV)} \\
& X1, X2, X3, X4, X5, X6, X7, X8 \geq 0 &
\end{array}
$$

	Decision Variable	Solution Value	Unit Cost or Profit c(j)	Total Contribution	Reduced Cost	Basis Status	Allowable Min. c(j)	Allowable Max. c(j)
1	LG W&R T	500	15.00	7,500.00	0	basic	7.00	17.00
2	SM W&R T	600	12.00	7,200.00	0	basic	10.00	18.00
3	LG OTHER T	200	20	4,000.00	0	basic	16	32
4	SM OTHER T	200	18	3,600.00	0	basic	12	22
5	LG W&R P	500	35	17,500.00	0	basic	33	43
6	SM W&R P	0	30	0.00	0	basic	22	32
7	LG OTHER P	0	50	0	12	at bound	38	M
8	SM OTHER P	0	40	0	4	at bound	36	M
	Objective	Function	(Min.) =	39,800.00				

Minimum Cost = $39,800; conduct survey as follows:

	Telephone	Personal
I	500	500
II	600	0
III	200	0
IV	200	0

CD4.5 See files ch4-5a.lpp and ch4-5b.lpp.

a. X1 = the number of trucks sent from East Warehouse to Northpark
X2 = the number of trucks sent from East Warehouse to Central City
X3 = the number of trucks sent from East Warehouse to Southgate
X4 = the number of trucks sent from West Warehouse to Northpark
X5 = the number of trucks sent from West Warehouse to Central City
X6 = the number of trucks sent from West Warehouse to Southgate

MIN $9X1 + 25X2 + 33X3 + 20X4 + 38X5 + 49X6$

S.T.

$X1 + X2 + X3 \leq 20$ (East)

$X4 + X5 + X6 \leq 20$ (West)

$X1 + X4 = 10$ (Northpark)

$X2 + X5 = 12$ (Central City)

$X3 + X6 = 15$ (Southgate)

$X1, X2, X3, X4, X5, X6 \geq 0$

	Decision Variable	Solution Value	Unit Cost or Profit c(j)	Total Contribution	Reduced Cost	Basis Status	Allowable Min. c(j)	Allowable Max. c(j)
1	EAST-NP	0	9.00	0.00	2	At bound	7.00	M
2	EAST-CC	5	25.00	125.00	0	basic	22.00	27.00
3	EAST-SG	15	33	495.00	0	basic	-M	36
4	WEST-NP	10	20	200.00	0	basic	-M	22
5	WEST-CC	7	38	266.00	0	basic	36	41

6	WEST-SG	0	49	0.00	3	At bound	46	M
	Objective	Function	(Min.) =	1,086.00				

Minimum mileage = 1086 total miles with the following trucking assignments:

East to Central City = 5 | West to Northpark = 10
to Southgate = 15 | to Central City = 7

b. The problem is now infeasible. 41 trucks are needed but only 40 are available.

CD4.6 See output for problem CD4.5.

a. Yes; the minimum value of the range of optimality is 46.

b. (i) Shadow price of -13 for trucks available at the East warehouse, means that for each extra truck made available to the East warehouse (up to 7 extra, 27 total) the total number of miles will decree by 13.

(ii) Shadow price of +20 for Northpark means that for each extra truck needed at this location (up to 3 more, 13 total) the total number of miles will increase by 20.

CD4.7 See file cd4-7.lpp.

a. X1 = the amount of time spent in large lectures
X2 = the amount of time spent in recitation sections
X3 = the amount of time spent with the individual instructor

MAX X1 + 2X2 + 10X3

S.T.
3X1 + 8X2 + 120X3 $\leq$ 2,160 (Teacher Time)
X1 + X2 + X3 = 180 (Total instruction)
X3 $\geq$ 10 (Min individual instruction)
.5X1 - .5X2 $\leq$ 0 (Max large lectures)
X1, X2, X3 $\geq$ 0

	Decision Variable	Solution Value	Unit Cost or Profit c(j)	Total Contribution	Reduced Cost	Basis Status	Allowable Min. c(j)	Allowable Max. c(j)
1	LECTURE	80.0000	1.0000	80.0000	0	basic	-M	1.6429
2	RECITATION	90.0000	2.0000	180.0000	0	basic	1.4576	M
3	INDIVIDUAL	10.0000	10.0000	100.0000	0	basic	-M	24.4000
	Objective	Function	(Max.) =	360.0000				

Schedule: 80 minutes in large lectures, 90 minutes in recitation sections, and 10 minutes of individual instruction.

b. Probably an 80-minute lecture should be scheduled once a week along with two 45-minute recitation sessions and 10 minutes of individual instruction.

CD4.8 See file cd4-8.lpp.

$X1$ = the number cut into 4 30-in. pieces
$X2$ = the number cut into 2 30-in. and 1 42-in. pieces
$X3$ = the number cut into 1 30-in. and 2 42-in. pieces
$X4$ = the number cut into 2 30-in. and 1 56-in. pieces
$X5$ = the number cut into 1 42-in. and 1 56-in. pieces
$X6$ = the number cut into 2 56-in. pieces

$$
\begin{array}{lllr}
\text{MIN} & 18X2 + 6X3 + 4X4 + 22X5 + 8X6 & & \\
\text{S.T.} & 4X1 + 2X2 + X3 + 2X4 & \geq 1{,}500 & \text{(30-in.)} \\
 & X2 + 2X3 + X5 & \geq 500 & \text{(42-in.)} \\
 & X4 + X5 + 2X6 & \geq 600 & \text{(56-in.)} \\
 & X1, X2, X3, X4, X5, X6 \geq 0 & &
\end{array}
$$

From file cd4-8.lpp, the following are alternate optimal solutions:

4 30-in. pieces	312.5 = 313	12.5 = 13
2 30-in. and 1 42-in. pieces		
1 30-in. and 2 42-in. pieces	250	250
2 30-in. and 1 56-in. pieces		600
1 42-in. and 1 56-in. pieces		
2 56-in. pieces	300	

Total 10-foot pipes = 863 -- total waste = 3900 inches.
Two 30-in. pieces will remain in inventory with rounded solution.

CD4.9 See file cd4-9.lpp.

a. This could violate the "no interaction" assumption of linear programming.

b. $X1$ = $ spent on television advertising
$X2$ = $ spent on radio advertising
$X3$ = $ spent on newspaper advertising
$X4$ = $ spent on its circulars

$$
\begin{array}{llllllll}
\text{MAX} & 28X1 + 18X2 + 20X3 + 15X4 & & & & & & \\
\text{S.T.} & X1 + & X2 + & X3 + & X4 & \leq & 700{,}000 & \text{(Budget)} \\
 & X1 + & X2 & & & \geq & 350{,}000 & \text{(TV/Radio)} \\
 & 10X1 + & 7X2 + & 8X3 + & 4X4 & \geq & 2{,}500{,}000 & \text{(Yuppies)} \\
 & 5X1 + & 2X2 + & 3X3 + & X4 & \geq & 1{,}200{,}000 & \text{(College)} \\
 & 5X1 + & 8X2 + & 6X3 + & 9X4 & \geq & 1{,}800{,}000 & \text{(Audiophiles)} \\
 & X1 & & & & \leq & 300{,}000 & \text{(Max TV)} \\
 & & X2 & & & \leq & 300{,}000 & \text{(Max Radio)} \\
 & & & X3 & & \leq & 300{,}000 & \text{(Max Newspapers)} \\
 & & & & X4 & \leq & 300{,}000 & \text{(Max Circulars)} \\
 & X1, X2, X3, X4 \geq 0 & & & & & &
\end{array}
$$

	Decision Variable	Solution Value	Unit Cost or Profit c(j)	Total Contribution	Reduced Cost	Basis Status	Allowable Min. c(j)	Allowable Max. c(j)
1	TV	300,000	28	8,400,000	0	basic	18	M
2	RADIO	100,000	18	1,800,000	0	basic	15	20
3	NEWS	300,000	20	6,000,000	0	basic	18	M
4	CIRCULAR	0	15	0	-3	at bound	-M	18
	Objective	Function	(Max.) =	16,200,000				

	Constraint	Left Hand Side	Direction	Right Hand Side	Slack or Surplus	Shadow Price	Allowable Min. RHS	Allowable Max. RHS
1	BUDGET	700,000	=	700,000	0	18	650,000	900,000
2	TV/RADIO	400,000	>=	350,000	50,000	0	-M	400,000
3	YUPPIE	6,100,000	>=	2,500,000	3,600,000	0	-M	6,100,000
4	COLLEGE	2,600,000	>=	1,200,000	1,400,000	0	-M	2,600,000
5	AUDIOPHILE	4,100,000	>=	1,800,000	2,300,000	0	-M	4,100,000
6	MAX TV	300,000	<=	300,000	0	10	100,000	400,000
7	MAX RAD	100,000	<=	300,000	200,000	0	100,000	M
8	MAX NEWS	300,000	<=	300,000	0	2	100,000	350,000
9	MAX CIRC	0	<=	300,000	300,000	0	0	M

Maximum exposure units = 16,200,000 by spending $300,000 on TV, $100,000 on radio, and $300,000 on newspaper ads.

c. It would not affect the optimal solution; it is a non-binding constraint (there is slack).

CD4.10 See file cd4-10.lpp.

X1 = the number of initial mailings sent out with no inducements
X2 = the number of initial mailings sent out with prenotification notices
X3 = the number of initial mailings sent out with return postage
X4 = the number of initial mailings sent out with both prenotification/return post.
Y1 = the number of follow-up mailings to those who received no inducements
Y2 = the number of follow-up mailings to those who received prenotification notices
Y3 = the number of follow-up mailings to those who received return postage
Y4 = the number of follow-up mailings to those who received both prenotification/return postage

$$\begin{aligned}
\text{MIN } & .35X1 + .70X2 + .67X3 + 1.02X4 + 1.15Y1 + 1.50Y2 + 1.47Y3 + 1.82\,Y4 \\
\text{S.T. } & X1 + X2 + X3 + X4 = 2000 \\
& .35X1 + .38X2 + .37X3 + .41X4 + .30Y1 + .32Y2 + .31Y3 + .33Y4 \geq 1100 \\
& -.65X1 + Y1 \leq 0 \\
& -.62X2 + Y2 \leq 0 \\
& -.63X3 + Y3 \leq 0 \\
& -.59X4 + Y4 \leq 0 \\
& X1, X2, X3, X4, Y1, Y2, Y3, Y4 \geq 0
\end{aligned}$$

	Decision Variable	Solution Value	Unit Cost or Profit c(j)	Total Contribution	Reduced Cost	Basis Status	Allowable Min. c(j)	Allowable Max. c(j)
1	NONE	1,700.60	0.35	595.21	0	Basic	0.2935	0.6983
2	PRENOT	299.40	0.7	209.58	0	Basic	0.3517	0.7249
3	RETURNP	0.00	0.67	0.00	0.175	at bound	0.495	M
4	BOTH	0	1.02	0	0.0445	at bound	0.9755	M
5	F-UP NONE	1,105.39	1.15	1,271.20	0	Basic	1.0631	1.6813
6	F-UP PREN	185.629	1.5	278.44	0	Basic	0.9382	1.5402
7	F-UP RET	0	1.47	0	0	Basic	1.1923	4.9424
8	F-UP BOTH	0	1.82	0	0	Basic	1.7446	5.2612
	Objective	Function	(Min.) =	2,354.43				

Mail 1700 initial pieces with no inducements and 300 with prenotification notices.
Mail 1105 follow-up pieces to people who were initially mailed no inducements.
Mail 186 follow-up pieces to people who initially received prenotification notices.
Total cost = $2,354.43.

Case 4.1 See file ca4-1.xls or ca4-1.lpp.

- X1 = the number of student economy desks produced in September
- X2 = the number of standard economy desks produced in September
- X3 = the number of executive economy desks produced in September
- X4 = the number of student basic desks produced in September
- X5 = the number of standard basic desks produced in September
- X6 = the number of executive basic desks produced in September
- X7 = the number of student hand crafted desks produced in September
- X8 = the number of standard hand crafted desks produced in September
- X9 = the number of executive hand crafted desks produced in September
- X10 = the total number of desks produced in September

MAX $20X1 + 30X2 + 40X3 + 50X4 + 80X5 + 125X6 + 100X7 + 250X8 + 325X9$

S.T. $X1 + X2 + X3 + X4 + X5 + X6 + X7 + X8 + X9 - X10 = 0$ (Def. X10)

$X1 \geq 750$ (Orders -- Student economy)

$X2 \geq 1500$ (Orders -- Standard economy)

$X3 \geq 100$ (Orders -- Executive economy)

$X4 \geq 400$ (Orders -- Student basic)

$X5 \geq 1500$ (Orders -- Standard basic)

$X6 \geq 100$ (Orders -- Executive basic)

$X7 \geq 25$ (Orders -- Student hand crafted)

$X8 \geq 150$ (Orders -- Standard hand crafted)

$X9 \geq 50$ (Orders -- Executive hand crafted)

$15X1 + 17X2 + 19X3 + 23X4 + 28X5 + 32X6 + 76X7 + 93X8 + 110X9 \leq 230{,}400$ (Estimated man-minutes)

$14X1 + 24X2 + 30X3 \leq 65{,}000$ (Aluminum)

$8X1 + 15X2 + 24X3 \leq 60{,}000$ (Particle board)

$22X4 + 40X5 + 55X6 + 25X7 + 45X8 + 60X9 \leq 175{,}000$ (Oak)

$1.5X1 + 2X2 + 2.5X3 \leq 9{,}600$ (Production line 1)

$X1 + X2 + X3 + X4 + X5 + X6 \leq 9{,}600$ (Production line 2)

$3X4 + 4X5 + 5X6 + 3X7 + 4X8 + 5X9 \leq 19{,}200$ (Production line 3)

$-X1 - X2 - X3 + .2X10 \leq 0$ (Minimum economy desks)

$X1 + X2 + X3 - .5X10 \leq 0$ (Maximum economy desks)

$-X4 - X5 - X6 + .4X10 \leq 0$ (Minimum basic desks)

$X4 + X5 + X6 - .6X10 \leq 0$ (Maximum basic desks)

$-X7 - X8 - X9 + .1X10 \leq 0$ (Minimum hand crafted desks)

$X7 + X8 + X9 - .2X10 \leq 0$ (Maximum hand crafted desks)

$-X1 - X4 - X7 + .2X10 \leq 0$ (Minimum student desks)

$X1 + X4 + X7 - .35X10 \leq 0$ (Maximum student desks)

$-X2 - X5 - X8 + .4X10 \leq 0$ (Minimum standard desks)

$X2 + X5 + X8 - .7X10 \leq 0$ (Maximum standard desks)

$-X3 - X6 - X9 + .05X10 \leq 0$ (Minimum executive desks)

$X3 + X6 + X9 - .15X10 \leq 0$ (Maximum executive desks)

$X1, X2, X3, X4, X5, X6, X7, X8, X9, X10 \geq 0$

	Decision Variable	Solution Value	Unit Cost or Profit c(j)	Total Contribution	Reduced Cost	Basis Status	Allowable Min. c(j)	Allowable Max. c(j)
1	ECONSTU	750.00	20	15,000.00	0	basic	-M	24.06
2	ECONSTD	1,500.00	30	45,000.00	0	basic	-M	42.05
3	ECONEXEC	100.00	40	4,000.00	0	basic	-M	78.34
4	BASIC STU	525.54	50	26,276.85	0	basic	45.4	61.67
5	BASICSTD	1,657.50	80	132,600.10	0	basic	72.03	83.06
6	BASICEXEC	825.4	125	103,175.30	0	basic	121.22	171.26
7	HCSTU	25	100	2,500.00	0	basic	-M	188.36
8	HCSTD	1,069.24	250	267,311.10	0	basic	246.89	270.17
9	HCEXEC	50	325	16,250.00	0	basic	-M	328.77
10	TOTAL	6,502.69	0	0.00	0	basic	-8.57	6.94
	Objective	Function	(Max.) =	612,113.40				
	Constraint	Left Hand Side	Direction	Right Hand Side	Slack or Surplus	Shadow Price	Allowable Min. RHS	Allowable Max. RHS
1	DEF-TOTAL	0.00	=	0.00	0.00	-2.11	-496.31	309.12
2	>750 E-STU	750.00	>=	750.00	0	-4.06	438.77	892.19
3	>1500E-STD	1,500.00	>=	1,500.00	0.00	-12.05	1,187.61	2,145.83
4	>100E-EXEC	100	>=	100	0	-38.34	0	616.67
5	>400B-STU	525.54	>=	400	125.54	0	-M	525.54
6	>1500B-STD	1,657.50	>=	1,500.00	157.5	0	-M	1,657.50
7	>100B-EXEC	825.4	>=	100	725.4	0	-M	825.4
8	>25HC-STU	25	>=	25	0	-88.36	0	151.28
9	>150HC-STD	1,069.24	>=	150	919.24	0	-M	1,069.24
10	>50 HC-EXE	50	>=	50	0	-3.77	0	778.39
11	MAN-MIN	230,400.00	<=	230,400.00	0	2.6	200,442.80	239,324.70
12	ALUMINUM	49,500.00	<=	65,000.00	15,500.00	0	49,500.00	M
13	PART.BD.	30,900.00	<=	60,000.00	29,100.00	0	30,900.00	M
14	OAK	175,000.00	<=	175,000.00	0	0.23	169,646.40	195,939.60
15	LINE1	4,375.00	<=	9,600.00	5,225.00	0	4,375.00	M
16	LINE2	5,358.44	<=	9,600.00	4,241.56	0	5,358.44	M
17	LINE3	16,935.61	<=	19,200.00	2,264.39	0	16,935.61	M
18	MIN-ECON	-1,049.46	<=	0	1,049.46	0	-1,049.46	M
19	MAX-ECON	-901.34	<=	0	901.34	0	-901.34	M
20	MIN-BAS	-407.37	<=	0	407.37	0	-407.37	M
21	MAX-BAS	-893.17	<=	0	893.17	0	-893.17	M
22	MIN-HC	-493.98	<=	0	493.98	0	-493.98	M
23	MAX-HC	-156.29	<=	0	156.29	0	-156.29	M
24	MIN-STU	0	<=	0	0	12.79	-281.91	114.49
25	MAX-STU	-975.4	<=	0	975.4	0	-975.4	M
26	MIN-STD	-1,625.67	<=	0	1,625.67	0	-1,625.67	M
27	MAX-STD	-325.13	<=	0	325.13	0	-325.13	M
28	MIN-EXEC	-650.27	<=	0	650.27	0	-650.27	M
29	MAX-EXEC	0	<=	0	0	31.09	-331.81	114.9

	Economy	Basic	Hand Crafted
Student	750	525.53	25
Standard	1500	1657.50	1069.24
Executive	100	825.40	50

Profit = $612,113.40

Case 4.2 See files ca4-2.lpp and ca4-2r.lpp (for relaxing the money market restriction)

X_j = amount invested in investment j where = 1 (Beekman Stock), 2 (Taco), 3 (Calton), 4 (Qube), 5 (LA Power), 6 (Beekman Bond), 7 (Metropolitan), 8 (Socal), 9 (T-bill), 10 (Money Market), 11 (C/D)

X12 = total invested in non-money investments

The average risk of the portfolio is the weighted average of the risk factors, i.e.:

$$33(X1/500{,}000) + 71(X2/500{,}000) + \ldots + 10(X10/500{,}000) + 0(X11/500{,}000)$$

The average liquidity of the portfolio is the weighted average of the liquidity factors, i.e.:

$$100(X1/500{,}000) + 100(X2/500{,}000) + \ldots + 100(X10/500{,}000) + 0(X11/500{,}000)$$

In the constraints below, the left and right side of the risk and liquidity constraints have been multiplied by 500,000.

MAX $.085X1 + .1X2 + .105X3 + .12X4 + .058X5 + .063X6 + .072X7 + .09X8 + .046X9 + .052X10 + .078X11$

S.T.

$X1 + X2 + X3 + X4 + X5 + X6 + X7 + X8 - X12 = 0$ (Definition X12)

$X1 + X2 + X3 + X4 + X5 + X6 + X7 + X8 + X9 + X10 + X11 = 500{,}000$ (Invest)

$62X1 + 71X2 + 78X3 + 95X4 + 19X5 + 33X6 + 23X7 + 50X8 + 10X10 \le 27{,}500{,}000$ (Risk)

$100X1 + 100X2 + 100X3 + 100X4 + 95X5 + 92X6 + 79X7 + 80X9 + 100X10 \ge 42{,}500{,}000$ (Liquidity)

$X1 + X6 \ge 10{,}000$ (Minimum Beekman)

$X1 + X2 + X3 + X4 - .2X12 \ge 0$ (Min. nonmoney -- stocks)

$X1 + X2 + X3 + X4 - .5X12 \le 0$ (Max. nonmoney -- stocks)

$X5 + X6 + X7 - .2X12 \ge 0$ (Min. nonmoney -- bonds)

$X5 + X6 + X7 - .5X12 \le 0$ (Max. nonmoney -- bonds)

$X3 + X8 + X9 - .2X12 \ge 0$ (Min. nonmoney -- real estate)

$X3 + X8 + X9 - .5X12 \le 0$ (Max. nonmoney -- real estate)

$X1 \le 100{,}000$ (Max. Beekman stock)

$X2 \le 100{,}000$ (Max. Taco Grande)

$X3 \le 100{,}000$ (Max. Calton)

$X4 \le 100{,}000$ (Max. Qube Electronics)

$X5 \le 100{,}000$ (Max. LA Power)

$X6 \le 100{,}000$ (Max. Beekman bonds)

$X7 \le 100{,}000$ (Max. Metropolitan)

$X8 \le 100{,}000$ (Max. Socal)

$X10 \ge 250{,}000$ (Min. Money market)

$X5 + X6 + X7 \ge 125{,}000$ (Min. bonds)

$.6X1 - .4X5 + .6X6 - .4X7 + .6X8 - .4X9 - .4X10 - .4X11 \le 0$ (Max. return < 10%)

$X1 + X2 + X3 + X4 + X10 \ge 250{,}000$ (Min. totally liquid)

$X1, X2, X3, X4, X5, X6, X7, X8, X9, X10, X11, X12 \ge 0$

Money market restriction included:

	Decision Variable	Solution Value	Unit Cost or Profit c(j)	Total Contribution	Reduced Cost	Basis Status	Allowable Min. c(j)	Allowable Max. c(j)
1	BEEK(S)	0.00	0.085	0.00	-0.0175	at bound	-M	0.1025
2	TACO G	37,500.00	0.1	3,750.00	0	basic	0.0825	0.105
3	CALTON	100,000.00	0.105	10,500.00	0	basic	0.1	M
4	QUBE	100,000.00	0.12	12,000.00	0	basic	0.1	M
5	LA POWER	32,500.00	0.058	1,885.00	0	basic	0.0217	0.0621
6	BEEK (B)	57,201.09	0.063	3,603.67	0	basic	0.059	0.0857
7	METRO TR	100,000.00	0.072	7,200.00	0	basic	0.0627	M
8	SOCAL	47,798.91	0.09	4,301.90	0	basic	0.063	0.1435
9	TBILL	0	0.046	0.00	-0.0373	at bound	-M	0.0833
10	MONMKT	25,000.00	0.052	1,300.00	0	basic	-M	0.08
11	C/D	0.00	0.078	0.00	-0.0288	at bound	-M	0.11
12	NONMON	475,000.00	0	0.00	0.00	basic	-0.0254	M
	Objective	Function	(Max.) =	44,540.57				

Money market restriction relaxed:

	Decision Variable	Solution Value	Unit Cost or Profit c(j)	Total Contribution	Reduced Cost	Basis Status	Allowable Min. c(j)	Allowable Max. c(j)
1	BEEK(S)	0.00	0.09	0.00	-0.02	at bound	-M	0.1
2	TACO G	50,000.00	0.1	5,000.00	0	basic	0.08	0.1
3	CALTON	100,000.00	0.1	10,500.00	0	basic	0.1	M
4	QUBE	100,000.00	0.12	12,000.00	0	basic	0.1	M
5	LA POWER	50,000.00	0.06	2,900.00	0	basic	0.02	0.06
6	BEEK (B)	52,717.39	0.06	3,321.20	0	basic	0.06	0.09
7	METRO TR	100,000.00	0.07	7,200.00	0	basic	0.06	M
8	SOCAL	47,282.61	0.09	4,255.44	0	basic	0.06	0.14
9	TBILL	0	0.05	0.00	-0.04	at bound	-M	0.08
10	MONMKT	0.00	0.05	0.00	-0.03	at bound	-M	0.08
11	C/D	0.00	0.08	0.00	-0.03	at bound	-M	0.11
12	NONMON	500,000.00	0	0.00	0.00	basic	-0.03	M
	Objective	Function	(Max.) =	45,176.63				

	Including Money Market Constraint	Excluding Money Market Constraint
Beekman Stock	$ 0	$ 0
Taco Grande	$ 37,500	$ 50,000
Calton REIT	$100,000	$100,000
Qube Electronics	$100,000	$100,000
LA Power	$ 32,500	$ 50,000
Beekman Bonds	$ 57,201	$ 52,717
Metropolitan Transit	$100,000	$100,000
Socal Apartments	$ 47,799	$ 47,283
T-bill Account	$ 0	$ 0
Money Market Account	$ 25,000	$ 0
Certificate of Deposit	$ 0	$ 0
TOTAL RETURN	$ 44,541 (8.908%)	$ 45,177 (9.035%)

Case CD4.1 See file cac4-1.lpp

For j = 1 (May), 2 (June), 3 (July), 4 (August):

Xj = the number of novices hired at the beginning of month j
Yj = the number of experienced temporary workers hired at the beginning of month j
Aj = the number of apprentices in month j
Ej = the total number of experienced temporary workers in month j
Fj = the number of novices terminated at the end of month j
Gj = the number of apprentices terminated at the end of month j
Hj = the number of experienced temporary workers terminated at the end of month j

MIN 3400X1 + 3400X2 + 3400X3 + 3400X4 + 1500Y1 + 1500Y2 + 1500Y3 + 1500Y4
+ 2800A2 + 2800A3 + 2800A4 + 3000E1 + 3000E2 + 3000E3 + 3000E4 + 250F1
+ 250F2 + 250F3 + 250F4 + 500G2 + 500G3 + 500G4 + 700H1 + 700H2 + 700H3
+ 700H4

S.T.

E1 - Y1 = 0	(Total experienced temporary workers -- May)
E2 - Y2 - E1 + H1 = 0	(Total experienced temporary workers -- June)
E3 - Y3 - E2 + H2 - A2 + G2 = 0	(Total experienced temporary workers -- July)
E4 - Y4 - E3 + H3 - A3 + G3 = 0	(Total experienced temporary workers -- Aug.)
E4 - H4 = 0	(Terminate all exp. temporary workers -- Aug.)
A2 - X1 + F1 = 0	(Total apprentices -- June)
A3 - X2 + F2 = 0	(Total apprentices -- July)
A4 - X3 + F3 = 0	(Total apprentices -- Aug.)
A4 - G4 = 0	(Terminate all apprentices -- Aug.)
X4 - F4 = 0	(Terminate all novices -- Aug.)
400X1 + 800E1 ≥ 200,000	(Production -- May)
400X2 + 600A2 + 800E2 ≥ 300,000	(Production -- June)
400X3 + 600A3 + 800E3 ≥ 270,000	(Production -- July)
400X4 + 600A4 + 800E4 ≥ 150,000	(Production -- Aug.)
F1 - X1 ≤ 0	(Max novices terminated -- May)
F2 - X2 ≤ 0	(Max novices terminated -- June)
F3 - X3 ≤ 0	(Max novices terminated -- July)
G2 - A2 ≤ 0	(Max apprentices terminated -- June)
G3 - A3 ≤ 0	(Max apprentices terminated -- July)
H1 - E1 ≤ 0	(Max exp. temporary workers terminated -- May)
H2 - E2 ≤ 0	(Max exp. temporary workers terminated -- June)
H3 - E3 ≤ 0	(Max exp. temporary workers terminated -- July)

All variables ≥ 0

	Decision Variable	Solution Value	Unit Cost or Profit c(j)	Total Contribution	Reduced Cost	Basis Status	Allowable Min. c(j)
1	NOV-MAY	0	3,400.00	0	3,650.00	at bound	-250
2	NOV-JUN	0	3,400.00	0	3,650.00	at bound	-250
3	NOV-JUL	0	3,400.00	0	2,150.00	at bound	1,250.00
4	NOV-AUG	0	3,400.00	0	0	basic	1,600.00
5	EXP-HIRE-MAY	250	1,500.00	375,000.00	0	basic	-1,500.00
6	EXP-HIRE-JUN	125	1,500.00	187,500.00	0	basic	800
7	EXP-HIRE-JUL	0	1,500.00	0	2,200.00	at bound	-700
8	EXP-HIRE-AUG	0	1,500.00	0	2,200.00	at bound	-700
9	APP-JUN	0	2,800.00	0	275	at bound	2,525.00
10	APP-JUL	0	2,800.00	0	800	at bound	2,000.00
11	APP-AUG	0	2,800.00	0	0	basic	2,000.00
12	TOT-EXP-MAY	250	3,000.00	750,000.00	0	basic	0
13	TOT-EXP-JUN	375	3,000.00	1,125,000.00	0	basic	-700
14	TOT-EXP-JUL	212.5	3,000.00	637,500.00	0	basic	0
15	TOT-EXP-AUG	125	3,000.00	375,000.00	0	basic	2,266.67
16	NOV-TERM-MAY	0	250	0	0	basic	-3,400.00
17	NOV-TERM-JUN	0	250	0	0	basic	-3,400.00
18	NOV-TERM-JUL	0	250	0	0	basic	-1,900.00
19	NOV-TERM-AUG	0	250	0	1,800.00	at bound	-1,550.00
20	APP-TERM-JUN	0	500	0	0	basic	225
21	APP-TERM-JUL	0	500	0	0	basic	-300
22	APP-TERM-AUG	0	500	0	800	at bound	-300
23	EXP-TERM-MAY	0	700	0	700	at bound	0
24	EXP-TERM-JUN	162.5	700	113,750.00	0	basic	500
25	EXP-TERM-JUL	87.5	700	61,250.00	0	basic	500
26	EXP-TERM-AUG	125	700	87,500.00	0	basic	-33.33
	Objective	Function	(Min.) =	3,712,500.00			
	Constraint	Left Hand Side	Direction	Right Hand Side	Slack or Surplus	Shadow Price	Allowable Min. RHS
1	EXP-MAY	0	=	0	0	0	-175
2	EXP-JUN	0	=	0	0	0	-250
3	EXP-JUL	0	=	0	0	700	-162.5
4	EXP-AUG	0	=	0	0	700	-87.5
5	TERM-EXP-AUG	0	=	0	0	-700	-M
6	APP-JUN	0	=	0	0	250	0
7	APP-JUL	0	=	0	0	250	0
8	APP-AUG	0	=	0	0	250	0
9	TERM-APP-AUG	0	=	0	0	300	0
10	TERM-NOV-AUG	0	=	0	0	1,550.00	0
11	PROD-MAY	200,000.00	>=	200,000.00	0	3.75	0
12	PROD-JUNE	300,000.00	>=	300,000.00	0	4.63	200,000.00
13	PROD-JULY	270,000.00	>=	270,000.00	0	3.75	200,000.00
14	PROD-AUG	150,000.00	>=	150,000.00	0	3.75	50,000.00
15	MAX NOV TERM-MAY	0	<=	0	0	0	0
16	MAX NOV TERM-JUN	0	<=	0	0	0	0
17	MAX NOV TERM-JUL	0	<=	0	0	0	0
18	MAX APP TERM-JUN	0	<=	0	0	-200	0
19	MAX APP TERM-JUL	0	<=	0	0	-200	0

20	MAX EXP TERM-MAY	-250	<=	0	250	0	-250
21	MAX EXP TERM-JUN	-212.5	<=	0	212.5	0	-212.5
22	MAX EXP TERM-JUL	-125	<=	0	125	0	-125

Unrounded solution:

Hire 250 experienced temporary workers in May and 125 experienced temporary workers in June. Terminate 162.5 of them at the end of June, 87.5 at the end of July and 125 at the end of August. Total Cost = $3,712,500.

Rounded solution:

Hire 250 experienced temporary workers in May and 125 experienced temporary workers in June. Terminate 162 of them at the end of June, 88 at the end of July and 125 at the end of August. Total Cost = $3,714,000.

Case CD4.2 See file cac4-2.lpp

Variable		Miller	Driller	Lathe Op.	on Lathe
X1 (Y1)	= number of dozen US (Ca) valves	Arnie	Arnie	Arnie	1
X2 (Y2)	= number of dozen US (Ca) valves	Arnie	Arnie	Arnie	2
X3 (Y3)	= number of dozen US (Ca) valves	Arnie	Arnie	Bruce	1
X4 (Y4)	= number of dozen US (Ca) valves	Arnie	Arnie	Bruce	2
X5 (Y5)	= number of dozen US (Ca) valves	Arnie	Bruce	Arnie	1
X6 (Y6)	= number of dozen US (Ca) valves	Arnie	Bruce	Arnie	2
X7 (Y7)	= number of dozen US (Ca) valves	Arnie	Bruce	Bruce	1
X8 (Y8)	= number of dozen US (Ca) valves	Arnie	Bruce	Bruce	2
X9 (Y9)	= number of dozen US (Ca) valves	Bruce	Arnie	Arnie	1
X10 (Y10)	= number of dozen US (Ca) valves	Bruce	Arnie	Arnie	2
X11 (Y11)	= number of dozen US (Ca) valves	Bruce	Arnie	Bruce	1
X12 (Y12)	= number of dozen US (Ca) valves	Bruce	Arnie	Bruce	2
X13 (Y13)	= number of dozen US (Ca) valves	Bruce	Bruce	Arnie	1
X14 (Y14)	= number of dozen US (Ca) valves	Bruce	Bruce	Arnie	2
X15 (Y15)	= number of dozen US (Ca) valves	Bruce	Bruce	Bruce	1
X16 (Y16)	= number of dozen US (Ca) valves	Bruce	Bruce	Bruce	2

REVENUE/COSTS/PROFIT PER DOZEN

Variable	Rev.	Mat'l	Drill	Mill	Lathe1	Lathe2	Arnie	Bruce	Chuck	Daryl	Unsk.	Profit
X1	88.00	16.50	3.15	2.80	9.00		5.89		2.00	3.00	1.50	44.16
X2	88.00	16.50	3.15	2.80		9.10	5.70		2.00	3.00	1.50	44.25
X3	88.00	16.50	3.15	2.80	8.40		3.04	2.94	2.00	3.00	1.50	44.67
X4	88.00	16.50	3.15	2.80		8.45	3.04	2.73	2.00	3.00	1.50	44.83
X5	88.00	16.50	3.15	2.00	9.00		4.56	1.05	2.00	3.00	1.50	45.24
X6	88.00	16.50	3.15	2.00		9.10	4.37	1.05	2.00	3.00	1.50	45.33
X7	88.00	16.50	3.15	2.00	8.40		1.71	3.99	2.00	3.00	1.50	45.75
X8	88.00	16.50	3.15	2.00		8.45	1.71	3.78	2.00	3.00	1.50	45.91
X9	88.00	16.50	3.50	2.80	9.00		4.18	2.10	2.00	3.00	1.50	43.42
X10	88.00	16.50	3.50	2.80		9.10	3.99	2.10	2.00	3.00	1.50	43.51
X11	88.00	16.50	3.50	2.80	8.40		1.33	5.04	2.00	3.00	1.50	43.93
X12	88.00	16.50	3.50	2.80		8.45	1.33	4.83	2.00	3.00	1.50	44.09
X13	88.00	16.50	3.50	2.00	9.00		2.85	3.15	2.00	3.00	1.50	44.50
X14	88.00	16.50	3.50	2.00		9.10	2.66	3.15	2.00	3.00	1.50	44.59
X15	88.00	16.50	3.50	2.00	8.40			6.09	2.00	3.00	1.50	45.01
X16	88.00	16.50	3.50	2.00		8.45		5.88	2.00	3.00	1.50	45.17
Y1	100.00	17.50	3.50	3.20	10.20		6.65		2.00	3.00	1.50	52.45
Y2	100.00	17.50	3.50	3.20		9.75	6.27		2.00	3.00	1.50	53.28
Y3	100.00	17.50	3.50	3.20	9.00		3.42	3.15	2.00	3.00	1.50	53.73
Y4	100.00	17.50	3.50	3.20		9.10	3.42	2.94	2.00	3.00	1.50	53.84
Y5	100.00	17.50	3.50	2.40	10.20		5.13	1.26	2.00	3.00	1.50	53.51
Y6	100.00	17.50	3.50	2.40		9.75	475	1.26	2.00	3.00	1.50	54.34
Y7	100.00	17.50	3.50	2.40	9.00		1.90	4.41	2.00	3.00	1.50	54.79
Y8	100.00	17.50	3.50	2.40		9.10	1.90	4.20	2.00	3.00	1.50	54.90
Y9	100.00	17.50	4.20	3.20	10.20		4.75	2.52	2.00	3.00	1.50	51.13
Y10	100.00	17.50	4.20	3.20		9.75	4.37	2.52	2.00	3.00	1.50	51.96
Y11	100.00	17.50	4.20	3.20	9.00		1.52	5.67	2.00	3.00	1.50	52.41
Y12	100.00	17.50	4.20	3.20		9.10	1.52	5.46	2.00	3.00	1.50	52.52
Y13	100.00	17.50	4.20	2.40	10.20		3.23	3.78	2.00	3.00	1.50	52.19
Y14	100.00	17.50	4.20	2.40		9.75	2.85	3.78	2.00	3.00	1.50	53.02
Y15	100.00	17.50	4.20	2.40	9.00			6.93	2.00	3.00	1.50	53.47
Y16	100.00	17.50	4.20	2.40		9.10		6.72	2.00	3.00	1.50	53.58

MAX 44.16X1 + 44.25X2 + 44.67X3 + 44.83X4 + 45.24X5 + 45.33X6 + 45.75X7 + 45.91X8 + 43.42X9 + 43.51X10 + 43.93X11 + 44.09X12 + 44.50X13 + 44.56X14 + 45.01X15 + 45.17X16 + 52.45Y1 + 53.28Y2 + 53.73Y3 + 53.84Y4 + 53.51Y5 + 54.34Y6 + 54.79Y7 + 54.90Y8 + 51.13Y9 + 51.96Y10 + 52.41Y11 + 52.52Y12 + 52.19Y13 + 53.02Y14 + 53.47Y15 + 53.58Y16

$$\text{S.T.}\ 9X1 + 9X2 + 9X3 + 9X4 + 9X5 + 9X6 + 9X7 + 9X8 + 10X9 + 10X10 + 10X11 + 10X12 + 10X13 + 10X14 + 10X15 + 10X16 + 10Y1 + 10Y2 + 10Y3 + 10Y4 + 10Y5 + 10Y6 + 10Y7 + 10Y8 + 12Y9 + 12Y10 + 12Y11 + 12Y12 + 12Y13 + 12Y14 + 12Y15 + 12Y16 \le 1{,}800 \quad \text{(Milling)}$$

$$7X1 + 7X2 + 7X3 + 7X4 + 5X5 + 5X6 + 5X7 + 5X8 + 7X9 + 7X10 + 7X11 + 7X12 + 5X13 + 5X14 + 5X15 + 5X16 + 8Y1 + 8Y2 + 8Y3 + 8Y4 + 6Y5 + 6Y6 + 6Y7 + 6Y8 + 8Y9 + 8Y10 + 8Y11 + 8Y12 + 6Y13 + 6Y14 + 6Y15 + 6Y16 \le 1{,}800 \quad \text{(Drilling)}$$

$$15X1 + 14X3 + 15X5 + 14X7 + 15X9 + 14X11 + 15X13 + 14X15 + 17Y1 + 15Y3 + 17Y5 + 15Y7 + 17Y9 + 15Y11 + 17Y13 + 15Y15 \le 1{,}800 \quad \text{(Lathe 1)}$$

$$14X2 + 13X4 + 14X6 + 13X8 + 14X10 + 13X12 + 14X14 + 13X16 + 15Y2 + 14Y4 + 15Y6 + 14Y8 + 15Y10 + 14Y12 + 15Y14 + 14Y16 \le 1{,}800 \quad \text{(Lathe 2)}$$

$$31X1 + 30X2 + 16X3 + 16X4 + 24X5 + 23X6 + 9X7 + 9X8 + 22X9 + 21X10 + 7X11 + 7X12 + 15X13 + 14X14 + 35Y1 + 33Y2 + 18Y3 + 18Y4 + 27Y5 + 25Y6 + 10Y7 + 10Y8 + 25Y9 + 23Y10 + 8Y11 + 8Y12 + 17Y13 + 15Y14 \le 2{,}160 \quad \text{(Max. Arnie)}$$

$$31X1 + 30X2 + 16X3 + 16X4 + 24X5 + 23X6 + 9X7 + 9X8 + 22X9 + 21X10 + 7X11 + 7X12 + 15X13 + 14X14 + 35Y1 + 33Y2 + 18Y3 + 18Y4 + 27Y5 + 25Y6 + 10Y7 + 10Y8 + 25Y9 + 23Y10 + 8Y11 + 8Y12 + 17Y13 + 15Y14 \ge 1{,}200 \quad \text{(Min. Arnie)}$$

$$14X3 + 13X4 + 5X5 + 5X6 + 19X7 + 18X8 + 10X9 + 10X10 + 24X11 + 23X12 + 15X13 + 15X14 + 29X15 + 28X16 + 15Y3 + 14Y4 + 6Y5 + 6Y6 + 21Y7 + 20Y8 + 12Y9 + 12Y10 + 27Y11 + 26Y12 + 18Y13 + 18Y14 + 33Y15 + 32Y16 \le 2{,}400 \quad \text{(Max. Bruce)}$$

$$14X3 + 13X4 + 5X5 + 5X6 + 19X7 + 18X8 + 10X9 + 10X10 + 24X11 + 23X12 + 15X13 + 15X14 + 29X15 + 28X16 + 15Y3 + 14Y4 + 6Y5 + 6Y6 + 21Y7 + 20Y8 + 12Y9 + 12Y10 + 27Y11 + 26Y12 + 18Y13 + 18Y14 + 33Y15 + 32Y16 \ge 1{,}200 \quad \text{(Min. Bruce)}$$

$$10X1 + 10X2 + 10X3 + 10X4 + 10X5 + 10X6 + 10X7 + 10X8 + 10X9 + 10X10 + 10X11 + 10X12 + 10X13 + 10X14 + 10X15 + 10X16 + 10Y1 + 10Y2 + 10Y3 + 10Y4 + 10Y5 + 10Y6 + 10Y7 + 10Y8 + 10Y9 + 10Y10 + 10Y11 + 10Y12 + 10Y13 + 10Y14 + 10Y15 + 10Y16 \le 1{,}920 \quad \text{(Max. Chuck)}$$

$$10X1 + 10X2 + 10X3 + 10X4 + 10X5 + 10X6 + 10X7 + 10X8 + 10X9 + 10X10 + 10X11 + 10X12 + 10X13 + 10X14 + 10X15 + 10X16 + 10Y1 + 10Y2 + 10Y3 + 10Y4 + 10Y5 + 10Y6 + 10Y7 + 10Y8 + 10Y9 + 10Y10 + 10Y11 + 10Y12 + 10Y13 + 10Y14 + 10Y15 + 10Y16 \geq 1{,}200 \quad \text{(Min. Chuck)}$$

$$12X1 + 12X2 + 12X3 + 12X4 + 12X5 + 12X6 + 12X7 + 12X8 + 12X9 + 12X10 + 12X11 + 12X12 + 12X13 + 12X14 + 12X15 + 12X16 + 12Y1 + 12Y2 + 12Y3 + 12Y4 + 12Y5 + 12Y6 + 12Y7 + 12Y8 + 12Y9 + 12Y10 + 12Y11 + 12Y12 + 12Y13 + 12Y14 + 12Y15 + 12Y16 \leq 2{,}100 \quad \text{(Max. Daryl)}$$

$$12X1 + 12X2 + 12X3 + 12X4 + 12X5 + 12X6 + 12X7 + 12X8 + 12X9 + 12X10 + 12X11 + 12X12 + 12X13 + 12X14 + 12X15 + 12X16 + 12Y1 + 12Y2 + 12Y3 + 12Y4 + 12Y5 + 12Y6 + 12Y7 + 12Y8 + 12Y9 + 12Y10 + 12Y11 + 12Y12 + 12Y13 + 12Y14 + 12Y15 + 12Y16 \geq 1{,}200 \quad \text{(Min. Daryl)}$$

$$X1 + X2 + X3 + X4 + X5 + X6 + X7 + X8 + X9 + X10 + X11 + X12 + X13 + X14 + X15 + X16 - 3Y1 - 3Y2 - 3Y3 - 3Y4 - 3Y5 - 3Y6 - 3Y7 - 3Y8 - 3Y9 - 3Y10 - 3Y11 - 3Y12 - 3Y13 - 3Y14 - 3Y15 - 3Y16 \geq 0 \quad \text{(Ratio -- US/Ca.)}$$

$$X1 + X2 + X3 + X4 + X5 + X6 + X7 + X8 + X9 + X10 + X11 + X12 + X13 + X14 + X15 + X16 \geq 50 \quad \text{(Orders -- US)}$$

$$Y1 + Y2 + Y3 + Y4 + Y5 + Y6 + Y7 + Y8 + Y9 + Y10 + Y11 + Y12 + Y13 + Y14 + Y15 + Y16 \geq 25 \quad \text{(Orders -- Ca.)}$$

All Xi's and Yi's ≥ 0

	Decision Variable	Solution Value	Unit Cost or Profit c(j)	Total Contribution	Reduced Cost	Basis Status	Allowable Min. c(j)	Allowable Max. c(j)
1	US-A/A/A1	0	44.16	0	-3.4047	at bound	-M	47.5647
2	US-A/A/A2	0	44.25	0	-3.5859	at bound	-M	47.8359
3	US-A/A/B1	0	44.67	0	-3.5859	at bound	-M	48.2559
4	US-A/A/B2	0	44.83	0	-3.4047	at bound	-M	48.2347
5	US-A/B/A1	27.6586	45.24	1,251.28	0	basic	45.2400	47.1376
6	US-A/B/A2	0	45.33	0	-0.1812	at bound	-M	45.5112
7	US-A/B/B1	0	45.75	0	-0.1812	at bound	-M	45.9312
8	US-A/B/B2	93.6408	45.91	4,299.05	0	basic	45.7296	45.91
9	US-B/A/A1	0	43.42	0	-7.3219	at bound	-M	50.7418
10	US-B/A/A2	0	43.51	0	-7.503	at bound	-M	51.013
11	US-B/A/B1	0	43.93	0	-7.503	at bound	-M	51.433
12	US-B/A/B2	0	44.09	0	-7.3218	at bound	-M	51.4118
13	US-B/B/A1	0	44.5	0	-3.9171	at bound	-M	48.4171
14	US-B/B/A2	0	44.59	0	-4.0983	at bound	-M	48.6883
15	US-B/B/B1	0	45.01	0	-4.0983	at bound	-M	49.1083
16	US-B/B/B2	0	45.17	0	-3.9171	at bound	-M	49.0871
17	CA-A/A/A1	0	52.45	0	-5.2165	at bound	-M	57.6665
18	CA-A/A/A2	0	53.28	0	-3.2236	at bound	-M	56.5036
19	CA-A/A/B1	0	53.73	0	-3.2236	at bound	-M	56.9536
20	CA-A/A/B2	0	53.84	0	-3.2236	at bound	-M	57.0636
21	CA-A/B/A1	0	53.51	0	-1.993	at bound	-M	55.503
22	CA-A/B/A2	16.6064	54.34	902.3901	0	basic	54.1461	54.34
23	CA-A/B/B1	0	54.79	0	0	at bound	-M	54.79
24	CA-A/B/B2	23.8268	54.9	1,308.09	0	basic	54.9	55.0941
25	CA-B/A/A1	0	51.13	0	-11.6013	at bound	-M	62.7313
26	CA-B/A/A2	0	51.96	0	-9.6083	at bound	-M	61.5683
27	CA-B/A/B1	0	52.41	0	-9.6083	at bound	-M	62.0183
28	CA-B/A/B2	0	52.52	0	-9.6083	at bound	-M	62.1283
29	CA-B/B/A1	0	52.19	0	-8.3777	at bound	-M	60.5677
30	CA-B/B/A2	0	53.02	0	-6.3848	at bound	-M	59.4048
31	CA-B/B/B1	0	53.47	0	-6.3848	at bound	-M	59.8548
32	CA-B/B/B2	0	53.58	0	-6.3848	at bound	-M	59.9648
	Objective	Function	(Max.) =	7,760.80				

	Constraint	Left Hand Side	Direction	Right Hand Side	Slack or Surplus	Shadow Price	Allowable Min. RHS	Allowable Max. RHS
1	MAX MILL	1,496.03	<=	1,800.00	303.9745	0	1,496.03	M
2	MAX DRILL	849.0955	<=	1,800.00	950.9045	0	849.0955	M
3	MAX LATHE1	414.879	<=	1,800.00	1,385.12	0	414.879	M
4	MAX LATHE2	1,800.00	<=	1,800.00	0	0.1312	1,554.19	2,156.04
5	MAX ARNIE	2,160.00	<=	2,160.00	0	1.5653	1,361.77	2,560.58
6	MIN ARNIE	2,160.00	>=	1,200.00	960	0	-M	2,160.00
7	MAX BRUCE	2,400.00	<=	2,400.00	0	1.7265	2,022.40	2,706.49
8	MIN BRUCE	2,400.00	>=	1,200.00	1,200.00	0	-M	2,400.00
9	MAX CHUCK	1,617.33	<=	1,920.00	302.6752	0	1,617.33	M
10	MIN CHUCK	1,617.33	>=	1,200.00	417.3248	0	-M	1,617.33
11	MAX DARYL	1,940.79	<=	2,100.00	159.2102	0	1,940.79	M
12	MIN DARYL	1,940.79	>=	1,200.00	740.7898	0	-M	1,940.79
13	RATIO US/CA	0	>=	0	0	-0.9601	-264.0094	63.4293
14	US ORDERS	121.2994	>=	50	71.2994	0	-M	121.2994
15	CAL ORDERS	40.4331	>=	25	15.4331	0	-M	40.4331

Maximum profit = $7,760.80 by producing the following:

121.30 dozen US valves

- 27.66 dozen milled by Arnie, drilled by Bruce, lathed by Arnie on Lathe 1
- 93.64 dozen milled by Arnie, drilled by Bruce, lathed by Bruce on Lathe 2

40.43 dozen California valves

- 16.60 dozen milled by Arnie, drilled by Bruce, lathed by Arnie on Lathe 2
- 23.83 dozen milled by Arnie, drilled by Bruce, lathed by Bruce on Lathe 2

Use of resources:

Mill	1,496 min.	(All by Arnie)
Drill	849 min.	(All by Bruce)
Lathe 1	415 min.	(All by Arnie producing US models)
Lathe 2	1,800 min.	(249 by Arnie making US models, 1551 by Bruce -- 1217 making US Models, 334 making California models)

Use of Personnel

Arnie	2,160 min.
Bruce	2,400 min.
Chuck	1,617 min.
Daryl	1,941 min.
Unskilled	2,426 min.

Chapter 13

Problem Summary

Prob. #	Concepts Covered	Level of Difficulty	Notes
13.1	Developing a relative frequency distribution, doing a fixed time simulation, comparing simulation and expected value results	2	
13.2	Doing a next event simulation of a queuing problem.	4	
13.3	Calculating a 95% confidence interval for Wq, comparing the simulated value for Wq with the steady state value	3	
13.4	Calculating a 95% confidence interval for Wq. Comparing two populations using independent samples	3	
13.5	Determining the critical path through a network, doing a critical path analysis using simulation, determining a 95% confidence interval	5	
13.6	Doing a fixed time simulation of an inventory problem.	5	
13.7	Using a paired t test for determining whether population means are different	3	
13.8	Calculating a 95% confidence interval for Wq, comparing the simulated value for Wq with the steady state value.	3	
13.9	Comparing multiple populations using analysis of variance.	3	
13.10	Conducting fixed time simulations of an inventory system. Comparing different policies using simulation	6	
13.11	Conducting a fixed time simulation of an inventory system	4	
13.12	Conducting a next event simulation of a queuing system	6	
13.13	Conducting a four week fixed time simulation of an inventory system. Calculating a 95% confidence interval	6	
13.14	Conducting a four week fixed time simulation of an inventory system. Comparing different policies using a paired t test.	6	
13.15	Conducting a next event simulation of a tandem queuing system.	6	Note that no columns are specified for the selection of the random numbers. The solution is given based on selecting random numbers from columns 1 through 5.

13.16	Conducting a next event simulation of a queuing system. Comparing the value for Wq calculated by the simulation to the steady state value.	5	
13.17	Conducting a next event simulation of a queuing system to calculate W.	4	
13.18	Conducting fixed time simulations of a gambling situation. Determining a 95% confidence interval.	6	The simulations were conducted using the RAND() function in Excel.
13.19	Conducting fixed time simulations to determine the shortest travel route	4	Note that no columns are specified for the selection of the random numbers. The solution is given based on selecting random numbers from columns 1 through 5.
13.20	Conducting a fixed time simulation of a queuing system to determine Lq.	5	
13.21	Conducting a next event simulation of a queuing system	5	Note that no columns were specified for the simulation. For this simulation random numbers from columns 1 through 3 were selected.
13.22	Conducting a next event simulation of a queuing system with different service characteristics	8	
13.23	Conducting a next event simulation of a tandem queuing system	6	
13.24	Conducting a fixed time simulation of an inventory system. Using simulation to compare policies.	5	
13.25	Conducting a fixed time simulation of a sales situation.	4	
CD13.1	Conducting next event simulations of two queuing systems to determine the best person to hire. Comparing the results from the simulation to that obtained based on using steady state results.	7	
CD13.2	Conducting simulations of an inventory system to determine the optimal policy.	6	
CD13.3	Conducting simulations of an inventory system using an Excel spreadsheet. Determining the optimal policy and a 95% confidence interval using ANOVA	9	
CD13.4	Conducting a fixed time simulation of an inventory system	7	
CD13.5	Conducting a fixed time simulation of an inventory system. Comparing policies using simulation	7	
CD13.6	Using goodness of fit analysis to determine whether phone number digits are uniformly distributed.	3	
CD13.7	Simulated a Markov Chain	3	

Case 13.1	Conducting a fixed time simulation of an inventory system, comparing different policies and determining the best policy from the set considered.	10	To make this problem easier for your students you may wish to provide them with the Excel spreadsheets given in files ca13-1a.xls and ca13-1b.xls.
Case 13.2	Conducting a next event simulation of a queuing system to determine whether a service improvement should be made	9	
Case CD13.1	Conducting a fixed time simulation of an inventory system, comparing different policies and determining the best policy from the set considered.	10	To make this problem easier for your students you may wish to provide them with the Excel spreadsheets given in files cac13-1a.xls and cac13-1b.xls.
Case CD13.2	Conducting a next event simulation of a queuing system to determine which server to hire.	9	

Problem Solutions

13.1 See file ch13-1.xls

a.

Price Change	Probability
-1/2	.06
-3/8	.08
-1/4	.10
-1/8	.12
0	.20
+1/8	.24
+1/4	.08
+3/8	.06
+1/2	.06

b. Expected Change = (.06)(-.5) + (.08)(-.375) + (.10)(-.25) + (.12)(-.125) + (.20)(0) + (.24)(.125) + (.08)(.25) + (.06)(.375) + (.06)(.5) = .0025. The price in 30 days would be 32 + 30*(.0025) = 32.075

c.

day	Rand #	Prob	Price Chg	Price 32.000
1	6506	0.6506	0.125	32.125
2	7761	0.7761	0.125	32.250
3	6170	0.617	0.125	32.375
4	8800	0.88	0.375	32.750
5	4211	0.4211	0	32.750
6	7452	0.7452	0.125	32.875
7	1182	0.1182	-0.375	32.500
8	4012	0.4012	0	32.500
9	0335	0.0335	-0.5	32.000
10	6299	0.6299	0.125	32.125
11	5482	0.5482	0	32.125
12	1085	0.1085	-0.375	31.750
13	1698	0.1698	-0.25	31.500
14	6969	0.6969	0.125	31.625
15	1696	0.1696	-0.25	31.375
16	0267	0.0267	-0.5	30.875
17	3175	0.3175	-0.125	30.750
18	7959	0.7959	0.125	30.875
19	4958	0.4958	0	30.875
20	4281	0.4281	0	30.875
21	2231	0.2231	-0.25	30.625
22	0002	0.0002	-0.5	30.125
23	8434	0.8434	0.25	30.375
24	8959	0.8959	0.375	30.750
25	8975	0.8975	0.375	31.125
26	1288	0.1288	-0.375	30.750
27	6412	0.6412	0.125	30.875
28	6463	0.6463	0.125	31.000
29	3856	0.3856	0	31.000
30	6117	0.6117	0.125	31.125

The price after 30 days is 31.125.

d. We would expect the simulation to be close to the expected value but not necessarily equal to the expected value.

13.2 See file ch13-2.xls

Cust. #	R #	IAT	completion time	R #	# Purchased	Total
1	0.651	1.577	1.577	0.334	0	0
2	0.776	2.245	3.822	0.987	5	5
3	0.617	1.44	5.262	0.263	0	5
4	0.88	3.18	8.442	0.914	3	8
5	0.421	0.82	9.262	0.965	5	13
6	0.745	2.051	11.31	0.488	1	14
7	0.118	0.189	11.5	0.826	2	16
8	0.401	0.769	12.27	0.278	0	16
9	0.034	0.051	12.32	0.964	5	21
10	0.63	1.491	13.81	0.47	1	22
11	0.548	1.192	15	0.202	0	22
12	0.109	0.172	15.18	0.443	0	22
13	0.17	0.279	15.46	0.846	2	24
14	0.697	1.791	17.25	0.942	4	28
15	0.17	0.279	17.53	0.669	1	29
16	0.027	0.041	17.57	0.632	1	30

The store will sell out of lottery tickets after 17.57 minutes

13.3. See files ch13-3.xls and ch13-3b.qaa

a. The 95% confidence interval for W = .6607 +/- 2.262*.0744/√10 = .6607 +/- .0532 = (.6075, .7139)

b. W_q = .6667 minutes

13.4 See files ch13-4a.xls and ch13-4b.xls

a. a. The 95% confidence interval for W is .58376 +/- 2.262*.07186/√10 = .5876 +/- .02272 = (.5610, .6064)

b. We can conclude that there is a significant difference in the average waiting times, t = 2.35 while the critical t value is 2.10. The p value of the test is .03.

13.5 See files ch13-5.xls, ch13-5.cpm, ch13-5R1.cpm, ch13-5R2.cpm, ch13-5R3.cpm, ch13-5R4.cpm, ch13-5R5.cpm,

a. D-E, 15 weeks

13.5 b. P(A is on critical path) = 1/5 = .2, P(B is on critical path) = 1/5 = .2, P(C is on critical path) = 1/5 = .2, P(D is on critical path) = 3/5 = .6, P(E is on critical path) = 3/5 = .6, P(F is on critical path) = 1/5 = .2, P(G is on critical path) = 1/5 = .2

c. The 95% confidence interval = 18 +/- 2.776*1.414/√5 = 18+/- 1.756 = (16.244, 19.756)

13.6 See file ch13-6.xls

Day	R #	# Arrive	Cust #	R #	Purchase	R #	color	white	almond	gold
								5	4	3
1	0.651	2	1	0.334	yes	0.22	white	4	4	3
			2	0.987	no			4	4	3
2	0.776	3	3	0.263	yes	0.983	gold	4	4	2
			4	0.914	no			4	4	2
			5	0.965	no			4	4	2
3	0.617	2	6	0.488	yes	0.627	almond	4	3	2
			7	0.826	no			4	3	2
4	0.88	3	8	0.278	yes	0.831	gold	4	3	1
			9	0.964	no			4	3	1
			10	0.47	yes	0.005	white	3	3	1
5	0.421	2	11	0.202	yes	0.562	almond	3	2	1
			12	0.443	yes	0.337	white	2	2	1
6	0.745	3	13	0.846	no			2	2	1
			14	0.942	no			2	2	1
			15	0.669	no			2	2	1
7	0.118	0								
8	0.401	2	16	0.632	no			2	2	1
			17	0.753	no			2	2	1
9	0.034	0								
10	0.63	2	18	0.052	yes	0.614	almond	2	1	1
			19	0.107	yes	0.165	white	1	1	1
11	0.548	2	20	0.188	yes	0.804	gold	1	1	0
			21	0.038	yes	0.702	gold	1	1	-1
12	0.109	0								
13	0.17	1	22	0.079	yes	0.499	almond	1	0	-1
14	0.697	2	23	0.724	no			1	0	-1
			24	0.589	yes	0.821	gold	1	0	-2
15	0.17	1	25	0.648	no			1	0	-2
16	0.027	0								
17	0.318	1	26	0.876	no			1	0	-2
18	0.796	3	27	0.014	yes	0.152	white	0	0	-2
			28	0.032	yes	0.953	gold	0	0	-3
			29	0.178	yes	0.105	white	-1	0	-3

It will take Taks 18 days to sell all 12 refrigerators.

13.7 See file ch13-7.xls

We cannot conclude that the estimated time is different between the two sets, $t = -1.797$ compared to a critical t value of 2.262. The p value for this test is .1059.

13.8 See files ch13-8.xls, ch13-8c.qaa

a. The 95% confidence interval in minutes is $11.695 +/- 2.262*6.782/\sqrt{10} = 11.695 +/- 4.851 =$ (6.844 ,16.546)

b. The 95% confidence interval in minutes is $12.787 +/- 2.262*8.496/\sqrt{10} = 12.878 +/- 6.078 =$ (6709 ,18.865)

c. $W_q = .3167$ hours = 19.002 minutes

13.9 See file ch13-9.xls

Yes, we can conclude that there is a difference. The F value = 32.98 compared to a critical F value of 3.354. The p value for the test is $5.65*10^{-8}$.

13.10 See files ch13-10a.xls and ch13-10b.xls

a.

Day	Rand. #	# Cust.	Cust. #	Rand #	Length	Day Ret.	Beg Day Invent.	End Day Invent.	Profit
1	0.632	2	1	0.109	1	2	3		
1			2	0.737	2	3		1	23
2	0.463	2	3	0.281	1	3	2		
2			4	0.515	1	3		0	36
3	0.866	3	5	0.912	3	6	3		
3			6	0.147	1	4			
3			7	0.648	2	5		0	36
4	0.003	0					1	1	23
5	0.562	2	8	0.245	1	6	2		
5			9	0.703	2	7		0	36
6	0.673	2	10	0.844	2	8	2		
6			11	0.739	2	8		0	36
7	0.593	2	12	0.179	1	8	1		
7			13	0.489	1	8		0	24
8	0.283	1	14	0.115	1	9	3	2	10
9	0.794	3	15	0.882	3	12	3		
9			16	0.945	3	12			
9			17	0.299	1	10		0	36
10	0.648	2	18	0.831	2	12	1		
			19	0.718	2	2		0	24
							Profit =		284

Profit over the ten days equals $284

b.

Day	R #	# Cust.	Cust. #	R #	Length	Day Ret.	Beg Day Invent.	End Day Invent.	Profit
1	0.632	2	1	0.109	1	2	4		
1			2	0.737	2	3		2	22
2	0.463	2	3	0.281	1	3	3		
2			4	0.515	1	3		1	35
3	0.866	3	5	0.912	3	6	4		
3			6	0.147	1	4			
3			7	0.648	2	5		1	35
4	0.003	0					2	2	22
5	0.562	2	8	0.245	1	6	3		
5			9	0.703	2	7		1	35
6	0.673	2	10	0.844	2	8	3		
6			11	0.739	2	8		1	48
7	0.593	2	12	0.179	1	8	2		
7			13	0.489	1	8		0	48
8	0.283	1	14	0.115	1	9	4	3	9
9	0.794	3	15	0.882	3	12	4		
9			16	0.945	3	12			
9			17	0.299	1	10		1	35
10	0.648	2	18	0.831	2	12	2		
			19	0.718	2	2		0	48
							Profit =		287

Profit over the ten days equals $287

13.11 See file ch13-11.xls

Day	R #	# Cust.	Cust. #	R #	System	Inventory Panas.	C.M.	R#	Action
						6	4		
1	0.22	1	1	0.893	ChanelM.		3		
2	0.983	6	2	0.885	ChanelM.		2		
			3	0.909	Other				
			4	0.261	Panas.	5			
			5	0.798	ChanelM.		1		
			6	0.554	ChanelM.		0		
			7	0.262	Panas.	4			
3	0.627	2	8	0.16	Panas.	3			
			9	0.406	ChanelM.			0.632	Back
4	0.831	3	10	0.652	ChanelM.			0.463	Back
			11	0.611	ChanelM.			0.866	Lost Sale
			12	0.681	ChanelM.	2		0.003	Panas.
5	0.005	0							
6	0.562	2	13	0.759	ChanelM.			0.562	Back
			14	0.659	ChanelM.			0.673	Back
7	0.337	1	15	0.088	Panas.	1			
8	0.614	2	16	0.319	Panas.	0			
			17	0.208	Panas.			0.593	Lost Sale

It will take 8 days for Dizzy Izzy to sell out of inventory, there will be 3 backorders and 2 lost sales.

13.12 See file ch13-12.xls

Cust #	RR# 1	IAT	AT	TSB	WT	RR#2	ST (1)	TSE	RR#3	ST (2)	TSE	Waiting Time
1	0.6506	1.58	1.58	1.58	0.00	0.3338	0.41	1.98	0.2197	0.68	2.67	1.09
2	0.7761	2.24	3.82	3.82	0.00	0.9874	4.37	8.20	0.9834	1.32	9.52	5.69
3	0.6170	1.44	5.26	8.20	2.93	0.2631	0.31	8.50	0.6268	1.02	9.52	4.26
4	0.8800	3.18	8.44	8.50	0.06	0.9139	2.45	10.95	0.8305	1.19	12.15	3.70
5	0.4211	0.82	9.26	10.95	1.69	0.9651	3.36	14.31	0.0051	0.50	14.81	5.55
6	0.7452	2.05	11.31	14.31	3.00	0.4883	0.67	14.98	0.5619	0.97	15.95	4.63
7	0.1182	0.19	11.50	14.98	3.48	0.8260	1.75	16.73	0.3367	0.78	17.51	6.01
8	0.4012	0.77	12.27	16.73	4.46	0.2781	0.33	17.05	0.6140	1.01	18.07	5.79
9	0.0335	0.05	12.32	17.05	4.73	0.9636	3.31	20.37	0.1645	0.64	21.00	8.68
10	0.6299	1.49	13.81	20.37	6.55	0.4696	0.63	21.00	0.8036	1.17	22.17	8.36
11	0.5482	1.19	15.00	21.00	6.00	0.2016	0.23	21.23	0.7017	1.08	22.31	7.31
12	0.1085	0.17	15.18	21.23	6.05	0.4426	0.58	21.81	0.4988	0.92	22.73	7.55
13	0.1698	0.28	15.46	21.81	6.35	0.8464	1.87	23.68	0.8208	1.18	24.87	9.41
14	0.6969	1.79	17.25	23.68	6.44	0.9424	2.85	26.54	0.1524	0.63	27.17	9.92
15	0.1696	0.28	17.53	26.54	9.01	0.6693	1.11	27.64	0.9525	1.29	28.94	11.41
16	0.0267	0.04	17.57	27.64	10.08	0.6317	1.00	28.64	0.1051	0.59	29.23	11.67
17	0.3175	0.57	18.14	28.64	10.50	0.7529	1.40	30.04	0.1759	0.65	30.69	12.55
18	0.7959	2.38	20.52	30.04	9.52	0.0515	0.05	30.09	0.8149	1.18	31.27	10.75
19	0.4958	1.03	21.55	30.09	8.54	0.1069	0.11	30.21	0.0318	0.53	30.73	9.18
20	0.4281	0.84	22.39	30.21	7.82	0.1877	0.21	30.42	0.8785	1.23	31.65	9.26
											W =	7.64

The average waiting time, W, equals 7.64 minutes

13.13 See file ch13-13.xls

Week	Day	R #	# Arriv.	R #	Cust #	Type	Inventory single	two	cost	profit
1							5	5		
	1	0.651	2	0.334	1	single	4			
				0.987	2	two		4		
	2	0.776	2	0.263	3	single	3			
				0.914	4	two		3		
	3	0.617	1	0.965	5	two		2		
	4	0.880	3	0.488	6	single	2			
				0.826	7	two		1		
				0.278	8	single	1			
	5	0.421	1	0.964	9	two		0		
	6	0.745	2	0.47	10	single	0			
				0.202	11	single			10	
	7	0.118	0							
									profit=	95

Week	Day	R #	# Arriv.	R #	Cust #	Type	Inventory single	two	cost	profit
							5	5		
2	1	0.401	1	0.443	1	single	4			
	2	0.034	0							
	3	0.630	1	0.846	2	two		4		
	4	0.548	1	0.942	3	two		3		
	5	0.109	0							
	6	0.170	0							
	7	0.697	2	0.669	4	two		2		
				0.632	5	two		1		
									profit=	66
							5	5		
3	1	0.170	0							
	2	0.027	0							
	3	0.318	1	0.753	1	two		4		
	4	0.796	2	0.052	2	single	4			
				0.107	3	single	3			
	5	0.496	1	0.188	4	single	2			
	6	0.428	1	0.038	5	single	1			
	7	0.223	0							
									profit=	39
							5	5		
4	1	0.000	0							
	2	0.843	2	0.079	1	single	4			
				0.724	2	two		4		
	3	0.896	3	0.589	3	single	3			
				0.648	4	two		3		
				0.876	5	two		2		
	4	0.898	3	0.014	6	single	2			
				0.032	7	single	1			
				0.178	8	single	0			
	5	0.129	0	0.409						
	6	0.641	1	0.251	9	single	1	1		
	7	0.646	1	0.171	10	single	0			
									profit=	84

The 95% confidence interval for Albright's mean weekly profit is \$71 +/- \$38.91 = (\$32.09, \$109.91)

13.14 See file ch13-14.xls

Week	Day	R #	# Arriv.	R #	Cust #	Type	Inventory single	two	cost	profit
1							6	4		
	1	0.651	2	0.334	1	single	5			
				0.987	2	two		3		
	2	0.776	2	0.263	3	single	4			
				0.914	4	two		2		
	3	0.617	1	0.965	5	two		1		
	4	0.880	3	0.488	6	single	3			
				0.826	7	two		0		
				0.278	8	single	2			
	5	0.421	1	0.964	9	two			10	
	6	0.745	2	0.47	10	single	1			
				0.202	11	single	0			
	7	0.118	0							
									profit=	86
							6	4		
2	1	0.401	1	0.443	1	single	5			
	2	0.034	0							
	3	0.630	1	0.846	2	two		3		
	4	0.548	1	0.942	3	two		2		
	5	0.109	0							
	6	0.170	0							
	7	0.697	2	0.669	4	two		1		
				0.632	5	two		0		
									profit=	66
							6	4		
3	1	0.170	0							
	2	0.027	0							
	3	0.318	1	0.753	1	two		3		
	4	0.796	2	0.052	2	single	5			
				0.107	3	single	4			
	5	0.496	1	0.188	4	single	3			
	6	0.428	1	0.038	5	single	2			
	7	0.223	0							
									profit=	39
							6	4		
4	1	0.000	0							
	2	0.843	2	0.079	1	single	5			
				0.724	2	two		3		
	3	0.896	3	0.589	3	single	4			
				0.648	4	two		2		
				0.876	5	two		1		
	4	0.898	3	0.014	6	single	3			
				0.032	7	single	2			
				0.178	8	single	1			
	5	0.129	0	0.409						
	6	0.641	1	0.251	9	single	0			
	7	0.646	1	0.171	10	single	1	1		
									profit=	85.5

We cannot claim that there is a difference in mean weekly profit between the two policies, t = .78 compared with a critical t value of 3.182. The p value for the test is .49.

13.15 See file ch13-15.xls

a. The twentieth customer completes registration at time 29.12 minutes

b. Average Waiting Time: Registration – 5.16 minutes, Manufacturing – 2 minutes, Retailing – 2 minutes, Import/Export – 3 minutes, Financial Services – 3 minutes.

13.16 See files ch13-16.xls and ch13-16.qaa

Cust #	RR# 1	IAT	AT	TSB	WT	ST	TSE
1	6320	1.499509	1.499509	1.499509	0	1.2	2.699509
2	4630	0.932636	2.432144	2.699509	0.267364	1.2	3.899509
3	8657	3.011519	5.443663	5.443663	0	1.2	6.643663
4	30	0.004507	5.44817	6.643663	1.195493	1.2	7.843663
5	5624	1.239675	6.687845	7.843663	1.155818	1.2	9.043663
6	6728	1.675776	8.36362	9.043663	0.680043	1.2	10.24366
7	5925	1.346572	9.710192	10.24366	0.533471	1.2	11.44366
8	2829	0.49881	10.209	11.44366	1.234661	1.2	12.64366
9	7939	2.369091	12.57809	12.64366	0.065571	1.2	13.84366
10	6476	1.564483	14.14258	14.14258	0	1.2	15.34258
11	3319	0.604976	14.74755	15.34258	0.595024	1.2	16.54258
12	8134	2.518182	17.26573	17.26573	0	1.2	18.46573
13	1712	0.281665	17.5474	18.46573	0.918335	1.2	19.66573
14	6317	1.498286	19.04568	19.66573	0.620049	1.2	20.86573
15	6605	1.620422	20.66611	20.86573	0.199627	1.2	22.06573
16	2734	0.479069	21.14517	22.06573	0.920558	1.2	23.26573
17	432	0.066241	21.21142	23.26573	2.054317	1.2	24.46573
18	3441	0.63262	21.84404	24.46573	2.621697	1.2	25.66573
19	726	0.113055	21.95709	25.66573	3.708641	1.2	26.86573
20	6969	1.790539	23.74763	26.86573	3.118103	1.2	28.06573

W_q from the simulation is .994 minutes. W_q from steady state is 2.4 minutes, one reason for the difference is the start up bias of the simulation.

13.17 See file ch13-17.xls

Copier #	R #	Arrival Time	Time Ser. Beg	service time	Completion Time	Waiting Time
1	0.109	0	0	0.097	0.097	0.097
2	0.737	1.5	1.500	1.113	2.613	1.113
3	0.281	3	3.000	0.275	3.275	0.275
4	0.515	4.5	4.500	0.603	5.103	0.603
5	0.912	6	6.000	2.021	8.021	2.021
6	0.147	7.5	8.021	0.132	8.153	0.653
7	0.648	9	9.000	0.870	9.870	0.870
8	0.245	10.5	10.500	0.234	10.734	0.234
9	0.703	12	12.000	1.012	13.012	1.012
10	0.844	13.5	13.500	1.549	15.049	1.549
11	0.739	15	15.049	1.118	16.168	1.168
12	0.179	16.5	16.500	0.164	16.664	0.164
13	0.489	18	18.000	0.560	18.560	0.560
14	0.115	19.5	19.500	0.102	19.602	0.102
15	0.882	21	21.000	1.782	22.782	1.782
16	0.945	22.5	22.782	2.419	25.201	2.701
17	0.299	24	25.201	0.296	25.497	1.497
18	0.831	25.5	25.500	1.484	26.984	1.484
19	0.718	27	27.000	1.054	28.054	1.054
20	0.277	28.5	28.500	0.270	28.770	0.270

The average waiting time, W, is .96 minutes.

13.18 See files ch13-18.xls and ch13-18a.xls

The playing times for the four simulations were: 21, 25, 5, and 7. The 95% confidence interval for the average playing time is 12.33 +/- 27.36 = (0, 39.69)

13.19 See file ch13-19.xls

R#1	Santa Ana	R#2	Orange	Total Time	R#3	Costa Mesa	R#4	River-side	Total Time	R#5	Garden Grove	Total Time
0.651	10	0.334	7	17	0.22	7	0.893	9	16	0.632	6	13
0.776	10	0.987	8	18	0.983	11	0.885	9	20	0.463	5	13
0.617	10	0.263	7	17	0.627	9	0.909	9	18	0.866	6	13
0.88	11	0.914	8	19	0.831	10	0.261	6	16	0.003	3	11
0.421	9	0.965	8	17	0.005	7	0.798	8	15	0.562	6	14
0.745	10	0.488	7	17	0.562	9	0.554	7	16	0.673	6	13
0.118	8	0.826	8	16	0.337	7	0.262	6	13	0.593	6	14
0.401	9	0.278	7	16	0.614	9	0.16	5	14	0.283	4	11
0.034	8	0.964	8	16	0.165	7	0.406	6	13	0.794	6	14
0.63	10	0.47	7	17	0.804	10	0.652	7	17	0.648	6	13

13.19 continued

The average time for the Santa Ana-Orange Freeway combination is 17 minuets, the average time for the Costa Mesa-Riverside Freeway combination is 15.8 minutes, and the average time for the Garden Grove-Orange Freeway combination is 12.9 minutes The Garden Grove-Orange Freeway combination appears to take the least amount of time based on these simulations.

13.20 See file ch13-20.xls

Int. #	R #	#arrive	R #	Unload Time	# Wait
1	0.109	0			0
2	0.737	1	0.200	0.5	0
3	0.281	0			0
4	0.515	0			0
5	0.912	2	0.797	0.5	
			0.355	0.5	1
6	0.147	0			0
7	0.648	1	0.630	0.5	0
8	0.245	0			0
9	0.703	1	0.550	0.5	0
10	0.844	1	0.517	0.5	0
11	0.739	1	0.934	1	0
12	0.179	0			0
13	0.489	0			0
14	0.115	0			0
15	0.882	2	0.700	0.5	
			0.144	0.5	1
16	0.945	2	0.357	0.5	
			0.015	0.5	2
17	0.299	0			1
18	0.831	1	0.093	0.5	1
19	0.718	1	0.708	0.5	1
20	0.277	0			0

Based on the simulation, $L_q = (1 + 1 + 2 + 1 + 1 + 1)/20 = .35$

13.21 See file ch13-21.xls

Cust	Customer Arrival			Number in	Customer Service				
#	R# 1	IAT	Arrival Time	System	R# 2	Dozen	R# 3	Ser. Time	Comp. Time
1	0.651	1.052	31.05	1	0.334	no	0.22	0.573	31.62
2	0.776	1.497	32.55	1	0.987	yes	0.983	6.148	38.7
3	0.617	0.96	33.51	2	0.263	no	0.627	0.709	39.4
4	0.88	2.12	35.63	3	0.914	yes	0.831	2.662	42.07
5	0.421	0.547	36.17	4	0.965	yes	0.005	0.008	42.07
6	0.745	1.367	37.54	5	0.488	no	0.562	0.687	42.76
7	0.118	0.126	37.67	6	0.826	yes	0.337	0.616	43.38
8	0.401	0.513	38.18	7	0.278	no	0.614	0.705	44.08
9	0.034	0.034	38.21	8	0.964	yes	0.165	0.27	44.35
10	0.63	0.994	39.21	8	0.47	no	0.804	0.768	45.12
11	0.548	0.795	40	8	0.202	no	0.702	0.734	45.85
12	0.109	0.115	40.12	9	0.443	no	0.499	0.666	46.52
13	0.17	0.186	40.3	10	0.846	yes	0.821	2.579	49.1
14	0.697	1.194	41.5	11	0.942	yes	0.152	0.248	49.35
15	0.17	0.186	41.68	12	0.669	no	0.953	0.818	50.16
16	0.027	0.027	41.71	13	0.632	no	0.105	0.535	50.7
17	0.318	0.382	42.09	12	0.753	yes	0.176	0.29	50.99
18	0.796	1.589	43.68	11	0.052	no	0.815	0.772	51.76
19	0.496	0.685	44.37	10	0.107	no	0.032	0.511	52.27
20	0.428	0.559	44.93	11	0.188	no	0.879	0.793	53.06

Customer 20 will complete service at time 6:53 a.m.. The maximum number of customers in the system is 13 during the period under study.

13.22 See file ch13-22.xls

For this simulation the average waiting time in line, W_q, appears to be much greater for full serve than for self serve (15.4 minutes versus 3.7 minutes).

13.23 See file ch13-23.xls

Cust #	R# 1	Start	Customs Insp. Ser. Time	End	cust #	Get Baggage R# 2	Ser. Time	End	cust #	Arrive Time	Baggage Insp. R# 3	Ser. Tim	End
1	0.651	0	52.53	52.53	1	0.334	73.11	125.6	1	125.6	0.22	0	0.22
2	0.776	52.53	58.81	111.3	2	0.987	787.3	898.7	3	217.1	0.983	180	397.1
3	0.617	111.3	50.85	162.2	3	0.263	54.95	217.1	6	445.1	0.627	60	505.1
4	0.88	162.2	64	226.2	4	0.914	441.4	667.6	8	449.1	0.831	60	565.1
5	0.421	226.2	41.06	267.2	5	0.965	603.9	871.2	11	551.6	0.005	0	565.1
6	0.745	267.2	57.26	324.5	6	0.488	120.6	445.1	10	577.8	0.562	60	637.8
7	0.118	324.5	25.91	350.4	7	0.826	314.8	665.2	12	641.7	0.337	0	641.7
8	0.401	350.4	40.06	390.5	8	0.278	58.66	449.1	7	665.2	0.614	60	725.2
9	0.034	390.5	21.68	412.1	9	0.964	596.4	1009	4	667.6	0.165	0	725.2
10	0.63	412.1	51.5	463.6	10	0.47	114.1	577.8	18	774.8	0.804	60	834.8
11	0.548	463.6	47.41	511.1	11	0.202	40.53	551.6	19	830.4	0.702	60	894.8
12	0.109	511.1	25.43	536.5	12	0.443	105.2	641.7	15	847.5	0.499	60	954.8
13	0.17	536.5	28.49	565	13	0.846	337.2	902.2	16	849.4	0.821	60	1015
14	0.697	565	54.85	619.8	14	0.942	513.8	1134	5	871.2	0.152	0	1015
15	0.17	619.8	28.48	648.3	15	0.669	199.2	847.5	20	888.9	0.953	180	1195
16	0.027	648.3	21.34	669.6	16	0.632	179.8	849.4	2	898.7	0.105	0	1195
17	0.318	669.6	35.88	705.5	17	0.753	251.6	957.1	13	902.2	0.176	0	1195
18	0.796	705.5	59.8	765.3	18	0.052	9.517	774.8	17	957.1	0.815	60	1255
19	0.496	765.3	44.79	810.1	19	0.107	20.35	830.4	9	1009	0.032	0	1255
20	0.428	810.1	41.41	851.5	20	0.188	37.42	888.9	14	1134	0.879	60	1315

Based on the simulation it will take 1315 seconds or 21.92 minutes for the twenty customers to get through customs.

13.24 See files ch13-24a.xls and ch13-24b.xls

a.

Week	R#	Demand	beg Invent	end Invent	Reorder	Q	R#	Lead Time	Arrival Date	hold cost	reorder cost	goodwill cost	total cost
				7		0							
1	0.22	0	7	7	0	0			0	14			14
2	0.98	5	7	2	1	13	0.893	3	5	9	50		59
3	0.63	3	2	0	1	13			5	1.333		40	41.33
4	0.83	4	0	0	1	13			5	0		160	160
5	0.01	0	0	0	1	13			5	0			0
6	0.56	3	13	10	0	5				23			23
7	0.34	1	10	9	0	5				19			19
8	0.61	3	9	6	0	5				15			15
9	0.16	0	6	6	0	5				12			12
10	0.80	4	6	2	1	13	0.885	3	5	8	50		58
												Total Cost =	401.3

The total cost over the ten weeks is $401.33

b.

Week	R #	Demand	beg Invent	end Invent	Reorder	hold cost	weekly cost	goodwill cost	total cost
				7					
1	0.22	0	7	7	0	14	5		19
2	0.98	5	7	2	0	9	5		14
3	0.63	3	7	4	1	11	5		16
4	0.83	4	4	0	0	4	5		9
5	0.01	0	5	5	1	10	5		15
6	0.56	3	5	2	0	7	5		12
7	0.34	1	7	6	1	13	5		18
8	0.61	3	6	3	0	9	5		14
9	0.16	0	8	8	1	16	5		21
10	0.80	4	8	4	0	12	5		17
							Total Cost =		155

The total cost over the ten weeks is $155.00

c. Based on the two simulations, Family should take the supplier's offer as it cost $246 less.

13.25 See file ch13-25.xls

b. According to the simulation, Marv's earnings over the 10-day period is $6055.73

CD13.1 See files cd13-1.xls, cd13-1a.qaa, and cd13-1b.qaa

Cust. #	R #	IAT	AT	5	Ann ST	Start	Finish	W	Jill ST	Start	Finish	W
1	8927	4	4	6320	3	4	7	3	2	4	6	2
2	8848	4	8	4630	2	8	10	2	2	8	10	2
3	9089	4	12	8657	4	12	16	4	4	12	16	4
4	2605	2	14	0030	1	16	17	3	1	16	17	3
5	7982	4	18	5624	2	18	20	2	2	18	20	2
6	5541	3	21	6728	3	21	24	3	2	21	23	2
7	2624	2	23	5925	2	24	26	3	2	23	25	2
8	1603	2	25	2829	1	26	27	2	1	25	26	1
9	4063	3	28	7939	3	28	31	3	3	28	31	3
10	6519	3	31	6476	3	31	34	3	2	31	33	2
11	6106	3	34	3319	2	34	36	2	1	34	35	1
12	6807	3	37	8134	4	37	41	4	4	37	41	4
13	7589	4	41	1712	1	41	42	1	1	41	42	1
14	6590	3	44	6317	3	44	47	3	2	44	46	2
15	0882	1	45	6605	3	47	50	5	2	46	48	3
16	3193	3	48	2734	1	50	51	3	1	48	49	1
17	2084	2	50	0432	1	51	52	2	1	50	51	1
18	4053	3	53	3441	2	53	55	2	1	53	54	1
19	9941	4	57	0726	1	57	58	1	1	57	58	1
20	0967	1	58	6969	3	58	61	3	2	58	60	2
21	7890	4	62	3434	2	62	64	2	1	62	63	1
22	1522	2	64	0059	1	64	65	1	1	64	65	1
23	8333	4	68	0617	1	68	69	1	1	68	69	1
24	7247	4	72	5586	2	72	74	2	2	72	74	2
25	5140	3	75	0996	1	75	76	1	1	75	76	1
26	0916	1	76	1114	1	76	77	1	1	76	77	1
27	7329	4	80	6063	3	80	83	3	2	80	82	2
28	3642	3	83	4400	2	83	85	2	2	83	85	2
29	1981	2	85	8971	4	85	89	4	4	85	89	4
30	2557	2	87	6612	3	89	92	5	2	89	91	4
31	9061	4	91									
						Wait	Time =	76		Wait	Time =	59
						Cost =	29.62			Cost =	31.43	

a. Based on the simulation, the cost using Ann is \$29.62

b. Based on the simulation, the cost using Jill is \$31.43

c. Ann appears to yield the lower average cost to the company.

d. For this process, $\lambda = 60/2.9 = 20.689655$. For Ann, $\mu = 60/2.3 = 26.086957$ and the average hourly cost is \$64.50. For Jill, $\mu = 60/2.1 = 28.571429$ and the average hourly cost is \$50.38. Hence, Jill should be hired.

CD13.2 See file cd13-2.xls

a. The average daily sales = .10*30 + .10*40 + .15*50 + .20*60 + .30*70 + .10*80 + .05*90 = 60.

b. The mean number of days required for the machine to run out of coffee = 300/60 = 5.

c.

			Fill Every 5 Days				Fill Every 4 Days			
			Inventory				Inventory			
Day	R#	Demand	Start	End	Labor	Profit	Start	End	Labor	Profit
1	3338	50	300	250	8	1	300	250	8	1
2	9874	90	250	160		16.2	250	160		16.2
3	2631	50	160	110		9	160	110		9
4	9139	80	110	30		14.4	110	30		14.4
5	9651	90	30	-60		-24.6	300	210	8	8.2
6	4883	60	300	240	8	2.8	210	150		10.8
7	8260	70	240	170		12.6	150	80		12.6
8	2781	50	170	120		9	80	30		9
9	9636	90	120	30		16.2	300	210	8	8.2
10	4696	60	30	-30		-9.6	210	150		10.8
11	2016	50	300	250	8	1	150	100		9
12	4426	60	250	190		10.8	100	40		10.8
13	8464	70	190	120		12.6	300	230	8	4.6
14	9424	80	120	40		14.4	230	150		14.4
15	6693	70	40	-30		-7.8	150	80		12.6
16	6317	70	300	230	8	4.6	80	10		12.6
17	7529	70	230	160		12.6	300	230	8	4.6
18	0515	30	160	130		5.4	230	200		5.4
19	1069	40	130	90		7.2	200	160		7.2
20	1877	40	90	50		7.2	160	120		7.2
				Profit =		115		Profit =		188.6

For this simulation the profit over the twenty days is $115.

d. For this simulation the profit over the twenty days is $188.60. On the basis of these simulations Jewel should fill the machine every five days.

CD13.3 See files cd13-3.xls and cd13-3a.xls

Note that multiple simulation runs can be made using file cd13-3.xls. To evaluate different policies one would change the value in cell G1. Based on the ten simulation runs for each policy (results given in see file cd13-3a.xls), the optimal decision is to fill the machine every 6 days. Basing the 95% confidence interval on the ANOVA analysis gives an interval of 5182.6 +/- $2.052*\sqrt{(66094.26/10)}$ = 5182.6 +/- 167.8 = (5,014.8, 5,350.4). If the 95% confidence interval is based only on the ten runs associated with filling the machine every 6 days the 95% confidence interval would be 5182.6 +/- $2.262*363.45/\sqrt{10}$ = 5182.6 +/- 260.00 = (4,922.60, 5,442.60).

CD13.4 See file cd13-4.xls

a. Because the demand and lead time is not stationary, the models presented in Chapter 10 may not give the optimal solution to this problem.

b. Over the 25 simulated days, the profit is $2,827.65.

CD13.5 See file cd13-5.xls

Over the 25 simulated days, the expected profit is $2,767.65. Hence, based on these results we would not be able to claim that this policy would not be an improvement over the policy of ordering 20 beds when the inventory reaches 10 beds.

CD13.6 See file cd13-6.xls

The chi square statistic for this data is 11.6. The critical chi square value based on 9 degrees of freedom and $\alpha = .05$ is 16.919. Since the calculated chi square statistic is not greater than the critical chi square value, we do not have sufficient evidence to claim that the data does not follow a uniform distribution.

CD13.7 See file cd13-7.xls

Based on the simulation, someone just promoted will stay with the company for one additional year.

Case 13.1 See files ca13-1a.xls, ca13-1b.xls, and ca13-1.its

File ca13-1a.xls is designed to do the simulation for an order point, order up to level inventory system under the current policy in which Office Central offers a $150 discount if they are out of stock of the copiers. Changing the values in cells I1 and I2 will enable different policies to be simulated. In particular, the value in cell I1 gives the order quantity upon which the order up to level policy is based, while the value in cell I2 gives the reorder point. File ca13-1b.xls is the simulation in which the discount policy is eliminated. Changing the holding cost can be accomplished in both simulations by changing the value in cell M1. File ca13-1.its can be used to a find reasonable starting point for the simulation analysis.

Case 13.2 See files ca13-2a.xls and ca13-2b.xls

File ca13-2a.xls is designed to do the simulation of Four Wheel Tire when the computerized balancing machine is not leased. File ca13-2b.xls is designed to do the simulation of Four Wheel Tire when the computerized balancing machine is leased. Running these simulations multiple times will enable the analysis to be done.

Case CD13.1 See file cac13-1.xls

File cac13-1.xls can be used to perform simulations of different order point, order quantity policies by changing the values in cells I1 and I2. In particular, the value in cell I1 gives the order quantity being considered, while the value in cell I2 gives the reorder point being considered.

Case CD13.2 See file cac13-2.xls

File cac13-2.xls can be used to perform simulations comparing the average hourly cost for hiring Harriet versus Larry. To change the arrival rate one would change the value in cell B2. To change the goodwill cost per minute one would change the value in cell K1.